LECTURES ON THE PHYSICS OF HIGHLY CORRELATED ELECTRON SYSTEMS VIII

Previous Proceedings in the Series of Lectures on the Physics of Highly Correlated Electron Systems

Year	Title	Publisher	ISBN
2002	Seventh Training Course	AIP Conf. Proceedings Vol. 678	0-7354-0147-0
2001	Sixth Training Course	AIP Conf. Proceedings Vol. 629	0-7354-0083-0
2000	Fifth Training Course	AIP Conf. Proceedings Vol. 580	0-7354-0019-9
1999	Fourth Training Course	AIP Conf. Proceedings Vol. 527	1-56396-950-5
1998	Third Training Course	*not published*	
1997	Second Training Course	AIP Conf. Proceedings Vol. 438	1-56396-789-8
1996	First Training Course	*not published*	

Other Related Titles from AIP Conference Proceedings

695 Highlights in Condensed Matter Physics
Edited by Adolfo Avella, Roberta Citro, Canio Noce, and Mario Salerno, December 2003, 0-7354-0167-5

677 Fundamental Physics of Ferroelectrics 2003
Edited by P. K. Davies and D. J. Singh, August 2003, 0-7354-0146-2

LECTURES ON THE PHYSICS OF HIGHLY CORRELATED ELECTRON SYSTEMS VIII

Eighth Training Course in the Physics of Correlated Electron Systems and High-Tc Superconductors

Salerno, Italy 6 – 17 October 2003

EDITORS
Adolfo Avella
Ferdinando Mancini
Università degli Studi di Salerno
Salerno, Italy

SPONSORING ORGANIZATIONS
European Commission
Università degli Studi di Salerno, Italy
International Institute for Advanced Scientific Studies
Vietri sul Mare (SA), Italy

AMERICAN INSTITUTE OF PHYSICS

Melville, New York, 2004

AIP CONFERENCE PROCEEDINGS ■ VOLUME 715

Editors:

Adolfo Avella
Ferdinando Mancini

Dipartimento di Fisica "E. R. Caianiello"
Università degli Studi di Salerno
Via S. Allende
I-84081 Baronissi (SA)
ITALY

E-mail: avella@sa.infn.it
mancini@sa.infn.it

L.C. Catalog Card No. 2004109621
ISBN 0-7354-0194-2
ISSN 0094-243X
Printed in the United States of America

Contents

PART II. PARTICIPANT CONTRIBUTIONS

Preface

The present volume contains the notes of the lectures delivered at the "Eighth Training Course in the Physics of Correlated Electron Systems and High-Tc Superconductors," held in Vietri sul Mare (Salerno, Italy) in October 2003. It also contains contributions by some of the participants who delivered seminars in the afternoon sessions.

Following the tradition of previous years, the meeting was devoted to the training of young scientists in one of the most intriguing fields of condensed matter physics: strongly correlated systems. The intent was to bring together for two weeks four senior and about thirty junior researchers in a close location with an informal atmosphere, paying special attention to foster the active participation of the young researchers.

The course consisted of four lectures every morning, held by Professors Antoine Georges, Masatoshi Imada, Miodrag L. Kulic, Alejandro Muramatsu, and of afternoon activities (seminars delivered by the junior researchers, solving of specific problems, round table on hot topics ...) aimed principally at promoting discussions between the attendees and the lecturers. The outcome of this type of course was a significant interchange of ideas among the participants thanks to both the enlightening morning lectures and the long afternoon sessions devoted to discussions.

It is hoped that both the meeting, which brought together leaders in the field as well as bright and eager beginners, and the present volume, which may be useful as an up-to-date book for researchers interested in the field, shall be considered as firm points by the scientific community.

We wish to acknowledge the support of those institutions that made the course possible. The main sponsor of the event has been the European Commission and funding was also granted by the University of Salerno. Finally, we wish to thank Professor Maria Marinaro, President of the International Institute of Advanced Scientific Studies, who hosted the event in the wonderful and warm venue of Vietri sul Mare.

Salerno, July 1, 2004

Adolfo Avella

Ferdinando Mancini

PART I

LECTURES

Strongly Correlated Electron Materials: Dynamical Mean-Field Theory and Electronic Structure

Antoine Georges

Centre de Physique Théorique, Ecole Polytechnique, 91128 Palaiseau Cedex, France

Abstract. These are introductory lectures to some aspects of the physics of strongly correlated electron systems. I first explain the main reasons for strong correlations in several classes of materials. The basic principles of dynamical mean-field theory (DMFT) are then briefly reviewed. I emphasize the formal analogies with classical mean-field theory and density functional theory, through the construction of free-energy functionals of a local observable. I review the application of DMFT to the Mott transition, and compare to recent spectroscopy and transport experiments. The key role of the quasiparticle coherence scale, and of transfers of spectral weight between low- and intermediate or high energies is emphasized. Above this scale, correlated metals enter an incoherent regime with unusual transport properties. The recent combinations of DMFT with electronic structure methods are also discussed, and illustrated by some applications to transition metal oxides and f-electron materials.

CONTENTS

CP715, *Lectures on the Physics of Highly Correlated Electron Systems VIII: Eighth Training Course*, edited by A. Avella and F. Mancini

1. INTRODUCTION: WHY STRONG CORRELATIONS ?

1.1. Hesitant electrons: delocalised waves or localised particles ?

The physical properties of electrons in many solids can be described, to a good approximation, by assuming an independent particle picture. This is particularly successful when one deals with broad energy bands, associated with a large value of the kinetic energy. In such cases, the (valence) electrons are highly *itinerant*: they are delocalised over the entire solid. The typical time spent near a specific atom in the crystal lattice is very short. In such a situation, valence electrons are well described using a *wave-like picture*, in which individual wavefunctions are calculated from an effective one-electron periodic potential.

For some materials however, this physical picture suffers from severe limitations and may fail altogether. This happens when valence electrons spend a larger time around a given atom in the crystal lattice, and hence have a tendency towards *localisation*. In such cases, electrons tend to "see each other" and the effects of statistical correlations between the motions of individual electrons become important. An independent particle description will not be appropriate, particularly at short or intermediate time scales (high to intermediate energies). A *particle-like picture* may in fact be more appropriate than a wave-like one over those time scales, involving wavefunctions localised around specific atomic sites. Materials in which electronic correlations are significant are generally associated with moderate values of the bandwidth (narrow bands). The small kinetic energy implies a longer time spent on a given atomic site. It also implies that the ratio of the Coulomb repulsion energy between electrons and the available kinetic energy becomes larger. As a result delocalising the valence electrons over the whole solid may become less favorable energetically. In some extreme cases, the balance may even become unfavorable, so that the corresponding electrons will remain localised. In a naive picture, these electrons sit on the atoms to which they belong and refuse to move. If this happens to all the electrons close to the Fermi level, the solid becomes an insulator. This insulator is difficult to understand in the wave-like language: it is not caused by the absence of available one-electron states caused by destructive interference in **k**-space, resulting in a band-gap, as in conventional band insulators. It is however very easy to understand in real space (thinking of the solid as made of individual atoms pulled closer to one another in order to form the crystal lattice). This mechanism was understood long ago [1, 2] by Mott (and Peierls), and such insulators are therefore called *Mott insulators* (Sec. 4. In other cases, such as f-electron materials, this electron localisation affects only part of the electrons in the solid (e.g the ones corresponding to the f-shell), so that the solid remains a (strongly correlated) metal.

The most interesting situation, which is also the one which is hardest to handle theoretically, is when the localised character on short time-scales and the itinerant character on long time-scales coexist. In such cases, the electrons "hesitate" between being itinerant and being localised. This gives rise to a number of physical phenomena, and also results in several possible instabilities of the electron gas which often compete, with very small energy differences between them. In order to handle such situations theoretically, it is necessary to think both in **k**-space and in real space, to handle both the particle-like and the wave-like character of the electrons and, importantly, to be able to describe physical phenomena on *intermediate energy scales*. For example, one needs to explain how long-lived (wave-like) quasiparticles may eventually emerge at low energy/temperature in a strongly correlated metal while at higher energy/temperature, only incoherent (particle-like) excitations are visible. It is the opinion of the author that, in many cases, understanding these intermediate energy scales and the associated coherent/incoherent crossover is the key to the intriguing physics often observed in correlated metals. In these lectures, we discuss a technique, the dynamical mean-field theory (DMFT), which is able to (at least partially) handle this problem. This technique has led to significant progress in our understanding of strong correlation physics, and allows for a quantitative description of many correlated materials[3, 4]. Extensions and generalisations of this technique are currently being developed in order to handle the most difficult/mysterious situations which cannot be tamed by the simplest version of DMFT.

1.2. Bare energy scales

Localised orbitals and narrow bands. In practice, strongly correlated materials are generally associated with partially filled d- or f- shells. Hence, the suspects are materials involving:

- Transition metal elements (particularly from the 3d-shell from Ti to Cu, and to a lesser extent 4d from Zr to Ag).
- Rare earth (4f from Ce to Yb) or actinide elements (5f from Th to Lw)

To this list, one should also add molecular (organic) conductors with large unit cell volumes in which the overlap between molecular orbitals is weak.

What is so special about d- and f- orbitals (particularly 3d and 4f) ? Consider the atomic wavefunctions of the 3d shell in a 3d transition metal atom (e.g Cu). There are no atomic wavefunctions with the same value $l = 2$ of the angular momentum quantum number, but lower principal quantum number n than $n = 3$ (since one must have $l \leq n-1$). Hence, the 3d wavefunctions are orthogonal to all the $n = 1$ and $n = 2$ orbitals just because of their angular dependence, and the radial part needs not have nodes or extend far away from the nucleus. As a result, the 3d-orbital wave functions are confined more closely to the nucleus than for s or p states of comparable energy. The same argument applies to the 4f shell in rare earths. It also implies that the 4d wavefunctions in the 4d transition metals or the 5f ones in actinides will be more extended (and hence that these materials are expected to display, on the whole, weaker correlation effects than 3d transition metals, or the rare earth, respectively).

Oversimplified as it may be, these qualitative arguments at least tell us that a key energy scale in the problem is the degree of overlap between orbitals on neighbouring atomic sites. This will control the bandwidth and the order of magnitude of the kinetic energy. A simple estimate of this overlap is the matrix element:

$$t_{\mathbf{RR}'}^{LL'} \sim \int d\mathbf{r}\, \chi_L^*(\mathbf{r}-\mathbf{R}) \frac{\hbar^2\nabla^2}{2m} \chi_{L'}(\mathbf{r}-\mathbf{R}') \tag{1}$$

In the solid, the wavefunction $\chi_L(\mathbf{r}-\mathbf{R})$ should be thought of as a Wannier-like wave function centered on atomic site $\mathbf{R}$. In narrow band systems, typical values of the bandwith are a few electron-volts.

Coulomb repulsion and the Hubbard U. Another key parameter is the typical strength of the Coulomb repulsion between electrons sitting in the most localized orbitals. The biggest repulsion is associated with electrons with opposite spins occupying the same orbital: this is the Hubbard repulsion which we can estimate as:

$$U \sim \int d\mathbf{r}d\mathbf{r}' |\chi_L(\mathbf{r}-\mathbf{R})|^2\, U_s(\mathbf{r}-\mathbf{r}')\, |\chi_L(\mathbf{r}'-\mathbf{R})|^2 \tag{2}$$

In this expression, U_s is the interaction between electrons *including screening effects* by other electrons in the solid. Screening is a very large effect: if we were to estimate (2) with the unscreened Coulomb interaction $U(\mathbf{r}-\mathbf{r}') = e^2/|\mathbf{r}-\mathbf{r}'|$, we would typically obtain values in the range of tens of electron-volts. Instead, the screened value of U

in correlated materials is typically a few electron-volts. This can be comparable to the kinetic energy for narrow bandwiths, hence the competition between localised and itinerant aspects. Naturally, other matrix elements (e.g between different orbitals, or between different sites) are important for a realistic description of materials (see the last section of these lectures).

In fact, a precise description of screening in solids is a rather difficult problem. An important point is, again, that this issue crucially *depends on energy scale*. At very low energy, one should observe the fully screened value, of order a few eV 's, while at high energies (say, above the plasmon energy in a metal) one should observe the unscreened value, tens of eV 's. Indeed, the screened effective interaction $W(\mathbf{r},\mathbf{r}';\omega)$ as estimated e.g from the RPA approximation, is a strong function of frequency (see e.g Ref. [5, 6] for an ab-initio GW treatment in the case of Nickel). As a result, using an energy-independent parametrization of the on-site matrix elements of the Coulomb interaction such as (2) can only be appropriate for a description restricted to low- enough energies [6]. The Hubbard interaction can only be given a precise meaning in a solid, over a large enery range, if it is made energy-dependent. I shall come back to this issue in the very last section of these lectures (Sec. 5.5).

The simplest model hamiltonian. From this discussion, it should be clear that the simplest model in which strong correlation physics can be discussed is that of a lattice of single-level "atoms", or equivalently of a single tight-binding band (associated with Wannier orbitals centered on the sites of the crystal lattice), retaining only the on-site interaction term between electrons with opposite spins:

$$H = -\sum_{\mathbf{RR}',\sigma} t_{\mathbf{RR}'} c^{+}_{\mathbf{R}\sigma} c_{\mathbf{R}'\sigma} + \varepsilon_0 \sum_{\mathbf{R}\sigma} n_{\mathbf{R}\sigma} + U \sum_{\mathbf{R}} n_{\mathbf{R}\uparrow} n_{\mathbf{R}\downarrow} \tag{3}$$

The kinetic energy term is diagonalized in a single-particle basis of Bloch's wavefunctions:

$$H_0 = \sum_{\mathbf{k}\sigma} \varepsilon_{\mathbf{k}} c^{+}_{\mathbf{k}\sigma} c_{\mathbf{k}\sigma} \; ; \; \varepsilon_{\mathbf{k}} \equiv \sum_{\mathbf{R}'} t_{\mathbf{RR}'} e^{i\mathbf{k}\cdot(\mathbf{R}-\mathbf{R}')} \tag{4}$$

with e.g for nearest-neighbour hopping on the simple cubic lattice in d-dimensions:

$$\varepsilon_{\mathbf{k}} = -2t \sum_{\mu=1}^{d} \cos(k_\mu a) \tag{5}$$

In the absence of hopping, we have, at each site, a single atomic level and hence four possible quantum states: $|0\rangle, |\uparrow\rangle, |\downarrow\rangle$ and $|\uparrow\downarrow\rangle$ with energies 0, ε_0 and $U+2\varepsilon_0$, respectively.

Eq. (3) is the famous Hubbard model [7, 8, 9]. It plays in this field the same role than that played by the Ising model in statistical mechanics: a laboratory for testing physical ideas, and theoretical methods alike. Simplified as it may be, and despite the fact that it already has a 40-year old history, we are far from having explored all the physical phenomena contained in this model, let alone of being able to reliably calculate with it in all parameter ranges !

1.3. Examples of strongly correlated materials

In this section, I give a few examples of strongly correlated materials. The discussion emphasizes a few key points but is otherwise very brief. There are many useful references related to this section, e.g [10, 11, 2, 12, 13].

1.3.1. Transition metals

In 3d transition metals, the 4s orbitals have lower energy than the 3d and are therefore filled first. The 4s orbitals extend much further from the nucleus, and thus overlap strongly. This holds the atoms sufficiently far apart so that the d- orbitals have a *small direct overlap*. Nevertheless, d-orbitals extend much further from the nucleus than the "core" electrons (corresponding to shells which are deep in energy below the Fermi level). As a result, throughout the 3d series of transition metals (and even more so in the 4d series), d-electrons do have an itinerant character, giving rise to quasiparticle bands. That this is the case is already clear from a very basic property of the material, namely how the equilibrium unit- cell volume depends on the element as one moves along the 3d series (Fig. 1). The unit-cell volume has a very characteristic, roughly parabolic, dependence. A simple model of a narrow band being gradually filled, introduced long ago by Friedel [14] accounts for this parabolic dependence (see also [15, 16]). Because the states at the bottom of the band are bonding-like while the states at the top of the band are anti-bonding like, the binding energy is maximal (and hence the equilibrium volume is minimal) for a half-filled shell. Instead, if the d-electrons were localised we would expect little contribution of the d-shell to the cohesive energy of the solid, and the equilibrium volume should not vary much along the series.

Screening is relatively efficient in transition metals because the 3d band is not too far in energy from the 4s band. The latter plays the dominant rôle in screening the Coulomb interaction (crudely speaking, one has to consider the following charge transfer process between two neighbouring atoms: $3d^n4s + 3d^n4s \to 3d^{n-1}4s^2 + 3d^{n+1}$, see e.g [17] for further discussion). For all these reasons (the band not being extremely narrow, screening being efficient), electron correlations do have important physical effects for 3d transition metals, but not extreme ones like localisation. Magnetism of these metals below the Curie temperature, but also the existence of fluctuating local moments in the paramagnetic phase are exemples of such correlation effects. Band structure calculations based on DFT-LDA methods overestimate the width of the occupied d-band (by about 30% in the case of nickel). Some features observed in spectroscopy experiments (such as the (in)famous 6 eV satellite in nickel) are also signatures of correlation effects, and are not reproduced by standard electronic structure calculations.

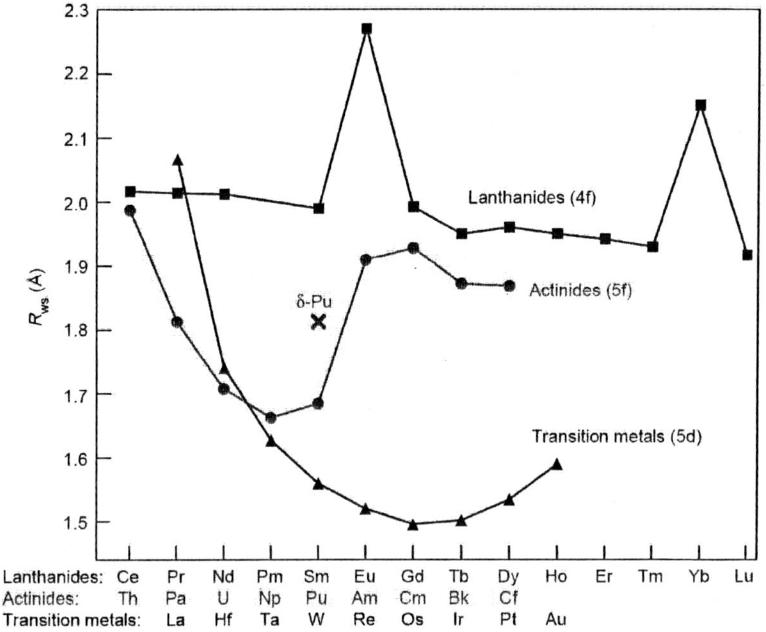

FIGURE 1. Experimental Wigner-Seitz Radius of Actinides, Lanthanides, and 5d Transition Metals. The equilibrium volume of the primitive unit cell is given by $V = 4\pi R_{WS}^3/3$. Elements that lie on top of each other have the same number of valence electrons. The volume of the transition metals has a roughly parabolic shape, indicating delocalised 5d electrons. The volumes of the lanthanides remain roughly constant, indicating localised 4f electrons. The volumes of the light actinides decrease with increasing atomic number, whereas the volumes of the late actinides behaves similarly to that of the lanthanides. From Ref. [16]

1.3.2. Transition metal oxides

In transition metal compounds (e.g oxides or chalcogenides), the direct overlap between d-orbitals is generally so small that d-electrons can only move through hybridisation with the ligand atoms (e.g oxygen 2p-bands). For example, in the cubic perovskite structure shown on Fig. 2, each transition-metal atom is "encaged" at the center of an octahedron made of six oxygen atoms. Hybridisation leads to the formation of bonding and antibonding orbitals. An important energy scale is the *charge-transfer energy* $\Delta = \varepsilon_d - \varepsilon_p$, i.e the energy difference between the average position of the oxygen and transition metal bands. When Δ is large as compared to the overlap integral t_{pd}, the bonding orbitals have mainly oxygen character and the antibonding ones mainly transition-metal character. In this case, the effective metal- to- metal hopping can be estimated as $t_{eff} \sim t_{pd}^2/\Delta$, and is therefore quite small.

The efficiency of screening in transition-metal oxides depends crucially on the relative position of the 4s and 3d band. For 3d transition metal monoxides MO with M to the right of Vanadium, the 4s level is much higher in energy than 3d, thus leading to poor screening and large values of U. This, in addition to the small bandwidth and relatively

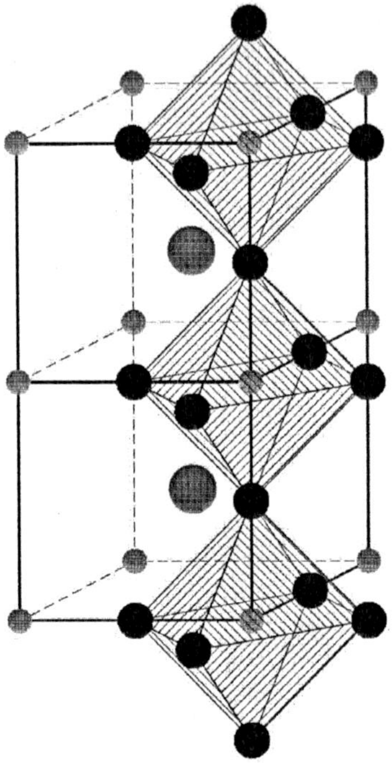

FIGURE 2. The cubic perovskite structure, e.g of the compound $SrVO_3$. Transition-metal atoms (V) - small grey spheres- are at the center of oxygen octahedra (dark spheres). Sr atoms are the larger spheres in-between planes. From Ref. [18].

large Δ, leads to dramatic correlation effects, turning the system into a Mott insulator (or rather, a charge- transfer insulator, see below), in spite of the incomplete filling of the d-band. The Mott phenomenon plays a key role in the physics of transition- metal oxides, as discussed in detail later in these lectures (see Sec. 4 and Fig. 6).

Crystal field splitting. The 5-fold (10-fold with spin) degeneracy of the d- orbitals in the atom is lifted in the solid, due to the influence of the electric field created by neighbouring atoms, i.e the ligand oxygen atoms in transition metal oxides. For a transition metal ion in an octahedral environment (as in Fig. 2), this results in a three-fold group of states (t_{2g}) which is lower in energy and a doublet (e_g) higher in energy. Indeed, the d_{xy}, d_{yz}, d_{zx} orbitals forming the t_{2g} multiplet do not point towards the ligand atoms, in contrast to the states in the e_g doublet ($d_{x^2-y^2}$, $d_{3z^2-r^2}$). The latter therefore lead to a higher cost in Coulomb repulsion energy. For a crystal with perfect cubic symmetry, the t_{2g} and e_g multiplets remain exactly degenerate, while a lower symmetry of the crystal lattice lifts the degeneracy further. For a tetrahedral environment of the transition-metal ion, the opposite situation is found, with t_{2g} higher in energy than e_g. In transition metals, the energy scale associated with crystal- field splitting is typically much smaller than the bandwith. This is not so in transition-metal oxides, for which these considerations become essential. In some materials, such as e.g $SrVO_3$ and the other d^1 oxides studied in Sec. 5.3 of these lectures, the energy bands emerging from the t_{2g} and e_g orbitals form two groups of bands well separated in energy.

Mott- and charge-transfer insulators. There are two important considerations, which are responsible for the different physical properties of the "early" (i.e involving Ti, V, Cr, ...) and "late" (Ni, Cu) transition- metal oxides:
- whether the Fermi level falls within the t_{2g} or e_g multiplets,
- what is the relative position of oxygen (ε_p) and transition-metal (ε_p) levels ?
For those compounds which correspond to an octahedral environment:

- In early transition-metal oxides, t_{2g} is partially filled, e_g is empty. Hence, the hybridisation with ligand is very weak (because t_{2g} orbitals point away from the 2p oxygen orbitals). Also, the d-orbitals are much higher in energy than the 2p orbitals of oxygen. As a result, the charge-transfer energy $\Delta = \varepsilon_d - \varepsilon_p$ is large, and the bandwidth is small. The local d-d Coulomb repulsion U_{dd} is a smaller scale than Δ but it can be larger than the bandwidth ($\sim t_{pd}^2/\Delta$): this leads to Mott insulators.
- For late transition-metal oxides, t_{2g} is completely filled, and the Fermi level lies within e_g. As a result, hybridisation with the ligand is stronger. Also, because of the greater electric charge on the nuclei, the attractive potential is stronger and as a result, the Fermi level moves closer to the energy of the 2p ligand orbitals. Hence, Δ is a smaller scale than U_{dd} and controls the energy cost of adding an extra electron. When this cost becomes larger than the bandwidth, insulating materials are obtained, often called "charge transfer insulators" [19, 20]. The mechanism is not qualitatively different than the Mott mechanism, but the insulating gap is set by the scale Δ rather than U_{dd} and separates the oxygen band from a d-band rather than a lower and upper Hubbard bands having both d-character.

The p-d model. The single-band Hubbard model is easily extended in order to take into account both transition-metal and oxygen orbitals in a simple modelisation of transition-metal oxides. The key terms to be retained are[1]:

$$H_{pd} = -\sum_{\mathbf{RR}',\sigma} t_{pd}\,(d^{+}_{\mathbf{R}\sigma} p_{\mathbf{R}'\sigma} + h.c) + \varepsilon_d \sum_{\mathbf{R}\sigma} n^d_{\mathbf{R}\sigma} + \varepsilon_p \sum_{\mathbf{R}'\sigma} n^p_{\mathbf{R}'\sigma} + U_{dd} \sum_{\mathbf{R}} n^d_{\mathbf{R}\uparrow} n^d_{\mathbf{R}\downarrow} \qquad (6)$$

to which one may want to add other terms, such as: Coulomb repulsions U_{pp} and U_{pd} or direct oxygen-oxygen hoppings t_{pp}.

1.3.3. f-electrons: rare earths, actinides and their compounds

A distinctive character of the physics of rare-earth metals (lanthanides) is that the 4f electrons tend to be localised rather than itinerant (at ambiant pressure). As a result, the f-electrons contribute little to the cohesive energy of the solid, and the unit-cell volume

---

[1] For simplicity, the hamiltonian is written in the case where only one d-band is relevant, as e.g for cuprates.

depends very weakly on the filling of the 4f shell (Fig. 1). Other electronic orbitals do form bands which cross the Fermi level however, hence the metallic character of the lanthanides. When pressure is applied, the f-electrons become increasingly itinerant. In fact, at some critical pressure, some rare-earth metals (mots notably Ce and Pr) undergo a sharp first-order transition which is accompanied by a discontinuous drop of the equilibrium unit-cell volume. Cerium is a particularly remarkable case, with a volume drop of as much as 15% and the same crystal symmetry (fcc) in the low-volume (α) and high-volume (γ) phase. In other cases, the transition corresponds to a change in crystal symmetry, from a lower symmetry phase at low pressure to a higher symmetry phase at high pressure. For a recent review on the volume-collapse transition of rare earth metals, see Ref. [15].

The equilibrium volume of actinide (5f) metals display behaviour which is intermediate between transition metals and rare earths. From the beginning of the series (Th) until Plutonium (Pu), the volume has an approximately parabolic dependence on the filling of the f-shell, indicating delocalised 5f electrons. From Americium onwards, the volume has a much weaker dependence on the number of f-electrons, suggesting localised behaviour. Interestingly, plutonium is right on the verge of this delocalisation to localisation transition. Not surprisingly then, plutonium is, among all actinide metals, the one which has the most complex phase diagram and which is also the most difficult to describe using conventional electronic structure methods (see [16, 21] for recent reviews). This will be discussed further in the last section of these lectures. This very brief discussion of rare-earths and actinide compounds is meant to illustrate the need for methods able to deal simultaneously with the itinerant and localised character of electronic degrees of freedom.

The physics of strong electronic correlations becomes even more apparent for f-electron materials which are compounds involving rare-earth (or actinide) ions and other atoms, such as e.g $CeAl_3$. A common aspect of such compounds is the formation of quasiparticle bands with extremely large effective masses (and hence large values of the low-temperature specific heat coefficient $\gamma = C/T$), up to a thousand time the bare electron mass ! Hence the term "heavy-fermion" given to these compounds: for reviews, see e.g [22, 23]. The origin of these large effective masses is the weak hybridization between the very localised f-orbitals and the rather broad conduction band associated with the metallic ion. At high temperature/energy, the f-electron have localised behaviour (yielding e.g local magnetic moments and a Curie law for the magnetic susceptibility). At low temperature/energy, the conduction electrons screen the local moments, leading to the formation of quasiparticle bands with mixed f- and conduction electron character (hence a large Fermi surface encompassing both f- and conduction electrons). The low-temperature susceptibility has a Pauli form and the low-energy physics is, apart from some specific compounds, well described by Fermi liquid theory. This screening process, the Kondo effect, is associated with a very low energy coherence scale, the (lattice [24]-) Kondo temperature, considerably renormalised as compared to the bare electronic energy scales.

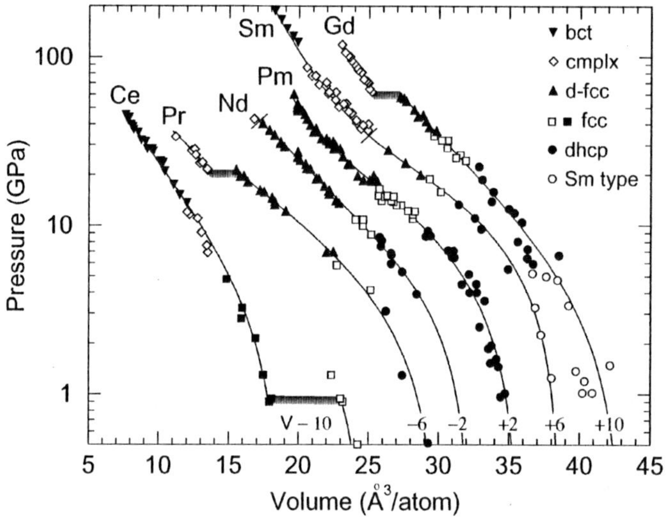

FIGURE 3. Pressure volume data for the rare earths. Structures are identified, with "cmplx" signifiying a number of complex, low-symmetry structures. The volume collapse transitions are marked by the wide hatched lines for Ce, Pr, and Gd, while lines perpendicular to the curves denote the d-fcc to hP3 symmetry change in Nd and Sm. The curves are guides to the eye. Note that the data and curves have been shifted in volume by the numbers (in Å^3/atom) shown at the bottom of the figure. Figure and caption reproduced from Ref. [15].

The periodic Anderson model. The simplest model hamiltonian appropriate for f-electron materials is the Anderson lattice or periodic Anderson model. It retains the f-orbitals associated with the rare-earth or actinide atoms at each lattice site, as well as the relevant conduction electron degrees of freedom which hybridise with those orbitals. In the simplest form, the hamiltonian reads:

$$H_{PAM} = \sum_{\mathbf{k}\sigma} \varepsilon_{\mathbf{k}} c^{\dagger}_{\mathbf{k}\sigma} c_{\mathbf{k}\sigma} + \sum_{\mathbf{k}\sigma m} (V_{\mathbf{k}} c^{\dagger}_{\mathbf{k}\sigma} f_{m\mathbf{k}\sigma} + h.c) + \varepsilon_f \sum_{\mathbf{R}\sigma m} n^f_{\mathbf{R}\sigma m} + U \sum_{\mathbf{R}} \left(\sum_{\sigma m} n^f_{\mathbf{R}\sigma m} \right)^2 \tag{7}$$

Depending on the material considered, other terms may be necessary for increased realism, e.g an orbital dependent f-level ε_{fm}, hybridisation $V_{\mathbf{k}m}$ or interaction matrix $U^{\sigma\sigma'}_{mm'}$ or a direct f-f hopping t_{ff}.

2. DYNAMICAL MEAN-FIELD THEORY AT A GLANCE

Dealing with strong electronic correlations is a notoriously difficult theoretical problem. From the physics point of view, the difficulties come mainly from the wide range of energy scales involved (from the bare electronic energies, on the scale of electron-Volts, to the low-energy physics on the scale of Kelvins) and from the many competing orderings and instabilities associated with small differences in energy.

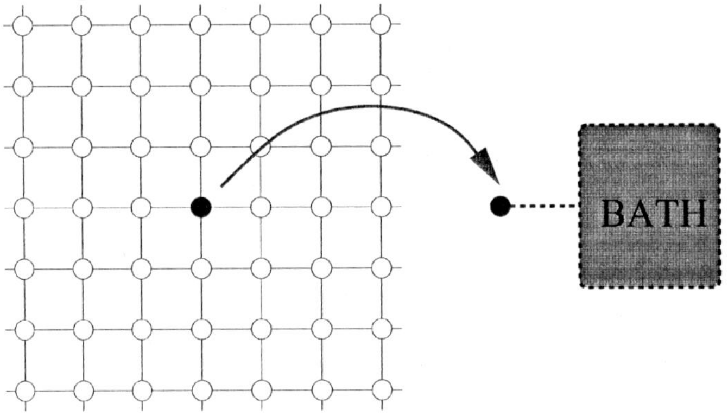

FIGURE 4. Mean-field theory replaces a lattice model by a single site coupled to a self-consistent bath.

It is the opinion of the author that, on top of the essential guidance from physical intuition and phenomenology, the development of quantitative techniques is essential in order to solve the key open questions in the field (and also in order to provide a deeper understanding of some "classic" problems, only partially understood to this day).

In this section, we explain the basic principles of Dynamical Mean-Field Theory (DMFT). This approach has been developed over the last fifteen years and has led to some significant advances in our understanding of strong correlations. In this section, we explain the basic principles of this approach in a concise manner. The Hubbard model is taken as an example. For a much more detailed presentation, the reader is referred to the available review articles [3, 4].

2.1. The mean-field concept, from classical to quantum

Mean-field theory approximates a lattice problem with many degrees of freedom by a *single-site effective problem* with less degrees of freedom. The underlying physical idea is that the dynamics at a given site can be thought of as the interaction of the local degrees of freedom at this site with an external bath created by all other degrees of freedom on other sites (Fig. 4).

Classical mean-field theory. The simplest illustration of this idea is for the Ising model:

$$H = -\sum_{(ij)} J_{ij} S_i S_j - h \sum_i S_i \tag{8}$$

Let us focus on the thermal average of the magnetization on each lattice site: $m_i = \langle S_i \rangle$. We consider an equivalent problem of *independent spins*:

$$H_{eff} = -\sum_i h_i^{eff} S_i \tag{9}$$

in which the (Weiss) effective field is chosen in such a way that the value of m_i is accurately reproduced. This requires:

$$\beta h_i^{eff} = \tanh^{-1} m_i \tag{10}$$

Let us consider, for definiteness, a ferromagnet with nearest-neighbour couplings $J_{ij} = J > 0$. The mean-field theory approximation (first put forward by Pierre Weiss, under the name of "molecular field theory") is that h_i^{eff} can be approximated by the thermal average of the local field seen by the spin at site i, namely:

$$h_i^{eff} \simeq h + \sum_j J_{ij} m_j = h + zJm \tag{11}$$

where z is the connectivity of the lattice, and translation invariance has been used ($J_{ij} = J$ for n.n sites, $m_i = m$). This leads to a self-consistent equation for the magnetization:

$$m = \tanh(\beta h + z\beta J m) \tag{12}$$

We emphasize that replacing the problem of interacting spins by a problem of non-interacting ones in a effective bath is not an approximation, as long as we use this equivalent model for the only purpose of calculating the local magnetizations. The approximation is made when relating the Weiss field to the degrees of freedom on neighbouring sites, i.e in the *self-consistency condition* (11). We shall elaborate further on this point of view in the next section, where exact energy functionals will be discussed. The mean-field approximation becomes *exact* in the limit where the connectivity z of the lattice becomes large. It is quite intuitive indeed that the neighbors of a given site can be treated globally as an external bath when their number becomes large, and that the spatial fluctuations of the local field become negligible.

Generalisation to the quantum case: dynamical mean-field theory. This construction can be extended to quantum many-body systems. Key steps leading to this quantum generalisation where: the introduction of the limit of large lattice coordination for interacting fermion models by Metzner and Vollhardt [25] and the mapping onto a self-consistent quantum impurity by Georges and Kotliar [26], which established the DMFT framework[2].

I explain here the DMFT construction on the simplest example of the Hubbard model[3]:

$$H = -\sum_{ij,\sigma} t_{ij} c_{i\sigma}^{\dagger} c_{j\sigma} + U \sum_i n_{i\uparrow} n_{i\downarrow} + \varepsilon_0 \sum_{i\sigma} n_{i\sigma} \tag{13}$$

As explained above, it describes a collection of single-orbital "atoms" placed at the nodes $\mathbf{R}_i$ of a periodic lattice. The orbitals overlap from site to site, so that the fermions

---

[2] See also the later work in Ref. [27], and Ref. [3] for an extensive list of references.

[3] The energy ε_0 of the single-electron atomic level has been introduced in this section for the sake of pedagogy. Naturally, in the single band case, everything depends only on the energy $\varepsilon_0 - \mu$ with respect to the global chemical potential so that one can set $\varepsilon_0 = 0$

can hop with an amplitude t_{ij}. In the absence of hopping, each "atom" has 4 eigenstates: $|0\rangle, |\uparrow\rangle, |\downarrow\rangle$ and $\uparrow\downarrow\rangle$ with energies 0, ε_0 and $U+2\varepsilon_0$, respectively.

The key quantity on which DMFT focuses is the *local* Green's function at a given lattice site:

$$G_{ii}^{\sigma}(\tau-\tau') \equiv -\langle T c_{i\sigma}(\tau) c_{i\sigma}^{\dagger}(\tau')\rangle \tag{14}$$

In classical mean-field theory, the local magnetization m_i is represented as that of a single spin on site i coupled to an effective Weiss field. In a completely analogous manner, we shall introduce a representation of the local Green's function as that of a *single atom coupled to an effective bath*. This can be described by the hamiltonian of an Anderson impurity model [4]:

$$H_{AIM} = H_{atom} + H_{bath} + H_{coupling} \tag{15}$$

in which:

$$\begin{aligned} H_{atom} &= U\, n_{\uparrow}^{c} n_{\downarrow}^{c} + (\varepsilon_0 - \mu)\,(n_{\uparrow}^{c} + n_{\downarrow}^{c}) \\ H_{bath} &= \textstyle\sum_{l\sigma} \widetilde{\varepsilon}_l\, a_{l\sigma}^{\dagger} a_{l\sigma} \\ H_{coupling} &= \textstyle\sum_{l\sigma} V_l\,(a_{l\sigma}^{\dagger} c_{\sigma} + c_{\sigma}^{\dagger} a_{l\sigma}) \end{aligned} \tag{16}$$

In these expressions, a set of non-interacting fermions (described by the $a_l^{\dagger}$'s) have been introduced, which are the degrees of freedom of the effective bath acting on site $\mathbf{R}_i$. The $\widetilde{\varepsilon}_l$ and V_l's are parameters which should be chosen in such a way that the c-orbital (i.e impurity) Green's function of (16) coincides with the local Green's function of the lattice Hubbard model under consideration. In fact, these parameters enter only through the hybridisation function:

$$\Delta(i\omega_n) = \sum_l \frac{|V_l|^2}{i\omega_n - \widetilde{\varepsilon}_l} \tag{17}$$

This is easily seen when the effective on-site problem is recast in a form which does not explicitly involves the effective bath degrees of freedom. However, this requires the use of an effective action functional integral formalism rather than a simple hamiltonian formalism. Integrating out the bath degrees of freedom one obtains the effective action for the impurity orbital only under the form:

$$S_{eff} = -\int_0^{\beta} d\tau \int_0^{\beta} d\tau' \sum_{\sigma} c_{\sigma}^{+}(\tau)\mathcal{G}_0^{-1}(\tau-\tau')c_{\sigma}(\tau') + U\int_0^{\beta} d\tau\, n_{\uparrow}(\tau) n_{\downarrow}(\tau) \tag{18}$$

in which:

$$\mathcal{G}_0^{-1}(i\omega_n) = i\omega_n + \mu - \varepsilon_0 - \Delta(i\omega_n) \tag{19}$$

[4] Strictly speaking, we have a collection of independent impurity models, one at each lattice site. In this section, for simplicity, we assume a phase with translation invariance and focus on a particular site of the lattice (we therefore drop the site index for the impurity orbital $c_{\sigma}^{\dagger}$). We also assume a paramagnetic phase. The formalism easily generalizes to phases with long-range order (i.e translational and/or spin-symmetry breaking) [3]

This local action represents the effective dynamics of the local site under consideration: a fermion is created on this site at time τ (coming from the "external bath", i.e from the other sites of the lattice) and is destroyed at time τ' (going back to the bath). Whenever two fermions (with opposite spins) are present at the same time, an energy cost U is included. Hence this effective action describes the fluctuations between the 4 atomic states $|0\rangle, |\uparrow\rangle, |\downarrow\rangle, \uparrow\downarrow\rangle$ induced by the coupling to the bath. We can interpret $\mathcal{G}_0(\tau - \tau')$ as the quantum generalisation of the Weiss effective field in the classical case. The main difference with the classical case is that this "dynamical mean-field" is a *function of energy* (or time) instead of a single number. This is required in order to take full account of local quantum fluctuations, which is the main purpose of DMFT. $\mathcal{G}_0$ also plays the role of a bare Green's function for the effective action S_{eff}, but it should *not be confused* with the non-interacting ($U = 0$) local Green's function of the original lattice model.

At this point, we have introduced the quantum generalisation of the Weiss effective field and have represented the local Green's function G_{ii} as that of a single atom coupled to an effective bath. This can be viewed as an *exact representation*, as further detailed in Sec. 3. We now have to generalise to the quantum case the mean-field *approximation* relating the Weiss function to G_{ii} (in the classical case, this is the self-consistency relation (12)). The simplest manner in which this can be explained - but perhaps not the more illuminating one conceptually (see Sec. 3 and [3, 28])- is to observe that, in the effective impurity model (18), we can define a local self-energy from the interacting Green's function $G(\tau - \tau') \equiv - < Tc(\tau)c^+(\tau') >_{S_{eff}}$ and the Weiss dynamical mean-field as:

$$\begin{aligned} \Sigma_{imp}(i\omega_n) &\equiv \mathcal{G}_0^{-1}(i\omega_n) - G^{-1}(i\omega_n) \\ &= i\omega_n + \mu - \varepsilon_0 - \Delta(i\omega_n) - G^{-1}(i\omega_n) \end{aligned} \tag{20}$$

Let us, on the other hand, consider the self-energy of the original lattice model, defined as usual from the full Green's function $G_{ij}(\tau - \tau') \equiv - < Tc_{i,\sigma}(\tau)c^+_{j,\sigma}(\tau') >$ by:

$$G(\mathbf{k}, i\omega_n) = \frac{1}{i\omega_n + \mu - \varepsilon_0 - \varepsilon_{\mathbf{k}} - \Sigma(\mathbf{k}, i\omega_n)} \tag{21}$$

in which $\varepsilon_{\mathbf{k}}$ is the Fourier transform of the hopping integral, i.e the dispersion relation of the non-interacting tight-binding band:

$$\varepsilon_{\mathbf{k}} \equiv \sum_j t_{ij} e^{i\mathbf{k}.(\mathbf{R}_i - \mathbf{R}_j)} \tag{22}$$

We then make the approximation that the lattice self-energy coincides with the impurity self-energy. In real-space, this means that we neglect all non-local components of Σ_{ij} and approximate the on-site one by Σ_{imp}:

$$\Sigma_{ii} \simeq \Sigma_{imp} \ , \ \ \Sigma_{i \neq j} \simeq 0 \tag{23}$$

We immediately see that this is a consistent approximation only provided it leads to a unique determination of the local (on-site) Green's function, which by construction is

TABLE 1. Correspondance between the mean-field theory of a classical system and the dynamical mean-field theory of a quantum system.

Quantum Case	Classical Case	
$-\sum_{ij\sigma} t_{ij} c^{+}_{i\sigma} c_{j\sigma} + \sum_i H_{atom}(i)$	$H = -\sum_{(ij)} J_{ij} S_i S_j - h \sum_i S_i$	Hamiltonian
$G_{ii}(i\omega_n) = - < c_i^{+}(i\omega_n) c_i(i\omega_n) >$	$m_i = < S_i >$	Local Observable
$H_{eff} = H_{atom} + \sum_{l\sigma} \tilde{\varepsilon}_l a^{+}_{l\sigma} a_{l\sigma} +$ $+ \sum_{l\sigma} V_l (a^{+}_{l\sigma} c_\sigma + h.c)$	$H_{eff} = -h_{eff} S$	Effective single-site Hamiltonian
$\Delta(i\omega_n) = \sum_l \frac{\lvert V_l\rvert^2}{i\omega_n - \tilde{\varepsilon}_l}$ $\mathcal{G}_0^{-1}(i\omega_n) \equiv i\omega_n + \mu - \Delta(i\omega_n)$	h_{eff}	Weiss function/Weiss field
$\sum_{\mathbf{k}} [\Delta(i\omega_n) + G(i\omega_n)^{-1} - \varepsilon_{\mathbf{k}}]^{-1} = G(i\omega_n)$	$h_{eff} = \sum_j J_{ij} m_j + h$	Self-consistency relation

the impurity-model Green's function. Summing (21) over $\mathbf{k}$ in order to obtain the on-site component G_{ii} of the the lattice Green's function, and using (20), we arrive at the self-consistency condition[5]:

$$\sum_{\mathbf{k}} \frac{1}{\Delta(i\omega_n) + G(i\omega_n)^{-1} - \varepsilon_{\mathbf{k}}} = G(i\omega_n) \tag{24}$$

Defining the non-interacting density of states:

$$D(\varepsilon) \equiv \sum_{\mathbf{k}} \delta(\varepsilon - \varepsilon_{\mathbf{k}}) \tag{25}$$

this can also be written as:

$$\int d\varepsilon \frac{D(\varepsilon)}{\Delta(i\omega_n) + G(i\omega_n)^{-1} - \varepsilon} = G(i\omega_n) \tag{26}$$

This *self-consistency condition* relates, for each frequency, the dynamical mean-field $\Delta(i\omega_n)$ and the local Green's function $G(i\omega_n)$. Furthermore, $G(i\omega_n)$ is the interacting Green's function of the effective impurity model (16) -or (18)-. Therefore, we have a closed set of equations that fully determine in principle the two functions Δ, G (or $\mathcal{G}_0, G$)). In practice, one will use an *iterative procedure*, as represented on Fig. 5. In many cases, this iterative procedure converges to a unique solution independently of the initial choice of $\Delta(i\omega_n)$. In some cases however, more than one stable solution can be found (e.g close to the Mott transition, see section below). The close analogy between the classical mean-field construction and its quantum (dynamical mean-field) counterpart is summarized in Table 1.

[5] Throughout these notes, the sums over momentum are normalized by the volume of the Brillouin zone, i.e $\sum_{\mathbf{k}} 1 = 1$

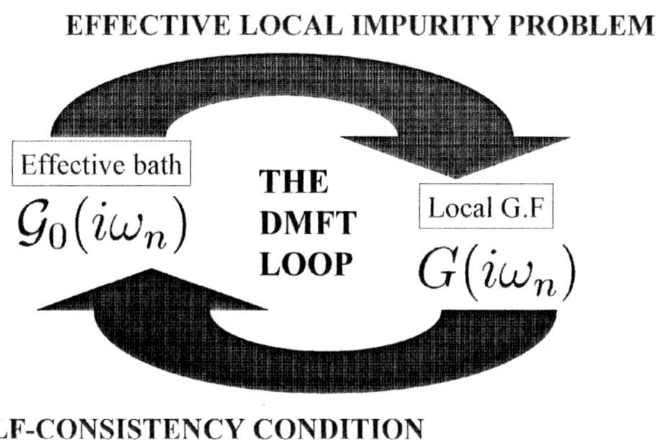

FIGURE 5. The DMFT iterative loop. The following procedure is generally used in practice: starting from an initial guess for $\mathcal{G}_0$, the impurity Green's function G_{imp} is calculated by using an appropriate solver for the impurity model (top arrow). The impurity self-energy is also calculated from $\Sigma_{imp} = \mathcal{G}_0^{-1}(i\omega_n) - G_{imp}^{-1}(i\omega_n)$. This is used in order to obtain the on-site Green's function of the lattice model by performing a **k**-summation (or integration over the free d.o.s): $G_{loc} = \sum_{\mathbf{k}}[i\omega_n + \mu - \varepsilon_{\mathbf{k}} - \Sigma_{imp}(i\omega_n)]^{-1}$. An updated Weiss function is then obtained as $\mathcal{G}_{0,new}^{-1} = G_{loc}^{-1} + \Sigma_{imp}$, which is injected again into the impurity solver (bottom arrow). The procedure is iterated until convergence is reached.

2.2. Limits in which DMFT becomes exact

Two simple limits: non-interacting band and isolated atoms. It is instructive to check that the DMFT equations yield the exact answer in two simple limits:

- In the *non-interacting limit* $U = 0$, solving (18) yields $G(i\omega_n) = \mathcal{G}_0(i\omega_n)$ and $\Sigma_{imp} = 0$. Hence, from (24), $G(i\omega_n) = \sum_{\mathbf{k}} 1/(i\omega_n + \mu - \varepsilon_0 - \varepsilon_{\mathbf{k}})$ reduces to the free on-site Green's function. DMFT is trivially exact in this limit since the self-energy is not only **k**-independent but vanishes altogether.
- In the *atomic limit* $t_{ij} = 0$, one just has a collection of independent atoms on each site and $\varepsilon_{\mathbf{k}} = 0$. Then (24) implies $\Delta(i\omega_n) = 0$: as expected, the dynamical mean-field vanishes since the atoms are isolated. Accordingly, the self-energy only has on-site components, and hence DMFT is again exact in this limit. The Weiss field reads $\mathcal{G}_0^{-1} = i\omega_n + \mu - \varepsilon_0$, which means that the action S_{eff} simply corresponds to the quantization of the atomic hamiltonian H_{atom}. This yields:

$$G(i\omega_n)_{atom} = \frac{1-n/2}{i\omega_n+\tilde{\mu}} + \frac{n/2}{i\omega_n+\tilde{\mu}-U}$$
$$\Sigma(i\omega_n)_{atom} = \frac{nU}{2} + \frac{n/2(1-n/2)U^2}{i\omega_n+\tilde{\mu}-(1-n/2)U} \tag{27}$$

with $\tilde{\mu} \equiv \mu - \varepsilon_0$ and $n/2 = (e^{\beta\tilde{\mu}} + e^{\beta(2\tilde{\mu}-U)})/(1+2e^{\beta\tilde{\mu}} + e^{\beta(2\tilde{\mu}-U)})$.

Hence, the dynamical mean-field approximation is exact in the two limits of the non-interacting band and of isolated atoms, and provides an interpolation in between. This interpolative aspect is a key to the success of this approach in the intermediate coupling regime.

Infinite coordination. The dynamical mean-field approximation becomes exact in the limit where the connectivity z of the lattice is taken to infinity. This is also true of the mean-field approximation in classical statistical mechanics. In that case, the exchange coupling between nearest-neighbour sites must be scaled as: $J_{ij} = J/z$ (for J_{ij}'s of uniform sign), so that the Weiss mean-field h_{eff} in (11) remains of order one. This also insures that the entropy and internal energy per site remain finite and hence preserves the competition which is essential to the physics of magnetic ordering. In the case of itinerant quantum systems [25], a similar scaling must be made on the hopping term in order to maintain the balance between the kinetic and interaction energy. The nearest-neighbour hopping amplitude must be scaled as: $t_{ij} = t/\sqrt{z}$. This insures that the non-interacting d.o.s $D(\varepsilon) = \sum_{\mathbf{k}} \delta(\varepsilon - \varepsilon_{\mathbf{k}})$ has a non-trivial limit as $z \to \infty$. Note that it also insures that the superexchange $J_{ij} \propto t_{ij}^2/U$ scales as $1/z$, so that magnetic ordering is preserved with transition temperatures of order unity. In practice, two lattices are often considered in the $z = \infty$ limit:

- The d-dimensional cubic lattice with $z = 2d \to \infty$ and $\varepsilon_{\mathbf{k}} = -2t \sum_{p=1}^{d} \cos(k_p)/\sqrt{z}$. In this case the non-interacting d.o.s becomes a Gaussian: $D(\varepsilon) = \frac{1}{t\sqrt{2\pi}} \exp -(\frac{\varepsilon^2}{2t^2})$
- The Bethe lattice (Cayley tree) with coordination $z \to \infty$ and nearest-neighbor hopping $t_{ij} = t/\sqrt{z}$. This corresponds to a semicircular d.o.s: $D(\varepsilon) = \frac{2}{\pi D}\sqrt{1-(\varepsilon/D)^2}$ with a half-bandwidth $D = 2t$. In this case, the self-consistency condition (24) can be inverted explicitly in order to relate the dynamical mean-field to the local Green's function as: $\Delta(i\omega_n) = t^2 G(i\omega_n)$.

Apart from the intrinsic interest of solving strongly correlated fermion models in the limit of infinite coordination, the fact that the DMFT equations become exact in this limit is important since it guarantees, for example, that exact constraints (such as causality of the self-energy, positivity of the spectral functions, sum rules such as the Luttinger theorem or the f-sum rule) are preserved by the DMFT approximation.

2.3. Important topics not reviewed here

There are several important topics related to the DMFT framework, which I have not included in these lecture notes. Some of them were covered in the lectures, but extensive review articles are available in which these topics are at least partially described.
This is a brief list of such topics:

DMFT for ordered phases. The DMFT equations can easily be extended to study phases with long-range order, calculate critical temperatures for ordering as well as phase diagrams, see e.g [3].

Response and correlation functions in DMFT. Response and correlation functions can be expressed in terms of the lattice Green's functions, and of the impurity model vertex functions, see e.g [3, 4]. Note that momentum-dependence enters, through the lattice Green's function.

Physics of the Anderson impurity model. Understanding the various possible fixed points of quantum impurity models is important for gaining physical intuition when solving lattice models within DMFT. See Ref. [23] for a review and references on the Anderson impurity model. It is important to keep in mind that, in contrast to the common situation in the physics of magnetic impurities or mesoscopics, the effective conduction electron bath in the DMFT context has significant energy-dependence. Also, the self-consistency condition can drive the effective impurity model from one kind of low-energy behaviour to another, depending on the range of parameters (e.g close to the Mott transition, see Sec. 4).

Impurity solvers. Using reliable methods for calculating the impurity Green's function and self-energy is a key step in solving the DMFT equations. A large numbers of "impurity solvers" have been implemented in the DMFT context[6], including: the quantum Monte Carlo (QMC) method [27] (see also [29, 30]), based on the Hirsch-Fye algorithm [31], adaptative exact diagonalisation or projective schemes (see [3] for a review and references), the Wilson numerical renormalisation group (NRG, see e.g [32] and references therein). Approximation schemes have also proven useful, when used in appropriate regimes, such as the "iterated perturbation theory" approximation (IPT, [26, 33]), the non-crossing approximation (NCA, see [4] for references) and various extensions [34], as well as schemes interpolating between high and low energies [35].

Beyond DMFT. DMFT does capture ordered phases, but does not take into account the coupling of short-range spatial correlations (let alone long-wavelength) to quasiparticle properties, in the absence of ordering. This is a key aspect of some strongly correlated materials (e.g cuprates, see the concluding section of these lectures), which requires an extension of the DMFT formalism. Two kinds of extensions have been explored:

- **k**-dependence of the self-energy can be reintroduced by considering cluster extensions of DMFT, i.e a small cluster of sites (or coupled atoms) into a self-consistent bath. Various embedding schemes have been discussed [3, 36, 37, 38, 39, 40, 41] and I will not attempt a review of this very interesting line of research here. One of the key questions is whether such schemes can account for a strong variation of the quasiparticle properties (e.g the coherence scale) along the Fermi surface.
- Extended DMFT (E-DMFT [42, 43, 44, 45]) focuses on two-particle local observables, such as the local spin or charge correlation functions, in addition to the local Green's function of usual DMFT. For applications to electronic structure, see Sec. 5.5.

---

[6] Some early versions of numerical codes are available at: http://www.lps.ens.fr/∼krauth

3. FUNCTIONALS, LOCAL OBSERVABLES, AND INTERACTING SYSTEMS

In this section[7], I would like to discuss a theoretical framework which applies quite generally to interacting systems. This framework reveals common concepts underlying different theories such as: the Weiss mean-field theory (MFT) of a classical magnet, the density functional theory (DFT) of the inhomogeneous electron gas in solids, and the dynamical mean-field theory (DMFT) of strongly correlated electron systems. The idea which is common to these diverse theories is the construction of *a functional of some local quantity* (effective action) by the Legendre transform method. Though exact in principle, it requires in practice that the exact functional is approximated in some manner. This method has a wide range of applicability in statistical mechanics, many-body physics and field-theory [46]. The discussion will be (hopefully) pedagogical, and for this reason I will begin with the example of a classical magnet. For a somewhat more detailed presentation, see Ref. [28].

There are common concepts underlying all these constructions (cf. Table), as will become clear below, namely:

- i) These theories focus on a specific *local quantity*: the local magnetization in MFT, the local electronic density in DFT, the local Green's function (or spectral density) in DMFT.
- ii) The original system of interest is replaced by an *equivalent system*, which is used to provide a representation of the selected quantity: a single spin in an effective field for a classical magnet, free electrons in an effective one-body potential in DFT, a single impurity Anderson model within DMFT. The effective parameters entering this equivalent problem define *generalized Weiss fields* (the Kohn-Sham potential in DFT, the effective hybridization within DMFT), which are self-consistently adjusted. I note that the associated equivalent system can be a non-interacting (one-body) problem, as in MFT and DFT, or a fully interacting many-body problem (albeit simpler than the original system) such as in DMFT and its extensions.
- iii) In order to pave the way between the real problem of interest and the equivalent model, the method of coupling constant integration will prove to be very useful in constructing (formally) the desired functional using the Legendre transform method. The coupling constant can be either the coefficient of the interacting part of the hamiltonian (which leads to a non-interacting equivalent problem, as in DFT), or in front of the non-local part of the hamiltonian (which leads in general to a local, but interacting, equivalent problem such as in DMFT).

Some issues and questions are associated with each of these points:

- i) While the theory and associated functional primarily aims at calculating the selected local quantity, it always come with the possibility of determining some more general object. For example, classical MFT aims primarily at calculating the local magnetization, but it can be used to derive the Ornstein-Zernike expression of the

---

[7] This section is based in part on Ref. [28]

TABLE 2. Comparison of theories based on functionals of a local observable

Theory	MFT	DFT	DMFT
Quantity	Local magnetization m_i	Local density $n(x)$	Local GF $G_{ii}(\omega)$
Equivalent system	Spin in effective field	Electrons in effective potential	Quantum impurity model
Generalised Weiss field	Effective local field	Kohn-Sham potential	Effective hybridisation

correlation function between different sites. Similarly, DFT aims at the local density, but Kohn-Sham orbitals can be *interpreted* (without a firm formal justification) as one-electron excitations. DMFT produces a local self-energy which one may interpret as the lattice self-energy from which the full k-dependent Green's function can be reconstructed. In each of these cases, the precise status and interpretation of these additional quantities can be questioned.

- ii) I emphasize that the choice of an equivalent representation of the local quantity has nothing to do with subsequent approximations made on the functional. The proposed equivalent system is in fact an exact representation of the problem under consideration (for the sake of calculating the selected local quantity). It does raise a *representability* issue, however: is it always possible to find values of the generalised Weiss field which will lead to a specified form of the local quantity, and in particular to the exact form associated with the specific system of interest? For example: given the local electronic density $n(x)$ of a specific solid, can one always find a Kohn-Sham effective potential such that the one-electron local density obtained by solving the Schrödinger equation in that potential coincides with $n(x)$? Or, in the context of DMFT: given the local Green's function of a specific model, can one find a hybridisation function such that it can be viewed as the local Green's function of the specified impurity problem ?
- iii) There is also a stability issue of the exact functional: is the equilibrium value of the local quantity a minimum ? More precisely, one would like to show that negative eigenvalues of the stability matrix correspond to true physical instabilities of the system. I will not seriously investigate this issue in this lecture (for a discussion within DMFT, where it is still quite open, see [47]).

3.1. The example of a classical magnet

For the sake of pedagogy, I will consider in this section the simplest example on which the above ideas can be made concrete: that of a classical Ising magnet with hamiltonian

$$H = -\sum_{ij} J_{ij} S_i S_j \qquad (28)$$

Construction of the effective action. We want to construct a functional $\Gamma[m_i]$ of a *preassigned* set of local magnetizations m_i, such that minimizing this functional yields the equilibrium state of the system. This functional is of course the Legendre transform of the free-energy with respect to a set of local magnetic fields. To make contact with the field-theory literature, I note that $\beta\Gamma$ is generally called the *effective action* in this context. I will give a formal construction of this functional, following a method due to Plefka [48] and Yedidia and myself [49]. Let us introduce a varying coupling constant $\alpha \in [0,1]$, and define:

$$H_\alpha \equiv \alpha H = \sum_{ij} \alpha J_{ij} S_i S_j \tag{29}$$

Introducing local Lagrange multipliers λ_i, we consider the functional:

$$\Omega[m_i, \lambda_i; \alpha] \equiv -\frac{1}{\beta} \ln \mathrm{Tr}\, e^{-\beta H_\alpha + \beta \sum_i \lambda_i (S_i - m_i)} = F[\lambda_i] + \sum_i \lambda_i m_i \tag{30}$$

Requesting stationarity of this functional with respect to the λ_i's amounts to impose that, *for all values of* α, $\langle S_i \rangle$ coincides with the preassigned local magnetization m_i. The equations $m_i = \langle S_i \rangle$ which expresses the magnetization as a function of the sources λ_i can then be inverted to yield the λ_i's as functions of the m_j's and of α:

$$\langle S_i \rangle_{\lambda,\alpha} = m_i \rightarrow \lambda_i = \lambda_i[m_j; \alpha] \tag{31}$$

(The average $\langle \cdots \rangle_{\lambda,\alpha}$ in this equation is with respect to the Boltzmann weight appearing in the above definition of Ω, including λ_i's and α). The Lagrange parameters can then be substituted into Ω to obtain the α-dependent Legendre transformed functional:

$$\Gamma_\alpha[m_i] = \Omega[m_i, \lambda_i[m_j, \alpha]] = F[\lambda_i[m]] + \sum_i \lambda_i[m] m_i \tag{32}$$

Of course, the functional we are really interested in is that of the original system with $\alpha = 1$, namely:

$$\Gamma[m_i] \equiv \Gamma_{\alpha=1}[m_i] \tag{33}$$

Let us first look at the non-interacting limit $\alpha = 0$ for which the explicit expression of Ω is easily obtained as:

$$\Omega_0 = \sum_i \left(-\frac{1}{\beta} \ln\cosh \beta\lambda_i + m_i \lambda_i \right) \tag{34}$$

Varying in the λ's yields:

$$\tanh \beta \lambda_i^{(\alpha=0)} = m_i \tag{35}$$

and finally:

$$\Gamma_{\alpha=0}[m_i] = \frac{1}{\beta} \sum_i \left(\frac{1+m_i}{2} \ln \frac{1+m_i}{2} + \frac{1-m_i}{2} \ln \frac{1-m_i}{2} \right) \tag{36}$$

The $\alpha = 0$ theory defines the *equivalent problem* that we want to use in order to deal with the original system. Here, it is just a theory of *independent spins in a local effective field*. The expression (36) is simply the entropy term corresponding to independent Ising spins for a given values of the local magnetizations.

The value taken by the Lagrange multiplier in the equivalent system, $\lambda_i^{\alpha=0}$ (denoted λ_i^0 in the following), must be interpreted as the *Weiss effective field*. We note that, in this simple example, there is an explicit and very simple relation (35) between the Weiss field and m_i, so that one can work *equivalently* in terms of either quantities. Also, because of the simple form of (35), *representability* is trivially satisfied: given the actual values of the magnetizations m_i's ($\in [-1,1]$) at equilibrium for the model under consideration, one can always represent them by the Weiss fields $\beta h_i^{eff} = \text{arctanh}\, m_i$.

To proceed with the construction of Γ, we use a coupling constant integration and write:

$$\Gamma[m_i;\alpha=1] = \Gamma_0[m_i] + \int_0^1 d\alpha \frac{d\Gamma_\alpha}{d\alpha}[m_i] \tag{37}$$

It is immediate that, because of the constraint $\langle (S_i - m_i) \rangle = 0$:

$$\frac{d\Gamma_\alpha}{d\alpha} = \langle H \rangle_{\alpha,\lambda[\alpha]} = -\sum_{ij} J_{ij} \langle S_i S_j \rangle_{\alpha,\lambda[\alpha,m]} \tag{38}$$

In this expression, the correlation must be viewed as a functional of the local magnetizations (thanks to the inversion formula (31)). Introducing the connected correlation function:

$$g_{ij}^c[\{m_k\};\alpha] \equiv \langle (S_i - m_i)(S_j - m_j) \rangle_{\alpha,\lambda[\alpha,m]} \tag{39}$$

we obtain:

$$\frac{d\Gamma_\alpha}{d\alpha} = -\sum_{ij} J_{ij} m_i m_j - \sum_{ij} J_{ij} g_{ij}^c[\{m_k\};\alpha] \tag{40}$$

So that finally, one obtains the formal expression for $\Gamma[m_i] \equiv \Gamma_{\alpha=1}[m_i]$:

$$\Gamma[m_i] = \Gamma_0[m_i] - \sum_{ij} J_{ij} m_i m_j - \sum_{ij} J_{ij} \int_0^1 d\alpha g_{ij}^c[m_k;\alpha] \equiv \Gamma_0 + E_{MF} + \Gamma_{corr} \tag{41}$$

In this expression, g^c denotes the connected correlation function for a given value of the coupling constant, *expressed as a functional of the local magnetisations*.

Hence, the *exact functional* Γ appears as a sum of three contributions:

- The part associated with the equivalent system (corresponding here to the entropy of constrained but otherwise free spins)
- The mean-field energy $\sum_{ij} J_{ij} m_i m_j$
- A contribution from correlations which contains all corrections beyond mean-field

As explained in the next section, there is a direct analogy between this and the various contributions to the density functional within DFT (kinetic energy, Hartree energy and exchange-correlation).

I note in passing that one can derive a closed equation for the exact functional, which reads (see [28] for a derivation):

$$\Gamma_\alpha[m_i] = \Gamma_0[m_i] - \alpha \sum_{ij} J_{ij} m_i m_j - \frac{1}{\beta} \sum_{ij} J_{ij} \int_0^\alpha d\alpha' \left[\frac{\delta^2 \Gamma_{\alpha'}}{\delta m_k \delta m_l} \right]^{-1}_{ij} \tag{42}$$

This equation fully determines in principle the effective action functional. However, in order to use it in practice, one generally has to start from a limit in which the functional is known explicitly, and expand around that limit. For example, an expansion around the high-temperature limit yields systematic corrections to mean-field theory [49, 28]. This equation is closely related [28] to the Wilson-Polchinsky equation [50] for the effective action (after a Legendre transformation: see also [51]), which can be taken as a starting point for a renormalisation group analysis by starting from the local limit and expanding in the "locality" (see e.g [52, 51].

Equilibrium condition and stability. The physical values of the magnetisations at equilibrium are obtained by minimising Γ, which yields:

$$m_i^* = \tanh\left(\beta \sum_j J_{ij} m_j^* - \beta \frac{\delta \Gamma_{corr}}{\delta m_i} \right) \tag{43}$$

and the Weiss field takes the following value:

$$h_i^{eff} \equiv (\lambda_i^0)^* = \sum_j J_{ij} m_j^* - \frac{\delta \Gamma_{corr}}{\delta m_i} |^* \tag{44}$$

This equation is a *self-consistency condition* which determines the Weiss field in terms of the local magnetizations on all other sites. Its physical interpretation is clear: h_i^{eff} is the true (average) local field seen by site i. It is equal to the sum of two terms: one in which all spins are treated as independent, and a correction due to correlations.

The stability of the functional around equilibrium is controlled by the fluctuation matrix:

$$\frac{\delta^2 \Gamma}{\delta m_i \delta m_j} = \frac{\delta \lambda_i^0}{\delta m_j} - J_{ij} + \frac{\delta^2 \Gamma_{corr}}{\delta m_i \delta m_j} \tag{45}$$

At equilibrium, this is nothing else than the inverse of the susceptibility (or correlation function) matrix:

$$\frac{\delta^2 \Gamma}{\delta m_i \delta m_j} \equiv (\chi^{-1})_{ij} = (\chi_0^{-1})_{ij} - J_{ij} + \frac{\delta^2 A_{corr}}{\delta m_i \delta m_j} \tag{46}$$

with:

$$(\chi_0^{-1})_{ij} = \frac{1}{\beta(1-m_i^2)} \delta_{ij} \tag{47}$$

Hence, our functional does satisfy a stability criterion as defined in the introduction: a negative eigenvalue of this matrix (i.e of $\chi(\vec{q})$) would correspond to a physical instability of the system. Note that at the simple mean-fied level, we recover the RPA formula for the susceptibility: $(\chi^{-1})_{ij} = (\chi_0^{-1})_{ij} - J_{ij}$.

Mean-field approximation and beyond. Obviously, this construction of the *exact* Legendre transformed free energy, and the exact equilibrium condition (43) has formal value, but concrete applications require some further approximations to be made on the correlation term Γ_{corr}. The simplest such approximation is just to neglect Γ_{corr} altogether. This is the familiar Weiss mean-field theory:

$$\Gamma_{MFT} = \frac{1}{\beta}\sum_i \left(\frac{1+m_i}{2}\ln\frac{1+m_i}{2} + \frac{1-m_i}{2}\ln\frac{1-m_i}{2}\right) - \sum_{ij} J_{ij} m_i m_j \tag{48}$$

For a ferromagnet (uniform positive J_{ij}'s), this approximation becomes *exact in the limit of infinite coordination* of the lattice.

The formal construction above is a useful guideline when trying to improve on the mean-field approximation. I emphasize that, within the present approach, *it is the self-consistency condition (44) (relating the Weiss field to the environment) that needs to be corrected*, while the equation $m_i = \tanh \beta h_i^{eff}$ is attached to our choice of equivalent system and will be always valid. For example, in [48, 49] it was shown how to construct Γ_{corr} by a systematic high-temperature expansion in β. This expansion can be conveniently generated by iterating the exact equation (42). It can also be turned into an expansion around the limit of infinite coordination [49]. The first contribution to Γ_{corr} in this expansion appears at order β (or α^2) and reads:

$$\Gamma^{(1)}_{corr} = -\frac{\beta}{2}\sum_{ij} J_{ij}^2 (1-m_i^2)(1-m_j^2) \tag{49}$$

This is a rather famous correction to mean-field theory, known as the "Onsager reaction term". For spin glass models (J_{ij}'s of random sign), it is crucial to include this term even in the large connectivity limit. The corresponding equations for the equilibrium magnetizations are those derived by Thouless, Anderson and Palmer [53].

3.2. Density functional theory

In this section, I explain how density-functional theory [8] (DFT) [55, 56] can be derived along very similar lines. This section borrows from the work of Fukuda et al. [57, 46] and of Valiev and Fernando [58]. For a recent pedagogical review emphasizing this point of view, see [59]. For detailed reviews of the DFT formalism, see e.g [60, 61].

Let us consider the inhomogeneous electron gas of a solid, with hamiltonian:

$$H = -\sum_i \frac{1}{2}\nabla_i^2 + \sum_i v(\mathbf{r}_i) + \frac{1}{2}\sum_{i\neq j} U(\mathbf{r}_i - \mathbf{r}_j) \tag{50}$$

in which $v(x)$ is the external potential due to the nuclei and $U(x-x')$ $(= e^2/|x-x'|)$ is the electron-electron interaction. (I use conventions in which $\hbar = m = 1$). Let us write

[8] I actually consider the finite-temperature extension of DFT [54]

this hamiltonian in second- quantized form, and again introduce a coupling-constant parameter α (the physical case is $\alpha = 1$):

$$H_\alpha = -\frac{1}{2}\int dx\, \psi^\dagger \nabla^2 \psi + \int dx\, v(x)\hat{n}(x) + \frac{\alpha}{2}\int dx dx'\, \hat{n}(x) U(x-x')\hat{n}(x') \qquad (51)$$

We want to construct the free energy functional of the system while constraining the average density to be equal to some specified function $n(x)$. In complete analogy with the previous section, we introduce a Lagrange multiplier function $\lambda(x)$, and consider [9]:

$$\Omega_\alpha[n(x), \lambda(x)] \equiv -\frac{1}{\beta}\ln \mathrm{Tr}\exp\left(-\beta H_\alpha + \beta\int dx\, \lambda(x)(n(x) - \hat{n}(x))\right) \qquad (52)$$

A functional of *both* $n(x)$ and $\lambda(x)$. As before, stationarity in λ insures that:

$$\langle \hat{n}(x)\rangle_{\lambda,\alpha} = n(x) \;\rightarrow\; \lambda(x) = \lambda_\alpha[n(x)] \qquad (53)$$

This will be used to eliminate $\lambda(x)$ in terms of $n(x)$ and construct the functional of $n(x)$ only:

$$\Gamma_\alpha[n(x)] \equiv \Omega_\alpha[n(x), \lambda_\alpha[n(x)]] \qquad (54)$$

3.2.1. Equivalent system: non-interacting electrons in an effective potential

Again, I first look at the non-interacting case $\alpha = 0$. Then we have to solve a one-particle problem in an x-dependent external potential. This yields:

$$\Omega_0[n[x], \lambda[x]] = -\mathrm{tr}\,\ln[i\omega_n - \hat{t} - \hat{v} - \hat{\lambda}] - \int dx\, \lambda(x) n(x) \qquad (55)$$

In this equation, tr denotes the trace over the degrees of freedom of a single electron, $i\omega_n$ is the usual Matsubara frequency, and $\hat{t} \equiv -\nabla^2/2$, $\hat{v}$, $\hat{\lambda}$ are the one-body operators corresponding to the kinetic energy, external potential and $\lambda(x)$ respectively. The identity $\ln det = \mathrm{tr}\,\ln$ has been used.

Minimisation with respect to $\lambda(x)$ yields the following relation between λ^0 and $n(x)$:

$$\frac{1}{\beta}\sum_n \langle x| \frac{1}{i\omega_n - \hat{t} - \hat{v} - \hat{\lambda}_0} |x\rangle = n(x) \qquad (56)$$

This defines the functional $\lambda_0[n(x)]$, albeit in a somewhat implicit manner. This is directly analogous to Eq.(35) defining the Weiss field in the Ising case (but in that case,

[9] Note that I chose in this expression a different sign convention for λ than in the previous section, and also that Tr denotes the full many-body trace over all N-electrons degrees of freedom.

this equation was easily invertible). If we want to be more explicit, what we have to do is solve the one-particle Schrodinger equation:

$$\left(-\frac{1}{2}\Delta + v_{KS}(x)\right)\phi_l(x) = \varepsilon_l \phi_l(x) \tag{57}$$

where the *effective one-body potential* (Kohn-Sham potential) is *defined* as:

$$v_{KS}(x) \equiv v(x) + \lambda^0(x) \tag{58}$$

It is convenient to construct the associated resolvent:

$$R(x,x';i\omega_n) = \sum_l \frac{\phi_l(x)\phi_l^*(x')}{i\omega_n - \varepsilon_l} \tag{59}$$

and the relation (56) now reads:

$$\sum_l |\phi_l(x)|^2 f_{FD}(\varepsilon_l) = n(x) \tag{60}$$

in which f_{FD} is the Fermi-Dirac distribution.

This relation expresses the local density in an interacting many-particle system as that of a one-electron problem *in an effective potential* defined by (56). In so doing, the effective one-particle wave functions and energies (Kohn-Sham orbitals) have been introduced, whose relation to the original system (and in particular their interpretation as excitation energies) is far from obvious (see e.g [61]). There is, for example, no fundamental justification in identifying the resolvent (59) with the true one-electron Green's function of the interacting system. The issue of *representability* (i.e whether an effective potential can always be found given a density profile $n(x)$) is far from being as obvious as in the previous section, but has been established on a rigorous basis [62, 63].

To summarize, the non-interacting functional $\Gamma_0[n(x)]$ reads:

$$\Gamma_0[n(x)] = -\text{tr}\ln[i\omega_n - \hat{t} - \hat{v} - \hat{\lambda}_0[n]] - \int dx\,\lambda^0[x;n]n(x) \tag{61}$$

which can be rewritten as:

$$\Gamma_0[n(x)] == -\frac{1}{\beta}\sum_l \ln\left[1 + e^{-\beta\varepsilon_l[n]}\right] - \int dx\,v_{KS}(x)n(x) + \int dx\,v(x)n(x) \tag{62}$$

in which λ_0 and v_{KS} are viewed as a functional of $n(x)$, as detailed above.

In the limit of zero temperature ($\beta \to \infty$), this reads:

$$\Gamma_0[n(x), T=0] = \sum_l^{l} \varepsilon_l - \int dx\,v_{KS}(x)n(x) + \int dx\,v(x)n(x) \tag{63}$$

in which the sum is over the N occupied Kohn-Sham states. We note that it contains extra terms beyond the ground-state energy of the KS equivalent system (see also Sec. 5.4).

We also note that Γ_0 is not a very explicit functional of $n(x)$. It is a somewhat more explicit functional of $\lambda_0(x)$ (or equivalently of the KS effective potential $v_{KS}(x)$) so that it is often more convenient to think in terms of this quantity directly. At any rate, in order to evaluate Γ_0 for a specific density profile or effective potential one must solve the Schrödinger equation for KS orbitals and eigenenergies. This is a time-consuming task for realistic three-dimensional potentials and practical calculations would be greatly facilitated if a more explicit accurate expression for $\Gamma[n(x)]$ would be available [10].

3.2.2. *The exchange-correlation functional*

We turn to the interacting theory, and use the coupling constant integration method (see [64] for its use in DFT):

$$\Gamma[n(x)] = \Gamma[n(x); \alpha = 0] + \int_0^1 d\alpha \frac{d\Gamma_\alpha}{d\alpha} \tag{64}$$

Similarly as before:

$$\frac{d\Gamma_\alpha}{d\alpha} = \langle \hat{U} \rangle_{\lambda,\alpha} = \frac{1}{2} \int dx dx' \, U(x - x') \langle \hat{n}(x) \hat{n}(x') \rangle_{\lambda,\alpha} \tag{65}$$

Separating again a Hartree (mean-field) term, we get:

$$\Gamma[n(x)] = \Gamma_0[n(x)] + E_{Hartree}[n(x)] + \Gamma_{xc}[n(x)] \tag{66}$$

with:

$$E_{Hartree}[n(x)] = \frac{1}{2} \int dx dx' \, U(x - x') n(x) n(x') \tag{67}$$

and Γ_{xc} is the correction-to mean field term (the exchange-correlation functional):

$$\Gamma_{xc}[n(x)] = \frac{1}{2} \int dx dx' \, U(x - x') \int_0^1 d\alpha g^c_\alpha[n; x, x'] \tag{68}$$

In which:

$$g^c_\alpha[n; x, x'] \equiv \langle (\hat{n}(x) - n(x))(\hat{n}(x') - n(x')) \rangle_{\lambda_\alpha[n],\alpha} \tag{69}$$

is the (connected) density-density correlation function, expressed as a functional of the local density, for a given value of the coupling α.

It should be emphasized that the exchange-correlation functional Γ_{xc} is *independent* of the specific form of the crystal potential $v(x)$: it is a *universal functional* which depends only on the form of the inter-particle interaction $U(x - x')$! To see this, we first observe that, because $\Gamma[n(x)]$ is the Legendre transform of the free energy with respect to the

---

[10] see e.g the lecture notes by K.Burke: http://dft.rutgers.edu/kieron/beta/index.html

one-body potential, we can easily relate the functional in the presence of the crystal potential $v(x)$ to that of the homogeneous electron gas (i.e with $v = 0$):

$$\Gamma[n(x)] = \Gamma_{HEG}[n(x)] + \int dx\, v(x) n(x) \tag{70}$$

Since this relation is also obeyed for the non-interacting system (see Eq. (61)), and using $\Gamma = \Gamma_0 + \Gamma_H + \Gamma_{xc}$, we see that the functional form of Γ_{xc} is independent of $v(x)$. It is the same for all solids, and also for the homogeneous electron gas.

I finally note that an exact relation can again be derived for the density functional (or alternatively the exchange-correlation functional) by noting that:

$$\beta g^c_\alpha[n;x,x'] = \left[\frac{\delta \Gamma_\alpha}{\delta n(x) \delta n(y)}\right]^{-1}_{xx'} \tag{71}$$

Inserting this relation into (66,68), one obtains:

$$\Gamma_\alpha[n] = \Gamma_0[n] + \alpha E_H[n] + \frac{1}{2}\int dx dx' U(x-x') \int_0^\alpha d\alpha' \left[\frac{\delta \Gamma_\alpha}{\delta n(x) \delta n(y)}\right]^{-1}_{xx'} \tag{72}$$

in complete analogy with (42). For applications of this exact functional equation, see e.g [65, 66]. Analogies with the exact renormalization group approach (see previous section) might suggest further use of this relation in the DFT context.

3.2.3. The Kohn-Sham equations

Let us now look at the condition for equilibrium. We vary $\Gamma[n(x)]$, and we note that, as before, the terms originating from the variation $\delta\lambda^0/\delta n(x)$ cancel because of the relation (56). We thus get:

$$\frac{\delta \Gamma}{\delta n(x)} = -\lambda_0(x) + \int dx' U(x-x') n(x') + \frac{\delta \Gamma_{xc}}{\delta n(x)} \tag{73}$$

so that the equilibrium density $n^*(x)$ is determined by:

$$\lambda^0(x)^* = \int dx' U(x-x') n^*(x') + \frac{\delta \Gamma_{xc}}{\delta n(x)}|_{n=n^*} \tag{74}$$

which equivalently specifies the KS potential at equilibrium as:

$$v^*_{KS}(x) = v(x) + \int dx' U(x-x') n^*(x') + \frac{\delta \Gamma_{xc}}{\delta n(x)}|_{n=n^*} \tag{75}$$

Equation (74) is the precise analog of Eq.(44) determining the Weiss field in the Ising case, and v^*_{KS} is the true effective potential seen by an electron at equilibrium, in a one-electron picture. Together with (57), it forms the fundamental (Kohn-Sham) equations of the DFT approach. To summarize, the expression of the total energy ($T=0$) reads:

$$\Gamma[n(x), T=0] = \sum_l^l \varepsilon_l - \int dx\, v_{KS}(x) n(x) + \int dx\, v(x) n(x) + \Gamma_{xc}[n(x)] \tag{76}$$

Concrete applications of the DFT formalism require an approximation to be made on the exchange-correlation term. The celebrated *local density approximation* (LDA) reads:

$$\Gamma_{xc}[n(x)]|_{LDA} = \int dx\, n(x)\, \varepsilon_{xc}^{HEG}[n(x)] \tag{77}$$

in which $\varepsilon_{xc}^{HEG}(n)$ is the exchange-correlation energy density of the *homogeneous* electron gas, for an electron density n. Discussing the reasons for the successes of this approximation (as well as its limitations) is quite beyond the scope of these lectures. The interested reader is referred e.g to [61, 59].

Finally, we observe that DFT satisfies the stability properties discussed in the introduction, since $\delta^2\Gamma/\delta n(x)\delta n(x')$ is the inverse of the density-density response function ($\mathbf{q}$-dependent compressibility). A negative eigenvalue would correspond to a charge ordering instability.

3.3. Exact functional of the local Green's function, and the Dynamical Mean-Field Theory approximation

In this section, I would like to explain how the concepts of the previous sections provide a broader perspective on the dynamical mean field approach to strongly correlated fermion systems. In contrast to DFT which focuses on ground-state properties (or thermodynamics), the goal of DMFT (see [3] for a review) is to address excited states by focusing on the *local Green's function* (or the *local spectral density*). Thus, it is natural to formulate this approach in terms of a functional of the local Green's function. This point of view has been recently emphasized by Chitra and Kotliar [67] and by the author in Ref. [28].

I describe below how such an *exact functional* can be formally constructed for a correlated electron model (irrespective, e.g of dimensionality), hence leading to a *local Green's function (or local spectral density) functional theory*. I will adopt a somewhat different viewpoint than in [67], by taking the *atomic limit* (instead of the non-interacting limit) as a reference system. This leads naturally to represent the exact local Green's function as that of a quantum impurity model, with a suitably chosen hybridisation function. There is no approximation involved in this mapping (only a representability assumption). This gives a general value to the impurity model mapping of Ref.[26]. Dynamical mean field theory as usually implemented can then be viewed as a *subsequent approximation* made on the non-local contributions to the exact functional (e.g. the kinetic energy).

For the sake of simplicity, I will take the Hubbard model as an example throughout this section. The hamiltonian is decomposed as:

$$H_\alpha = U \sum_i n_{i\uparrow} n_{i\downarrow} - \alpha \sum_{ij,\sigma} t_{ij} c_{i\sigma}^{+} c_{j\sigma} \tag{78}$$

I emphasize that the varying coupling constant $\alpha \in [0,1]$ has been introduced in front of the hopping term, which is the non-local term of this hamiltonian, and *not* in front of the interaction. When dealing with a more general hamiltonian, we would similarly decompose $H = H_{\rm loc} + \alpha H_{\rm non\text{-}loc}$.

3.3.1. Representing the local Green's function by a quantum impurity model

In order to constrain the local Green's function $\langle c_i(\tau)c_i^{+}(\tau')\rangle$ to take a specified value $G(\tau-\tau')$, we introduce conjugate sources (or Lagrange multipliers) $\Delta(\tau-\tau')$ and consider [11]:

$$\Omega_\alpha[G(\omega),\Delta(\omega)] \equiv -\frac{1}{N_s\beta}\ln\int DcDc^+\exp\{\int_0^\beta d\tau(\sum_{i\sigma}c_{i\sigma}^+(-\partial_\tau+\mu)c_{i\sigma}-H_\alpha[c,c^+])+ + \int_0^\beta\int_0^\beta d\tau d\tau'\sum_{i\sigma}\Delta(\tau-\tau')[G(\tau-\tau')-c_{i\sigma}^+(\tau)c_{i\sigma}(\tau')]\} \tag{79}$$

Inverting the relation $G = G_\alpha[\Delta]$ yields $\Delta = \Delta_\alpha[G]$, and a functional of the local Green's function is obtained as $\Gamma_\alpha[G] = \Omega_\alpha[G,\Delta_\alpha[G]]$. This is the Legendre transform of the free energy with respect to the local source Δ.

I would like to emphasize that this construction is quite different from the Baym-Kadanoff formalism, which considers a functional of all the components of the lattice Green's function G_{ij}, not only of its local part G_{ii}. The Baym-Kadanoff approach also gives interesting insights into the DMFT construction [3, 47], and will be considered at a later stage in these lectures.

Consider first the $\alpha = 0$ case, in which the hamiltonian is purely local (atomic limit). Then, we have to consider a local problem defined by the action:

$$S_{imp} = -\int_0^\beta d\tau\int_0^\beta d\tau'\sum_\sigma c_\sigma^+(\tau)\left[(-\partial_\tau+\mu)\delta(\tau-\tau')-\Delta_0(\tau-\tau')\right]c_\sigma(\tau') + U\int_0^\beta d\tau\, n_\uparrow(\tau)n_\downarrow(\tau) \tag{80}$$

Hence, the local Green's function $G(i\omega_n)$ is represented as that of a quantum impurity problem (an Anderson impurity problem in the context of the Hubbard model):

$$G = G_{imp}[\Delta_0] \tag{81}$$

As before, Δ_0 plays the role of a Weiss field (analogous to the effective field for a magnet, or to the KS effective potential in DFT). Formally, this Weiss field specifies [26] the

[11] In this section, I will divide the free energy functional by the number N_s of lattice sites (restricting myself for simplicity to an homogeneous system)

effective bare Green's function of the impurity action (80):

$$\mathcal{G}_0^{-1}(i\omega_n) = i\omega_n + \mu - \Delta_0(i\omega_n) \tag{82}$$

There are however two important new aspects here:

- i) The Weiss function Δ_0 is a *dynamical* (i.e frequency dependent) object. As a result the local equivalent problem (80) is not in Hamiltonian form but involves retardation
- ii) The equivalent local problem is not a one-body problem, but involves local interactions.

We note that, as in DFT, the explicit inversion of (81) is not possible in general. In practice, one needs a (numerical or approximate) technique to solve the quantum impurity problem (an *"impurity solver"*), and one can use an iterative procedure. Starting from some initial condition for Δ_0 (or $\mathcal{G}0$) , one computes the interacting Green's function G_{imp}, and the associated self-energy $\Sigma_{imp} \equiv \mathcal{G}_0^{-1} - G_{imp}^{-1}$. One then updates $\mathcal{G}0$ as: $\mathcal{G}0^{new} = [\Sigma_{imp} + G^{-1}]^{-1}$, where G is the specified value of the local Green's function.

3.3.2. Exact functional of the local Green's function

We proceed with the construction of the exact functional of the local Green's function, by coupling constant integration (starting from the atomic limit).

At $\alpha = 0$ (decoupled sites, or infinitely separated atoms), we have [12] :

$$\Omega_0[\Delta_0, G] = F_{imp}[\Delta_0] - \mathrm{Tr}\,(G\Delta_0) \tag{83}$$

where F_{imp} is the free energy of the local quantum impurity model viewed as a functional of the hybridisation function. By formal inversion $\Delta_0 = \Delta_0[G]$:

$$\Gamma_0[G] = F_{imp}\,[\Delta_0[G]] - \mathrm{Tr}\,\,(G\Delta_0[G]) \tag{84}$$

We then observe that (since the α-derivatives of the Lagrange multipliers do not contribute because of the stationarity of Ω):

$$\frac{d\Gamma_\alpha}{d\alpha} = \langle H_{\text{non-loc}} \rangle \tag{85}$$

which, for the Hubbard model, reduces to the kinetic energy:

$$\frac{d\Gamma_\alpha}{d\alpha} = \langle \hat{T} \rangle = -\frac{1}{N_s}\sum_{ij} t_{ij} \langle c_i^+ c_j \rangle|_G = \mathrm{Tr}\,\frac{1}{N_s}\sum_{\mathbf{k}} \varepsilon_{\mathbf{k}}\, G_\alpha(\mathbf{k}, i\omega_n)|_G \tag{86}$$

In this expression, the lattice Green's function $G_\alpha(\mathbf{k}, i\omega_n)$ should be expressed, for a given α, as a functional of the local Green's function G.

---

[12] In this formula and everywhere below, Tr denotes $\frac{1}{\beta}\sum_n$, with possibly a convergence factor $e^{i\omega_n 0+}$.

This leads to the following formal expression of the exact functional $\Gamma[G] = \Gamma_{\alpha=1}[G]$:

$$\Gamma[G] = F_{imp}[\Delta_0[G]] - \text{Tr}\,(G\Delta_0[G]) + \mathscr{T}[G] \tag{87}$$

in which $\mathscr{T}[G]$ is the kinetic energy functional (evaluated while keeping $G_{ii} = G$ fixed):

$$\mathscr{T}[G] = \int_0^1 d\alpha \frac{1}{N_s} \sum_{ij} t_{ij} \langle c_i^+ c_j \rangle |_G = \int_0^1 d\alpha \text{Tr} \frac{1}{N_s} \sum_{\mathbf{k}} \varepsilon_{\mathbf{k}} G_\alpha(\mathbf{k}, i\omega_n)|_G \tag{88}$$

The condition $\delta\Gamma/\delta G = 0$ determines the actual value of the local Green's function at equilibrium as (using $\delta\Gamma_0/\delta G = -\Delta_0$):

$$\Delta_0[G(i\omega_n)] = \frac{\delta \mathscr{T}[G]}{\delta G(i\omega_n)} \tag{89}$$

We recall that the generalized Weiss function (hybridization) and G are, by construction, related by (81):

$$G = G_{imp}[\Delta_0] \tag{90}$$

Equations (89,90) (together with the definition of the impurity model, Eq. 80)) are the key equations of dynamical mean-field theory, viewed as an exact approach. The cornerstone of this approach [26] is that, in order to obtain the local Green's function, one has to solve an impurity model (80), submitted to the self-consistency condition (89) relating the hybridization function Δ_0 to $G(i\omega_n)$ itself. I emphasize that, since $\Gamma[G]$ is an exact functional, this construction is completely general: it is valid for the Hubbard model in arbitrary dimensions and on an arbitrary lattice.

Naturally, using it in practice requires a concrete approximation to the kinetic energy functional $\mathscr{T}[G]$ (similarly, the DFT framework is only practical once an approximation to Γ_{xc} is used, for example the LDA). The DMFT *approximation* usually employed is described below. In fact, it might be useful to employ a different terminology and call "local spectral density functional theory" (or "local impurity functional theory") the exact framework, and DMFT the subsequent approximation commonly made in $\mathscr{T}[G]$.

3.3.3. A simple case: the infinite connectivity Bethe lattice

It is straightforward to see that the formal expression for the kinetic energy functional $\mathscr{T}[G]$ simplifies into a simple closed expression for the Bethe lattice with connectivity z, in the limit $z \to \infty$. In fact, a closed form can be given on an arbitrary lattice in the limit of large dimensions, but this is a bit more tedious and we postpone it to the next section.

In the limit of large connectivity, the hopping must be scaled as: $t_{ij} = t/\sqrt{z}$ [25]. Expanding the kinetic energy functional in (87) in powers of α, one sees that only the term of order α remains in the $z = \infty$ limit thanks to the tree-like geometry, namely:

$$\alpha \sum_{ijkl} t_{ij} t_{kl} \langle c_i^+ c_j c_k^+ c_l \rangle_{\alpha=0} = \alpha \sum_{ij} t_{ij}^2 \text{Tr}\, G^2 = \alpha (zN_s) \frac{t^2}{z} \text{Tr}\, G^2 \tag{91}$$

So that, integrating over α, one obtains $\mathscr{T}[G] = t^2 \mathrm{Tr}\, G^2/2$ and finally:

$$\Gamma_{Bethe,z=\infty}[G] = F_{imp}[\Delta_0[G]] - \mathrm{Tr}\,(G\Delta_0[G]) + \frac{t^2}{2}\mathrm{Tr}\, G^2 \tag{92}$$

This functional is similar (although different in details) to the one recently used by Kotliar [68] in a Landau analysis of the Mott transition within DMFT.

The self-consistency condition (89) that finally determines both the local Green's function and the Weiss field (through an iterative solution of the impurity model) thus reads in this case:

$$\Delta_0[G, i\omega_n] = t^2\, G(i\omega_n) \tag{93}$$

3.3.4. DMFT as an approximation to the kinetic energy functional.

Now, I will show that the usual form of DMFT [3] (for a general non-interacting dispersion ε_k) corresponds to a very simple approximation of the kinetic energy term $\mathscr{T}[G]$ in the exact functional $\Gamma[G]$. Consider the one-particle Green's function $G_\alpha(\mathbf{k}, i\omega_n)$ associated with the action (79) of the Hubbard model, in the presence of the source term Δ_α and for an arbitrary coupling constant. We can define a self-energy associated with this Green's function:

$$G_\alpha(\mathbf{k}, i\omega_n) = \frac{1}{i\omega_n + \mu - \Delta_\alpha[i\omega_n] - \alpha\varepsilon_\mathbf{k} - \Sigma_\alpha[\mathbf{k}, i\omega_n]} \tag{94}$$

The self-energy Σ_a is in general a $\mathbf{k}$-dependent object, except obviously for $\alpha = 0$ in which all sites are decoupled into independent impurity models. The DMFT approximation consists in replacing Σ_a for arbitrary α by the impurity model self-energy Σ_0 (hence depending only on frequency), at least for the purpose of calculating $\mathscr{T}[G]$. Hence:

$$G_\alpha(\mathbf{k}, i\omega_n)|_{DMFT} = \frac{1}{i\omega_n + \mu - \Delta_\alpha[i\omega_n; G] - \alpha\varepsilon_\mathbf{k} - \Sigma_{\alpha=0}[i\omega_n, G]} \tag{95}$$

With:

$$\Sigma_{\alpha=0}[G; i\omega_n] \equiv \mathscr{G}0^{-1} - G^{-1} = i\omega_n + \mu - \Delta_0[i\omega_n, G] - G^{-1} \tag{96}$$

Summing over $\mathbf{k}$, one then expresses the local Green's function in terms of the hybridisation as:

$$G(i\omega_n) = \int d\varepsilon \frac{D(\varepsilon)}{\zeta - \alpha\varepsilon} = \frac{1}{\alpha}\widetilde{D}\left(\frac{\zeta}{\alpha}\right) \tag{97}$$

With $\zeta \equiv i\omega_n + \mu - \Delta_\alpha - \Sigma_0 = \Delta_0 - \Delta_a + G^{-1}$. In this expression, $D(\varepsilon) = \frac{1}{N_s}\sum_\mathbf{k} \delta(\varepsilon - \varepsilon_\mathbf{k})$ is the non-interacting density of states, and $\widetilde{D}(z) = \int d\varepsilon \frac{D(\varepsilon)}{z-\varepsilon}$ its Hilbert transform. Introducing the inverse function such that $\widetilde{D}[R(g)] = g$, we can invert the relation above to obtain the hybridisation function as a functional of the local G for $U = 0$:

$$\Delta_\alpha[i\omega_n; G] = G^{-1} + \Delta_0[G] - \alpha R[\alpha G] \tag{98}$$

So that the lattice Green's function is also expressed as a functional of G as:

$$G_\alpha(\mathbf{k}, i\omega_n) = \frac{1}{\alpha R(\alpha G) - \alpha\varepsilon_\mathbf{k}} \tag{99}$$

Inserting this into (87), we can evaluate the kinetic energy:

$$\frac{1}{N_s}\sum_\mathbf{k} \varepsilon_\mathbf{k} G_\alpha(\mathbf{k}) = \frac{1}{\alpha}\int d\varepsilon \frac{\varepsilon D(\varepsilon)}{R(\alpha G) - \varepsilon} = \frac{1}{\alpha}[-1 + \alpha G R(\alpha G)] \tag{100}$$

and hence the DMFT approximation to $\mathscr{T}[G]$:

$$\mathscr{T}_{DMFT}[G] = \int_0^1 d\alpha \mathrm{Tr}\left[G(i\omega_n) R(\alpha G(i\omega_n)) - \frac{1}{\alpha}\right] \tag{101}$$

So that the total functional reads, in the DMFT approximation:

$$\begin{aligned}\Gamma_{DMFT}[G] = \quad & F_{imp}[\Delta_0[G]] - \mathrm{Tr}\ (G\Delta_0[G]) + \\ & + \int_0^1 d\alpha \mathrm{Tr}\left[G(i\omega_n) R(\alpha G(i\omega_n)) - \frac{1}{\alpha}\right]\end{aligned} \tag{102}$$

In the case of an infinite-connectivity Bethe lattice, corresponding to a semi-circular d.o.s of width $4t$, one has: $R[g] = t^2 g + 1/g$, so that the result (92) is recovered from this general expression. I note that the DMFT approximation to the functional $\mathscr{T}[G]$ is completely independent of the interaction strength U.

The equilibrium condition (89) $\delta\Gamma/\delta G = 0$ thus reads [13], in the DMFT approximation [3]:

$$\Delta_0[i\omega_n, G]|_{DMFT} = R[G(i\omega_n)] - \frac{1}{G(i\omega_n)} \tag{103}$$

This can be rewritten in a more familiar form, using (96):

$$G(i\omega_n) = \int d\varepsilon \frac{D(\varepsilon)}{i\omega_n + \mu - \Sigma_{imp}(i\omega_n)}, \quad \text{with: } \Sigma_{imp} = \mathscr{G}0^{-1} - G^{-1} \tag{104}$$

The self-consistency condition is equivalent to the condition $\Delta_{\alpha=1}[G] = 0$, as expected from the fact that $\Delta_{\alpha=1} = \delta\Gamma/\delta G$. Hence, within the DMFT approximation, the lattice Green's function is obtained by setting $\alpha = 1$ into (95):

$$G(\mathbf{k}, i\omega_n)|_{DMFT} = \frac{1}{i\omega_n + \mu \quad \varepsilon_\mathbf{k} \quad \Sigma_{imp}(i\omega_n)} \tag{105}$$

3.4. The Baym-Kadanoff viewpoint

Finally, let me briefly mention that the DMFT approximation can also be formulated using the more familiar Baym-Kadanoff functional. In contrast to the previous section,

---

[13] When deriving this equation, it is useful to note that $R(\alpha G) + \alpha G R'(\alpha G) = \partial_\alpha[\alpha R(\alpha G)]$.

this is a functional of *all components* G_{ij} of the lattice Green's function, not only of the local one G_{ii}. The Baym-Kadanoff functional is defined as:

$$\Omega_{BK}[G_{ij},\Sigma_{ij}] = -\mathrm{tr}\ln\left[(i\omega_n+\mu)\delta_{ij}-t_{ij}-\Sigma_{ij}(i\omega_n)\right] - \mathrm{tr}[\Sigma\cdot G] + \Phi_{LW}[\{G_{ij}\}] \quad (106)$$

Variation with respect to Σ_{ij} yields the usual Dyson's equation relating the Green's function and the self-energy. The Luttinger-Ward functional Φ_{LW} has a simple diagrammatic definition as the sum of all skeleton diagrams in the free-energy. Variation with respect to G_{ij} express the self-energy as a total derivative of this functional:

$$\Sigma_{ij}(i\omega_n) = \frac{\delta\Phi}{\delta G_{ij}(i\omega_n)} \quad (107)$$

The DMFT approximation amounts to approximate the Luttinger-Ward functional by a functional which is the sum of that of *independent atoms*, retaining only the dependence over the local Green's function, namely:

$$\Phi_{LW}^{DMFT} = \sum_i \Phi_{imp}[G_{ii}] \quad (108)$$

An obvious consequence is that the self-energy is site-diagonal:

$$\Sigma_{ij}(i\omega_n) = \delta_{ij}\Sigma(i\omega_n) \quad (109)$$

Eliminating Σ_{ii} amounts to do a Legendre transformation with respect to G_{ii}, and therfore leads to a different expression of the local DMFT functional introduced in the previous section [67]:

$$\Gamma_{DMFT}[G_{ii}] = -\mathrm{tr}\ln\left[(i\omega_n+\mu-\frac{\delta\Phi_{imp}}{\delta G_{ii}})\delta_{ij}-t_{ij}\right] - \mathrm{tr}[\frac{\delta\Phi}{\delta G_{ii}}\cdot G_{ii}] + \sum_i \Phi_{imp}[G_{ii}] \quad (110)$$

The Baym-Kadanoff formalism is useful for total energy calculations, and will be used in Sec. 5.4.

4. THE MOTT METAL-INSULATOR TRANSITION

4.1. Materials on the verge of the Mott transition

Interactions between electrons can be responsible for the insulating character of a material, as realized early on by Mott [1, 2]. The Mott mechanism plays a key role in the physics of strongly correlated electron materials. Outstanding examples [2, 11] are transition-metal oxides (e.g superconducting cuprates), fullerene compounds, as well as organic conductors[14]. Fig. 6 illustrates this in the case of transition metal oxides with perovskite structure ABO_3 [74].

[14] The Mott phenomenon may also be partly responsible for the localization of f-electrons in some rare earth and actinides *metals*, see [69, 70, 71, 72, 73] and [16, 21] for recent reviews.

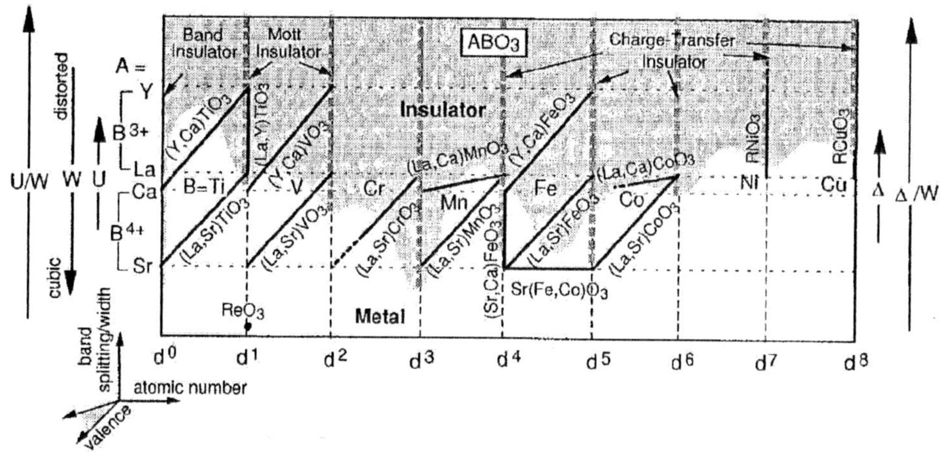

FIGURE 6. This diagram (due to A.Fujimori [74], see also [11]) can be viewed as a map of the vast territory of transition- metal compounds with perovskite structure ABO_3. Varying the transition metal ion B corresponds to gradual filling of the 3d-shell. Different substitutions on the A-site can be made (A= Sr,Ca and A=La,Y are mainly considered in this diagram). This allows to change either the valence of the transition metal ion (doping), or the structural parameters in an isoelectronic manner. The shaded region corresponds to insulating compounds, while the unshaded one corresponds to metals. This illustrates the key role of the Mott phenomenon in the physics of transition-metal oxides.

A limited number of materials are poised right on the verge of this electronic instability. This is the case, for example, of V_2O_3, $NiS_{2-x}Se_x$ and of quasi two-dimensional organic conductors of the κ-BEDT family. These materials are particularly interesting for the fundamental investigation of the Mott transition, since they offer the possibility of going from one phase to the other by varying some external parameter (e.g chemical composition,temperature, pressure,...). Varying external pressure is definitely a tool of choice since it allows to sweep continuously from the insulating phase to the metallic phase (and back). The phase diagrams of $(V_{1-x}Cr_x)_2O_3$ and of κ-(BEDT-TTF)$_2$Cu[N(CN)$_2$]Cl under pressure are displayed in Fig. 7. There is a great similarity between the high-temperature part of the phase diagrams of these materials, despite very different energy scales. At low-pressure they are *paramagnetic* Mott insulators, which are turned into metals as pressure is increased. Above a critical temperature T_c (of order ~ 450K for the oxide compound and ~ 40K for the organic one), this corresponds to a smooth crossover. In contrast, for $T < T_c$ a first-order transition is observed, with a discontinuity of all physical observables (e.g resistivity). The first order transition line ends in a second order critical endpoint at (T_c, P_c). We observe that in both cases, the critical temperature is a very small fraction of the bare electronic energy scales (for V_2O_3 the half-bandwidth is of order $0.5-1$ eV, while it is of order 2000 K for the organics).

There are also some common features between the low-temperature part of the phase diagram of these compounds, such as the fact that the paramagnetic Mott insulator orders into an antiferromagnet as temperature is lowered. However, there are also striking differences: the metallic phase has a superconducting instability for the organics, while

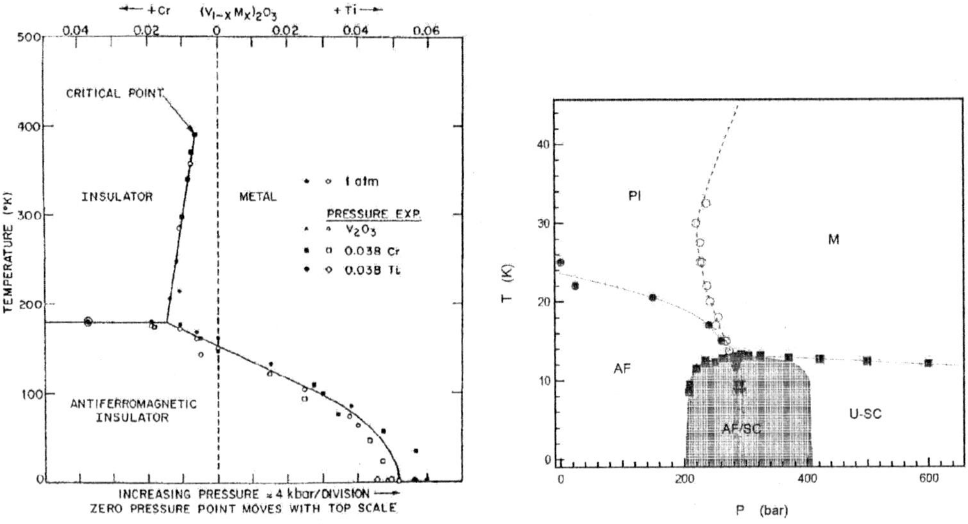

FIGURE 7. Left: Phase diagram of $(V_{1-x}Cr_x)_2O_3$ as a function of either Cr-concentration x or pressure (after[75]). Increasing x by 1% produces similar effects than *decreasing* pressure by $\sim$ 4kbar, for this material. Right: Phase diagram of κ-(BEDT-TTF)$_2$Cu[N(CN)$_2$]Cl as a function of pressure (after [76]).

this is not the case for V_2O_3. Also, the magnetic transition is only superficially similar: in the case of V_2O_3, it is widely believed to be accompanied (or even triggered) by orbital ordering[77] (in contrast to $NiS_{2-x}Se_x$[78]), and as a result the transition is first-order. In general, there is a higher degree of universality associated with the vicinity of the Mott critical endpoint than in the low-temperature region, in which long-range order takes place in a material- specific manner.

Mott localization into a paramagnetic insulator implies a high spin entropy, which must therefore be quenched in some way as temperature is lowered. An obvious possibility is magnetic ordering, as in these two materials. In fact, a Mott transition between a paramagnetic Mott insulator and a metallic phase is only observed in those compounds where magnetism is sufficiently *frustrated* so that the transition is not preempted by magnetic ordering. This is indeed the case in both compounds discussed here: V_2O_3 has competing ferromagnetic and antiferromagnetic exchange constants, while the two-dimensional layers in the organics have a triangular structure. Another possibility is that the entropy is quenched through a Peierls instability (dimerization), in which case the Mott insulator can remain paramagnetic (this is the case, for example, of VO_2). Whether it is possible to stabilize a paramagnetic Mott insulator down to $T = 0$ without breaking spin or translational symmetries is a fascinating problem, both theoretically and from the materials point of view (for a recent review on resonating valence bond phases in frustrated quantum magnets, see e.g [79] and [80]). The compound κ-(BEDT-TTF)$_2$Cu$_2$(CN)$_3$ may offer [81] a realization of such a spin-liquid state (presumably through a combination of strong frustration and strong charge fluctuations [82]), but this behaviour is certainly more the exception than the rule.

4.2. Dynamical mean-field theory of the Mott transition

Over the last decade, a detailed theory of the strongly correlated metallic state, and of the Mott transition itself has emerged, based on the *dynamical mean-field theory* (DMFT). We refer to [3] for a review and an extensive list of original references [26, 29, 30, 83, 84, 85] We now review some key features of this theory.

Quasiparticle coherence scale. In the metallic state, Fermi-liquid theory applies below a low energy scale ε_F^*, which can be interpreted as the coherence-scale for quasiparticles (i.e long-lived quasiparticles exist only for energies and temperature smaller than ε_F^*). This low-energy coherence scale is given by $\varepsilon_F^* \sim ZD$ (with D the half-bandwith, also equal to the Fermi energy of the non-interacting system at half-filling) where Z is the quasiparticle weight. In the strongly correlated metal close to the transition, $Z \ll 1$, so that ε_F^* is strongly reduced as compared to the bare Fermi energy.

Three peaks in the d.o.s: Hubbard bands and quasiparticles. In addition to low-energy quasiparticles (carrying a fraction Z of the spectral weight), the one-particle spectrum of the strongly correlated metal contains high-energy excitations carrying a spectral weight $1-Z$. These are associated to the atomic-like transitions corresponding to the addition or removal of one electron on an atomic site, which broaden into Hubbard bands in the solid. As a result, the **k**-integrated spectral function $A(\omega) = \sum_{\mathbf{k}} A(\mathbf{k}, \omega)$ (density of states d.o.s) of the strongly correlated metal is predicted [26] to display a three-peak structure, made of a quasiparticle band close to the Fermi energy surrounded by lower and upper Hubbard bands (Fig. 8 and inset of Fig. 14). The quasiparticle part of the d.o.s has a reduced width of order $ZD \sim \varepsilon_F^*$. The lower and upper Hubbard bands are separated by an energy scale Δ.

The insulating phase: local moments, magnetism and frustration. At strong enough coupling (see below), the paramagnetic solution of the DMFT equations is a Mott insulator, with a gap Δ in the one-particle spectrum. This phase is characterized by unscreened local moments, associated with a Curie law for the local susceptibility $\sum_q \chi_q \propto 1/T$, and an extensive entropy. Note however that the uniform susceptiblity $\chi_{q=0}$ is finite, of order $1/J \sim U/D^2$. As temperature is lowered, these local moments order into an antiferromagnetic phase [27, 83]. The Néel temperature is however strongly dependent on frustration [3] (e.g on the ratio t'/t between the next nearest-neighbour and nearest-neighbour hoppings) and can be made vanishingly small for fully frustrated models.

Separation of energy scales, spinodals and transition line. Within DMFT, a separation of energy scales holds close to the Mott transition. The mean-field solution corresponding to the paramagnetic metal at $T = 0$ disappears at a critical coupling U_{c2}. At this

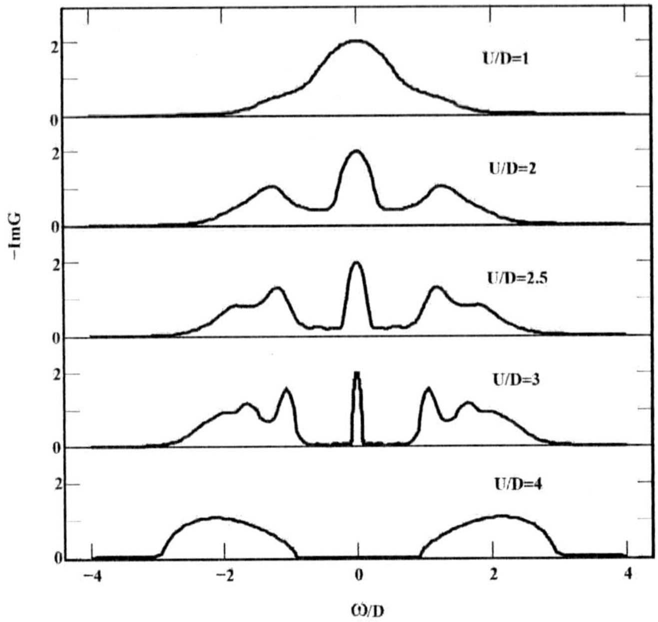

FIGURE 8. Local spectral function for several values of the interaction strength in DMFT. These results have been obtained using the IPT approximation, for the half-filled Hubbard model with a semi-circular d.o.s (from Ref. [3]). Close to the transition, the separation of scales between the quasiparticle coherence energy (ε_F^*) and the distance between Hubbard bands (Δ) is clearly seen.

point, the quasiparticle weight vanishes ($Z \propto 1 - U/U_{c2}$) as in Brinkman-Rice theory[15] On the other hand, a mean-field insulating solution is found for $U > U_{c1}$, with the Mott gap Δ opening up at this critical coupling (Mott-Hubbard transition). As a result, Δ is a finite energy scale for $U = U_{c2}$ and the quasiparticle peak in the d.o.s is well separated from the Hubbard bands in the strongly correlated metal.

These two critical couplings extend at finite temperature into two spinodal lines $U_{c1}(T)$ and $U_{c2}(T)$, which delimit a region of the $(U/D, T/D)$ parameter space in which two mean-field solutions (insulating and metallic) are found (Fig. 9). Hence, within DMFT, a first-order Mott transition occurs at finite temperature even in a purely electronic model. The corresponding critical temperature T_c^{el} is of order $T_c^{el} \sim \Delta E/\Delta S$, with ΔE and $\Delta S \sim \ln(2S+1)$ the energy and entropy differences between the metal and the insulator. Because the energy difference is small ($\Delta E \sim (U_{c2} - U_{c1})^2/D$), the critical temperature is much lower than D and U_c (by almost two orders of magnitude).

[15] Since the self-energy only depends on frequency within DMFT, this also implies that quasiparticles become heavy close to the transition, with $m^*/m = 1/Z$. In real materials, we expect however that magnetic exchange will quench out the spin entropy associated with local moments, resulting in a saturation of the effective mass close to the Mott transition. In the regime where $\varepsilon_F^* \ll J$, the effective mass is then expected to be of order J, as found e.g in slave-boson theories. Describing this effect requires extensions of DMFT in order to deal with short-range spatial correlations

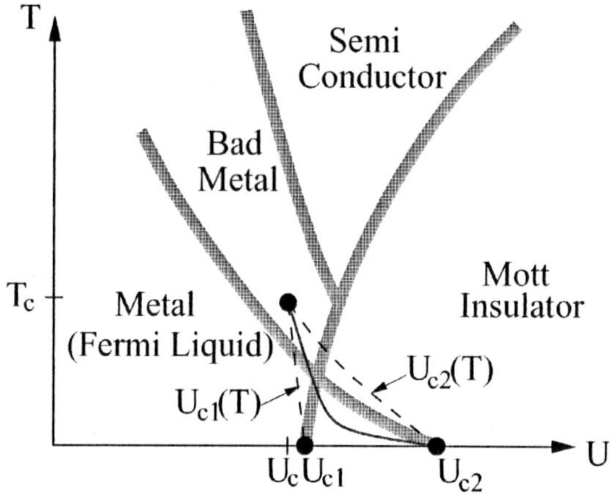

FIGURE 9. Paramagnetic phases of the Hubbard model within DMFT, displaying schematically the spinodal lines of the Mott insulating and metallic mean-field solutions (dashed), the first-order transition line (plain) and the critical endpoint. The shaded crossover lines separating the different transport regimes discussed in Sec.3 are also shown. The Fermi-liquid to "bad metal" crossover line corresponds to the quasiparticle coherence scale and is a continuation of the spinodal $U_{c2}(T)$ above T_c. The crossover into the insulating state corresponds to the continuation of the U_{c1} spinodal. Magnetic phases are not displayed and depend on the degree of frustration. Figure from Refs. [86] and [87].

Indeed, in V_2O_3 as well as in the organics, the critical temperature corresponding to the endpoint of the first-order Mott transition line is a factor of 50 to 100 smaller than the bare electronic bandwith.

4.3. Physical properties of the correlated metallic state: DMFT confronts experiments

4.3.1. Three peaks: evidence from photoemission

In Fig. 10, we reproduce the early photoemission spectra of some d^1 transition metal oxides, from the pioneering work of Fujimori and coworkers [88]. This work established experimentally, more than ten years ago, the existence of well-formed (lower) Hubbard bands in correlated metals, in addition to low-energy quasiparticles. This experimental study and the theoretical prediction of a 3-peak structure from DMFT [26] came independently around the same time. However, back in 1992, the existence of a narrow quasiparticle peak in $A(\omega)$ resembling the DMFT results was, to say the least, not obvious from these early data. Further studies [89] on $Ca_{1-x}Sr_xVO_3$ therefore aimed at studying the dependence of low-energy quasiparticle spectral features upon the degree of correlations. One of the main difficulty raised by these photoemission results is that the weight Z of the low-energy quasiparticle peak estimated from these early data is quite small (particularly for $CaVO_3$), while specific heat measurements do not reveal a dramatic mass enhancement. This triggered some discussion [89, 90, 11] about the

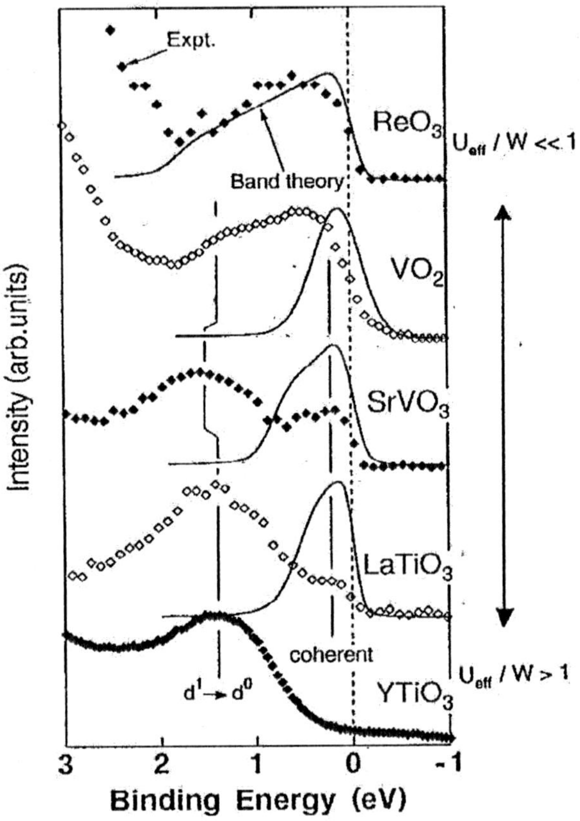

FIGURE 10. Photoemission spectra of several d^1 transition metal oxides, reproduced from Ref. [88]. The effects of correlations increases from ReO_3 (a weakly correlated metal) to $YTiO_3$ (a Mott insulator). The plain lines are the d.o.s obtained from band structure calculations. A lower Hubbard band around -1.5 eV is clearly visible in the most correlated materials, both in the metallic and insulating case.

possibility of a strong k-dependence of the self-energy. A decisive insight into this question came from further experimental developments by Maiti and coworkers [91, 18] in which it was demonstrated that the photoemission spectra are actually quite sensitive to the photon energy. Studies at different photon energies allowed these authors to extract the estimated spectra corresponding to the bulk and the surface of the material. Surface and bulk spectra were found to be very different indeed: the surface of $CaVO_3$ being apparently insulating-like while the bulk spectrum did show a much more pronounced quasiparticle peak. Very recently, high resolution, high-photon energy photoemission studies [92, 93] clarified considerably this issue. The high photon-energy spectrum reproduced on Fig. 11 displays a clear quasiparticle d.o.s at low-energy (with a weight in good agreement with $m/m^{\star}$ and a height comparable to the LDA d.o.s), as well as a lower Hubbard band carrying the rest of the spectral weight. Moreover, recent calculations [94, 95, 93] combining electronic structure methods and DMFT (see next section) compare favorably to the experimental spectra, on a quantitative level.

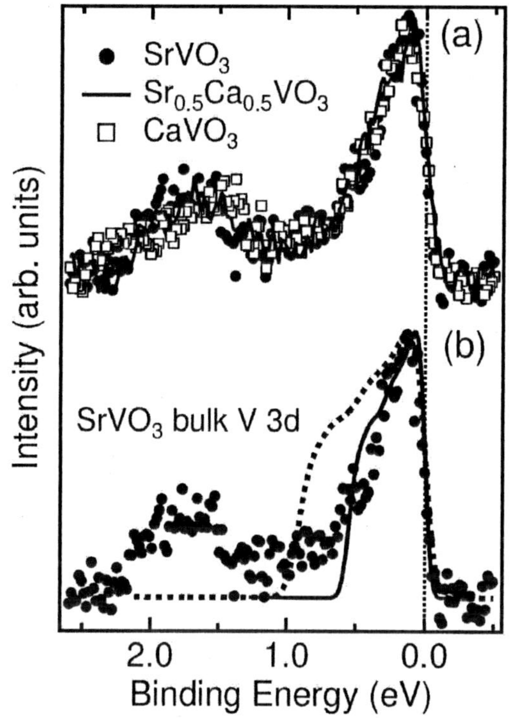

FIGURE 11. (a) Bulk V 3*d* spectral functions of $SrVO_3$ (closed circles), $Sr_{0.5}Ca_{0.5}VO_3$ (solid line) and $CaVO_3$ (open squares). (b) Comparison of the experimentally obtained bulk V 3*d* spectral function of $SrVO_3$ (closed circles) to the V 3*d* partial density of states for $SrVO_3$ (dashed curve) obtained from the band-structure calculation, which has been broadened by the experimental resolution of 140 meV. The solid curve shows the same V 3*d* partial density of states but the energy is scaled down by a factor of 0.6. Figure and caption from Ref. [92] (see also [93]).

In the case of $NiS_{2-x}Se_x$, angular resolved photoemission have revealed a clear quasiparticle peak, with strong spectral weight redistributions as a function of temperature [96]. For the metallic phase of V_2O_3, high photon energy photoemission proved to be an essential tool in the recent experimental finding of the quasiparticle peak (Fig. 12) by Mo et al. [97].

4.3.2. Spectral weight transfers

The quasiparticle peak in the d.o.s is characterized by an extreme sensitivity to changes of temperature, as shown in the inset of Fig. 14. Its height is strongly reduced as T is increased, and the peak disappears altogether as T reaches ε_F^*, leaving a pseudogap at the Fermi energy. Indeed, above ε_F^*, long-lived coherent quasiparticles no longer exist. The corresponding spectral weight is redistributed over a very large range of energies, of order U (hence much larger than temperature itself). This is reminiscent

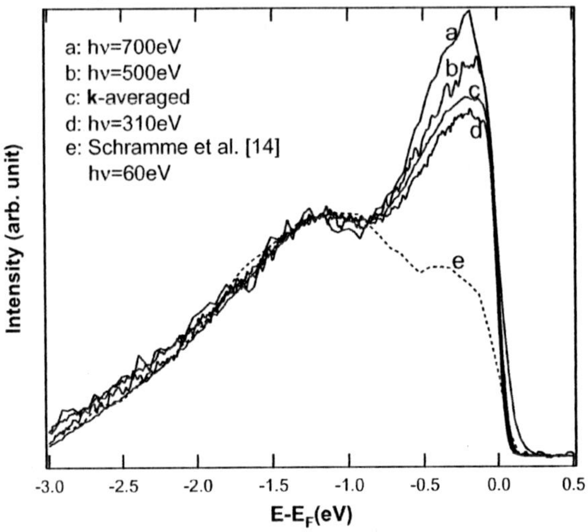

FIGURE 12. Photoemission spectra of V_2O_3, for various photon energies, from Ref. [97]. The highest photon energy spectrum, corresponding to the greatest bulk sensitivity, reveals a prominent quasiparticle peak.

of Kondo systems [98], and indeed DMFT establishes a formal and physical connection [26] between a metal close to the Mott transition and the Kondo problem. The local moment present at short time-scales is screened through a self-consistent Kondo process involving the low-energy part of the (single- component) electronic fluid itself.

These spectral weight transfers and redistributions are a distinctive feature of strongly correlated systems. As already mentioned, they have been observed in the photoemission spectra of $NiS_{2-x}Se_x$. They are also commonly observed in optical spectroscopy of correlated materials, as shown on Fig. 13 for metallic V_2O_3 [99] and the κ-BEDT organics [100]. DMFT calculations give a good description of the optical spectral weight transfers for these materials, at least on a qualitative level [99, 101].

4.3.3. Transport regimes and crossovers

The disappearance of coherent quasiparticles, and associated spectral weight transfers, results in three distinct transport regimes [99, 102, 101, 103, 87] for a correlated metal close to the Mott transition, within DMFT (Figs. 9 and 14):

- In the *Fermi-liquid regime* $T \ll \varepsilon_F^*$, the resistivity obeys a T^2 law with an enhanced prefactor: $\rho = \rho_M (T/\varepsilon_F^*)^2$. In this expression, ρ_M is the Mott-Ioffe-Regel resistivity $\rho_M \propto ha/e^2$ corresponding to a mean-free path of the order of a single lattice spacing in a Drude picture.
- For $T \sim \varepsilon_F^*$, an *"incoherent" (or "bad") metal* regime is entered. The quasiparticle lifetime shortens dramatically, and the quasiparticle peak is strongly suppressed (but still present). In this regime, the resistivity is metallic-like (i.e increases with

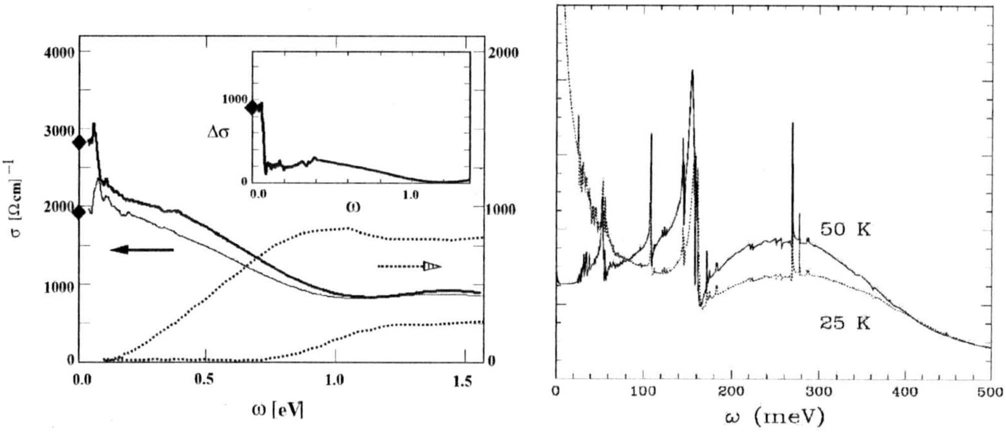

FIGURE 13. Left: Optical conductivity of metallic V_2O_3 [99] at $T = 170K$ (thick line) and $T = 300K$ (thin line)). The inset contains the difference of the two spectra $\Delta\sigma(\omega) = \sigma_{170K}(\omega) - \sigma_{300K}(\omega)$. Diamonds indicate the measured dc conductivity σ_{dc}. Dotted lines are for the insulating compounds $V_{2-y}O_3$ with $y = .013$ at $10K$ (upper) and $y = 0$ at $70K$ (lower). Right: Optical conductivity of κ-(BEDT-TTF)$_2$Cu[N(CN)$_2$]Br at ambiant pressure [100], for $T = 25$K and $T = 50$K. For both materials, transfer of spectral weight from high energies to the Drude peak is clearly visible as temperature is lowered.

T) but reaches values considerably larger than the Mott "limit" ρ_M. A Drude description is no longer applicable in this regime.

- Finally, for $\varepsilon_F^* \ll T \ll \Delta$, quasiparticles are gone altogether and the d.o.s displays a pseudogap associated with the scale Δ and filled with thermal excitations. This yields an insulating-like regime of transport, with the resistivity decreasing upon heating ($d\rho/dT < 0$). At very low temperature, the resistivity follows an activated behaviour, but deviations from a pure activation law are observed at higher temperature (these two regimes are depicted as the "insulating" and "semi-conducting" ones on Fig. 9).

These three regimes, and the overall temperature dependence of the resistivity obtained within DMFT are illustrated by Fig. 14. A distinctive feature is the resistivity maximum, which occurs close to the Mott transition. This behaviour is indeed observed experimentally in both Cr-doped V_2O_3 and the organics. In the latter case, the transport data obtained recently in the Orsay group are depicted on Fig. 15, and compared to DMFT model calculations [103, 87].

Within DMFT, the conductivity can be simply obtained from a calculation of the one-particle self-energy since vertex corrections are absent [104, 3]. However, a precise determination of both the real and imaginary part of the real-frequency self-energy is required. This is a challenge for most "impurity solvers". In practice, early calculations[102, 99, 101] used the iterated perturbation theory (IPT) approximation[26]. The results displayed in Fig. 14 have been obtained with this technique, and the overall shape of the resistivity curves are qualitatively reasonable.

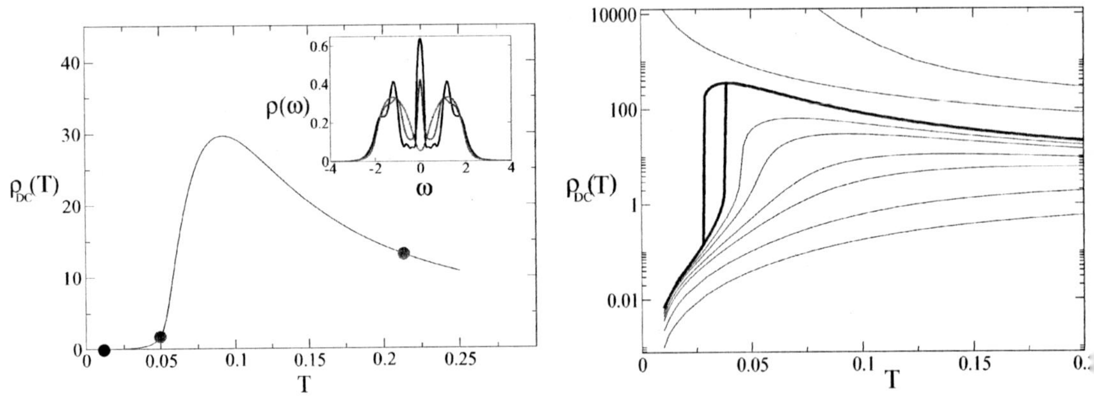

FIGURE 14. Left: Resistivity in the metallic phase close to the Mott transition ($U = 2.4D$), as a function of temperature, calculated within DMFT using the IPT approximation. For three selected temperatures, corresponding to the three regimes discussed in the text, the corresponding spectral density is displayed in the inset. Right: IPT results for the resistivity for values of U in the metallic regime (lower curves), the coexistence region (bold curve) and the insulating regime (upper two curves). From Ref. [86, 87].

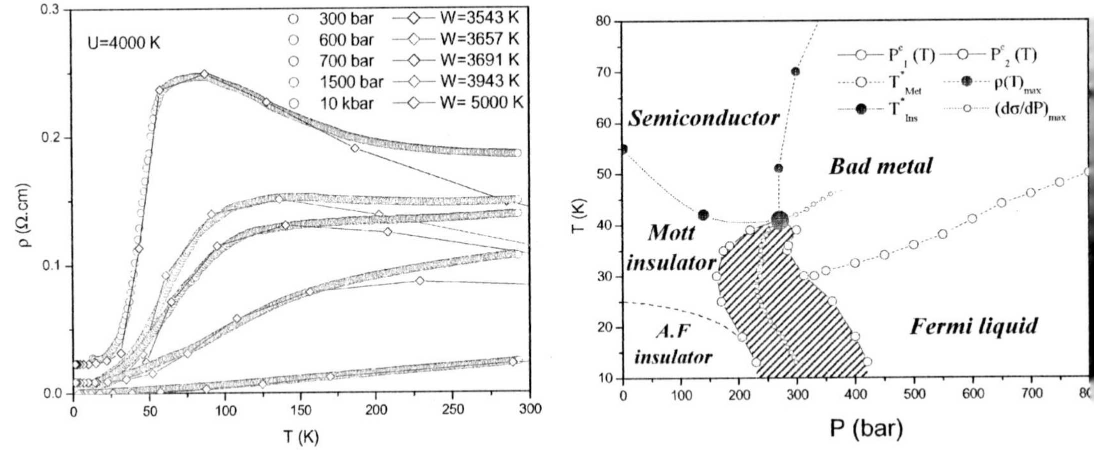

FIGURE 15. Left: Temperature-dependence of the resistivity at different pressures, for κ-(BEDT-TTF)$_2$Cu[N(CN)$_2$]Cl. The data (circles) are compared to a DMFT-NRG calculation (diamonds), with a pressure dependence of the bandwidth as indicated. The measured residual resistivity ρ_0 has been added to the theoretical curves. Right: Transport regimes and crossovers for this compound. Figures reproduced from Limelette *et al.* [103].

However, the IPT approximation does a poor job on the quasiparticle lifetime in the low-temperature regime, as shown on Fig. 16. Indeed, we expect on general grounds that, close to the transition, $D\text{Im}\Sigma$ becomes a scaling function [105] of ω/ε_F^* and T/ε_F^*, so that for $T \ll \varepsilon_F^*$ it behaves as: $\text{Im}\Sigma(\omega = 0) \propto D(T/\varepsilon_F^*)^2 \propto T^2/(Z^2 D)$ which leads to an enhancement of the T^2 coefficient of the resistivity by $1/Z^2$ as mentioned above. The IPT approximation does not capture this enhancement and yields the in-

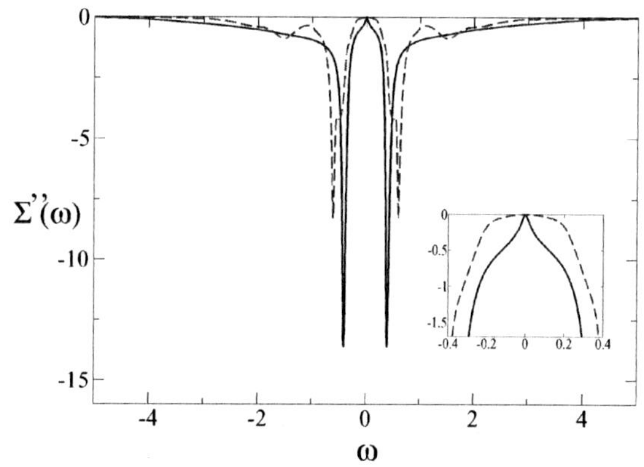

FIGURE 16. Comparison between the IPT (dashed lines) and NRG methods (plain lines), reproduced from Ref. [87]. The low-frequency behaviour of the inverse lifetime ImΣ clearly displays a critically enhanced curvature, which is not reproduced by IPT.

correct result $\text{Im}\Sigma_{IPT}(\omega = 0) \propto U^2T^2/D^3$, as illustrated For this reason, the numerical renormalization group (NRG) has been used recently [103, 87] in order to perform accurate transport calculations within DMFT. This method is very appropriate in this context, since it is highly accurate at low energies and yields real-frequency data[32]. DMFT-NRG calculations compare favorably to transport data on organics, as shown on Fig. 15.

The crossovers described here in electrical transport also have consequences for thermal transport. The thermopower, in particular, displays a saturation in the incoherent metal regime [106, 101]. This is presumably relevant for the cobalt- based thermoelectric oxides such as Na_xCoO_2 . Finally, let me emphasize that an interesting experimental investigation of the correlations between transport crossovers (both ab-plane and c-axis) and the loss of quasiparticle coherence observed in photoemission has been performed by Valla *et al.* [107] for several layered materials. This study raises intriguing questions in connection with DMFT, and particularly its **k**-dependent extensions.

4.4. Critical behaviour: a liquid-gas transition

Progress has been made recently in identifying the critical behaviour at the Mott critical endpoint, both from a theoretical and experimental standpoint. It was been pointed early on by Castellani *et al.*[108] (see also [109]) that an analogy exists with the liquid-gas transition in a classical fluid. This is based on a qualitative picture illustrated on Fig. 17. The Mott insulating phase has few double occupancies (or holes) and corresponds to a low-density "gas", while the metallic phase corresponds to a high-density "liquid" with many double-occupancies and holes (so that the electrons can be itinerant).

FIGURE 17. Cartoon of a typical real-space configuration of electrons in the Mott insulator (left) and metallic (right) phase. The insulator has few double-occupancies or holes, and corresponds to a gas of these excitations. Fluctuating local moments exist in this phase. The metal has many double-occupancies and holes, corresponding to a dense "liquid". Electrons are itinerant in the metallic phase, and the local moments are quenched. Within DMFT this quenching is akin to a (self-consistent) local Kondo effect.

Recently, this analogy has been given firm theoretical foundations within the framework of a Landau theory [68, 110, 111] derived from DMFT by Kotliar and coworkers. In this framework, a scalar order parameter ϕ is associated with the low-energy electronic degrees of freedom which build up the quasiparticle resonance in the strongly correlated metallic phase close to the transition. This order parameter couples to the singular part of the double occupancy (hence providing a connection to the qualitative picture above), as well as to other observables such as the Drude weight or the dc-conductivity. Because of the scalar nature of the order parameter, the transition falls in the Ising universality class. In Table 1, the correspondence between the Ising model quantities, and the physical observables of the liquid-gas transition and of the Mott metal-insulator transition is summarized.

In Fig. 18, the dc-conductivity obtained from DMFT in the half-filled Hubbard model (using IPT) is plotted as a function of the half-bandwith D, for several different temperatures. The curves qualitatively resemble those of the Ising model order parameter as a function of magnetic field (in fact, $D-D_c$ is a linear combination of the field h and of the mass term r in the Ising model field theory). Close to the critical point, scaling implies that the whole data set can be mapped onto a universal form of the equation of state:

$$\langle\phi\rangle = h^{1/\delta} f_{\pm}\left(h/|r|^{\gamma\delta/(\delta-1)}\right) \qquad (111)$$

In this expression, γ and δ are critical exponents associated with the order parameter and susceptibility, respectively: $\langle\phi\rangle \sim h^{1/\delta}$ at $T=T_c$ and $\chi = d\langle\phi\rangle/dh \sim |T-T_c|^{-\gamma}$. $f_{\pm}$ are

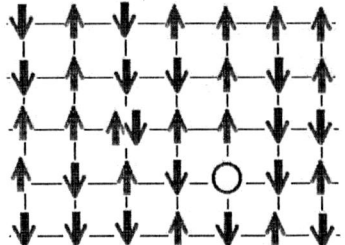

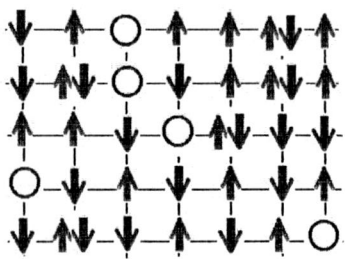

TABLE 3. Liquid-gas description of the Mott critical endpoint. The associated Landau free-energy density reads $r\phi^2 + u\phi^4 - h\phi$ (a possible ϕ^3 can be eliminated by an appropriate change of variables and a shift of ϕ).

Hubbard model	Mott MIT	Liquid-gas	Ising model
$D - D_c$	$p - p_c$	$p - p_c$	Field h (w/ some admixture of r)
$T - T_c$	$T - T_c$	$T - T_c$	Distance to critical point r (w/ some admixture of h)
Low-ω spectral weight	Low-ω spectral weight	$v_g - v_L$	Order parameter (scalar field ϕ)

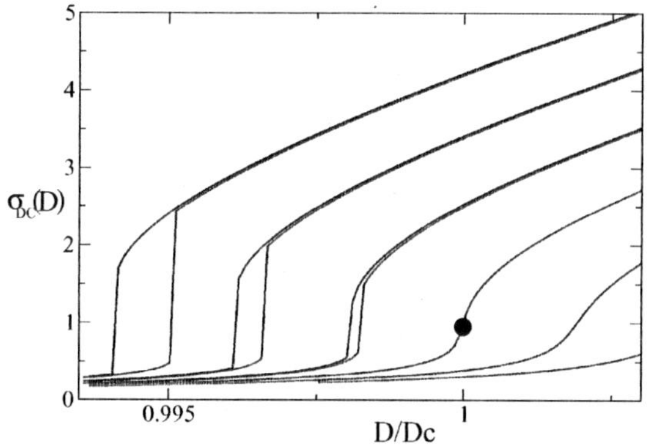

FIGURE 18. IPT calculation of the dc-conductivity as a function of the half-bandwith for the half-filled Hubbard model within DMFT, for several different temperatures. Increasing D drives the system more metallic. The curve at $T = T_c$ displays a singularity (vertical slope: dot), analogous to the non-linear dependence of the order parameter upon the magnetic field at a second-order magnetic transition. Hysteretic behaviour is found for $T < T_c$.

universal scaling functions associated with $T > T_c$ (resp. $T < T_c$). A quantitative study of the critical behaviour of the double occupancy within DMFT was made in Ref. [110], with the expected mean-field values of the exponents $\gamma = 1, \delta = 3$.

Precise experimental studies of the critical behaviour at the Mott critical endpoint have been performed very recently, using a variable pressure technique, for Cr-doped V_2O_3 by Limelette *et al.* [112] (Fig. 19) and also for the κ-BEDT organic compounds by Kagawa *et al.* [113]. These studies provide the first experimental demonstration of the liquid-gas critical behaviour associated with the Mott critical endpoint, including a a full scaling [112] onto the universal equation of state (111).

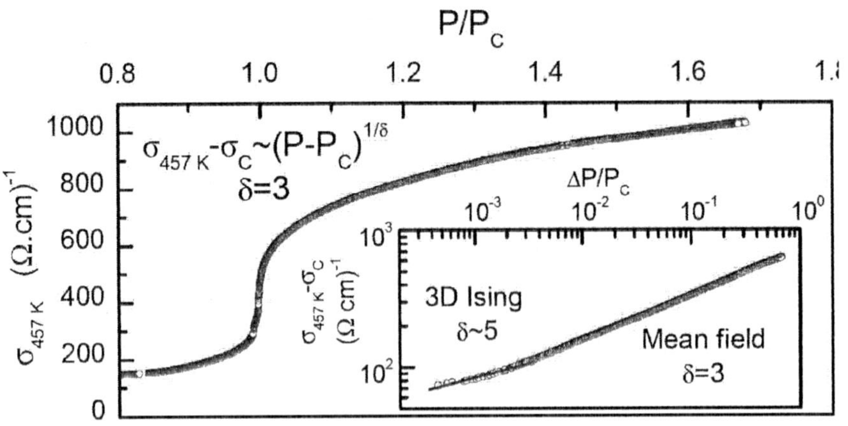

FIGURE 19. Conductivity of Cr-doped V_2O_3, at the critical endpoint $T = T_c$, measured as a function of pressure P/P_c (Limelette *et al.* [112]). A characteristic sigmoidal form is found, which is well fitted by $\sigma - \sigma_c \sim |P - P_c|^{1/\delta}$ (plain line). Inset: log-log scale. See Ref. [112] for a full experimental study of the critical behaviour, including scaling onto the universal equation of state.

4.5. Coupling to lattice degrees of freedom

Lattice degrees of freedom do play a role at the Mott transition in real materials, e.g the lattice spacing changes discontinuously through the first-order transition line in $(V_{1-x}Cr_x)_2O_3$, as displayed in Fig. 20. In the metallic phase, the d-electrons participate in the cohesion of the solid, hence leading to a smaller lattice spacing than in the insulating phase.

Both the electronic degrees of freedom and the ionic positions must be retained in order to describe these effects. In Ref. [114] (see also [115]), such a model was treated in the simplest approximation where all phonon excitations are neglected. The free energy then reads:

$$F = \frac{1}{2} B_0 \frac{(v - v_0)^2}{v_0} + F_{el}\left[D(v)\right] \tag{112}$$

In this expression, v is the unit-cell volume, B_0 is a reference elastic modulus and the electronic part of the free-energy F_{el} depends on v through the volume-dependence of the bandwith. In such a model, the critical endpoint is reached when the electronic response function:

$$\chi = -\frac{\partial^2 F_{el}}{\partial D^2} \tag{113}$$

is large enough (but not infinite), and hence the critical temperature T_c of the compressible model is larger than T_c^{el} (at which χ diverges in the Hubbard model). The compressibility $\kappa = \left(v\partial^2 F/\partial v^2\right)^{-1}$ diverges at T_c. This implies an anomalous lowering of the sound-velocity at the transition [116, 117], an effect that has been experimentally observed in the κ-BEDT compounds recently [118], as shown on Fig. 21.

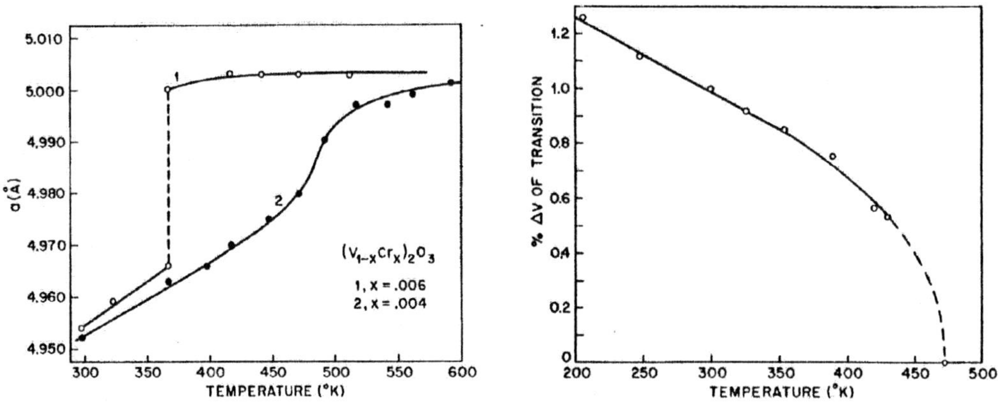

FIGURE 20. Left: Change of the lattice constant as a function of temperature for two samples of $(V_{1-x}Cr_x)_2O_3$ with different Cr-concentrations. The discontinuous change in the lattice constant through the first-order transition transition line is clearly seen for $x = .006$, while the sample with $x = .004$ is slightly to the right of the critical point. Right: Percentage volume change of the unit-cell volume close to the critical line, reflecting the critical behaviour of the order parameter. Reproduced from Ref. [109]

We emphasize that, within DMFT, an unambiguous answer is given to the "chicken and egg" question: is the first-order Mott transition driven by electronic or lattice degrees of freedom ? Within DMFT, the transition is described as an electronic one, with lattice degrees of freedom following up. In fact, it is aremarkable finding of DMFT that a purely electronic model can display a first-order Mott transition and a finite-T critical endpoint (associated with a diverging χ), provided that magnetism is frustrated enough so that ordering does not preempt the transition. Whether this also holds for the finite-dimensional Hubbard model beyond DMFT is to a large extent an open question (see [41] for indications supporting this conclusion in the 2D case).

4.6. The frontier: k-dependent coherence scale, cold and hot spots

A key question, still largely open, in our theoretical understanding of the Mott transition is the role of spatial correlations (inadequately treated by DMFT). This is essential in materials like cuprates, in which short-range spatial correlations play a key role (in particular magnetic correlations due to superexchange, leading to a strong tendency towards the formation of singlet bonds, as well as pair correlations). In the regime where the quasiparticle coherence scale ε_F^* is small as compared to the (effective strength of the) superexchange J, the DMFT picture is certainly deeply modified. There is compelling experimental evidence that the quasiparticle coherence scale then has a strong variation as the momentum **k** is varied along the Fermi surface, leading to the formation of "cold spots" and "hot regions". Such effects have been found in recent studies using cluster extensions of the DMFT framework ([119], see also [39, 120]).

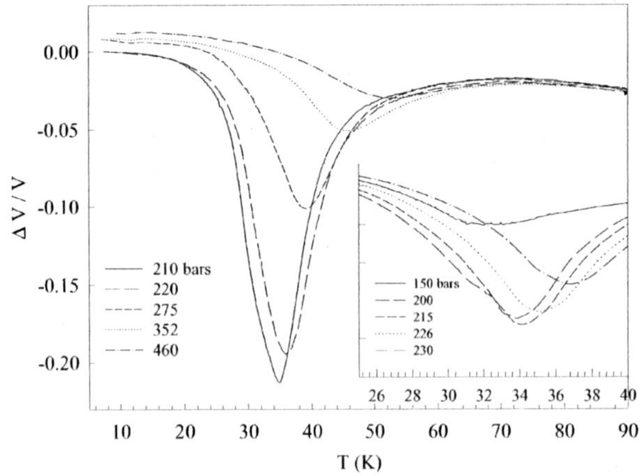

FIGURE 21. Relative change in the sound velocity of κ-(BEDT-TTF)$_2$Cu[N(CN)$_2$]Cl as a function of temperature, at various pressures. The velocity variation is relative to the value at 90 K. Inset: position and amplitude of the anomaly below 230 bars. (Figure and caption from Ref. [118]).

5. ELECTRONIC STRUCTURE AND DYNAMICAL MEAN-FIELD THEORY

The possibility of using DMFT in combination with electronic structure calculation methods, in order to overcome some of the limitations of DFT-LDA for strongly correlated materials, was pointed out early on [3]. In the last few years, very exciting developments have taken place, in which theorists from the electronic structure and many-body communities joined forces and achieved concrete implementations of DMFT within electronic structure calculations. The first papers [121, 122] implementing this combination appeared in 1997-1998, and the field has been extremely active since then. For reviews of the early developments in this field[16], see Refs. [123, 124, 125, 126]. For on-line material presented at recent workshops, see Refs. [127, 128, 129]

5.1. Limitations of DFT-LDA for strongly correlated systems

In Sec. 3.2, I briefly presented the basic principles of density functional theory (DFT). In practice, the local density approximation (LDA) to the exchange-correlation energy, and its extensions (such as the generalised gradient approximation) have been remarkably successful at describing ground-state properties of many solids from first principles. This is also the state of the art method for band structure calculations, with the additional assumption that Kohn-Sham eigenvalues can be interpreted as single-particle excitations. For strongly correlated materials however, DFT-LDA has severe limitations,

[16] This section is merely a brief introduction to the field and certainly not as an exhaustive review.

which we now briefly review.

Issues about ground-state properties. Ground-state properties, such as equilibrium unit-cell volume, are not accurately predicted from LDA (or even GGA) for the most strongly correlated materials. This is particularly true of materials in which some electrons are very localized, such as the 4f electrons of rare-earth elements at ambiant or low pressure (Sec. 1.3.3). If these orbitals are treated as valence orbitals, the LDA leads to a much too itinerant character, and therefore overestimates the contribution of these orbitals to the cohesive energy of the solids, hence leading to a too small unit-cell volume. If instead the f-orbitals are treated as core states, the equilibrium volume is then overestimated (albeit closer to experimental value, in the case of rare earth), since binding is underestimated. Phenomena such as the volume-collapse transitions, associated with the partial delocalization of the f-electrons, (and associated structural changes) under pressure [15] are simply out of reach of standard methods. At high pressures however, the f-electrons recover itinerant character and DFT-LDA(GGA) does better, as expected. In some particular cases, the electrons are just on the verge of the itinerant/localized behaviour. In such cases, standard electronic structure methods perform very poorly. A spectacular example is the δ-phase of metallic plutonium in which the unit-cell volume is underestimated (compared to the experimental value) by as much as 35% by standard electronic structure methods (Fig. 24) ! All these examples illustrate the need of a method which is able to handle intermediate situations between fully localized and fully itinerant electrons. I emphasize that this issue may depend crucially on energy scales, with localized character most pronounced at high-energy (short time) scales, and itinerant quasiparticles forming at low-energy (long time scales).

Excitation spectra. Even though the Kohn-Sham eigenvalues and wavefunctions are, strictly speaking, auxiliary quantities in the DFT formalism used to represent the local density, they are commonly interpreted as energy bands in electronic structure calculations. This is very successful in many solids, but does fail badly in strongly correlated ones. The most spectacular difficulty is that Mott insulators are found to have metallic Kohn-Sham spectra. This is documented, e.g by Fig. 23, in which the LDA density of states of two Mott insulators, $LaTiO_3$ and $YTiO_3$ are shown. I emphasize that, in both compounds (as well as in many other Mott insulators), the Mott insulating gap has nothing to do with the magnetic ordering in the ground-state. Even though magnetic long-range order is found at low-enough temperatures in both materials (below $T_N \simeq 140$ K in $LaTiO_3$ and $T_C \simeq 30$ K in $YTiO_3$), the insulating behaviour and Mott gap ($\simeq 1$ eV for $YTiO_3$) are maintained well above the ordering temperature. In other cases (such as VO_2), the insulating phase is a paramagnet and the LDA spectrum is again metallic.

In strongly correlated metals, e.g close to Mott insulators, the LDA bandstructure is also in disagreement with experimental observations. The two main discrepancies are the following. (i) LDA single-particle bands are generally too broad. Correlation effects lead to band-narrowing, corresponding to a (Brinkman-Rice) enhancement of the effective masses of quasiparticles. This becomes dramatic in f-electron materials, where the large effective mass is due to the Kondo effect, a many-body process which is beyond the reach of single-particle theories. (ii) The spectral weight Z associated with quasiparticles is reduced by correlations, and the corresponding missing spectral weight $1-Z$ is found in intermediate or high-energy incoherent excitations. In correlated metals, as well as in Mott insulators, lower and upper Hubbard bands are observed, which are absent in the LDA density of states (e.g for $SrVO_3$ and $CaVO_3$ in Fig. 11 and Fig 23).

Related correlation effects are observed also for pure transition metals, such as nickel, in which the LDA spectrum is unable to account for: the $\sim -6\,\mathrm{eV}$ photoemission satellite, and for the correct values of the occupied bandwidth and exchange splitting between the majority and minority band in the ferromagnetic ground-state.

5.2. Marrying DMFT and DFT-LDA

In this section, I briefly describe the (happy) marriage of electronic structure methods and dynamical mean-field theory. I first give a simple practical formulation in terms of a realistic many-body hamiltonian, and keep for the next section the construction of energy functionals.

The first issue to be discussed is the choice of the basis set for the valence electrons. Since DMFT emphasizes local correlations, we need a localised basis set, i.e basis functions which are centered on the atomic positions $\mathbf{R}$ in the crystal lattice. Up to now, most implementations have used basis sets based on linear muffin-tin orbitals [130, 131] (LMTOs) $\chi_{L\mathbf{R}}(\mathbf{r}) = \chi_L(\mathbf{r}-\mathbf{R})$ (in which $L = \{l,m\}$ stands for the angular momentum quantum number of the valence electrons). These basis sets offer the advantage to carry over the physical intuition of atomic orbitals from the isolated atoms to the solid. In the words of their creator, O.K. Andersen, LMTO- based electronic structure methods are "intelligible" because they are based on a minimal and flexible basis set of short-range orbitals [132]. There are several possible choices of basis even within the LMTO method. Basically, a compromise has to be made between the degree of localisation and the orthogonality of the basis set. The most localised basis set (the so-called "screened" or α-basis) is not orthogonal and will therefore involve[17] an overlap matrix $O_{LL'} = \langle \chi_L | \chi_{L'} \rangle$. Since DMFT neglects non-local correlations, they may be the best one to choose. However, a non-orthogonal basis set may not be simple to implement, for technical reasons, when using some impurity solvers (e.g QMC). Orthogonal LMTOs basis sets are somewhat more extended.

Another possibility is to use basis sets made of Wannier functions. This has been little explored yet in combination with DMFT. Wannier functions can in fact be constructed starting from the LMTO formalism by using the "downfolding" procedure (the so-called third-generation LMTO [132, 133]). Recently, DMFT has been implemented within a downfolded (NMTO) Wannier basis, and successfully applied to transition metal oxides with non-cubic structures. Other routes to Wannier functions (such as the Marzari-Vanderbilt construction of maximally localised Wannier functions [134]) might be worth pursuing. Given a basis set, the electron creation operator at a point $\mathbf{r}$ in the solid can be

[17] In the following, we assume an orthogonal basis set to simplify the formalism. The overlap matrix can be easily reintroduced where it is appropriate

decomposed as:

$$\psi^\dagger(\mathbf{r}) = \sum_{\mathbf{R},L} \chi^*_{L\mathbf{R}}(\mathbf{r}) c^\dagger_{L\mathbf{R}} \tag{114}$$

The decomposition of the full Green's function in the solid: $G(\mathbf{r},\mathbf{r}',\tau-\tau') \equiv -\langle T\psi(\mathbf{r},\tau)\psi^\dagger(\mathbf{r}',\tau')\rangle$ (as well as of any other one-particle quantity) thus reads:

$$G(\mathbf{r},\mathbf{r}',i\omega) = \sum_{\mathbf{R}\mathbf{R}'} \sum_{LL'} \chi_{L\mathbf{R}}(\mathbf{r}) G_{LL'}(\mathbf{R}-\mathbf{R}',i\omega) \chi_{L'\mathbf{R}'}(\mathbf{r}')^* \tag{115}$$

The simplest combination of DMFT and electronic structure methods uses a starting point which is similar to that of the LDA+U approach [135, 136]. Namely, one first separates the valence electrons into two groups: those for which standard electronic structure methods are sufficient on one hand (e.g $l = s, p$ in an oxide or $l = s, p, d$ in rare-earth compounds), and on the other hand the subset of orbitals which will feel strong correlations (e.g $l = d$ or $l = f$). This separation refers, of course, to the specific choice of basis set which has been made. In the following, I denote the orbitals with l in the correlated subset by the index $a \equiv \{m, \sigma\}$ (and $b, \cdots$). Let us then consider the one-particle hamiltonian:

$$H_{KS} = \sum_\lambda \varepsilon_\lambda^{KS} |\lambda\rangle\langle\lambda| = \sum_{\mathbf{k}L} h^{KS}_{LL'}(\mathbf{k}) c^\dagger_{\mathbf{k}L} c_{\mathbf{k}L'} \tag{116}$$

obtained from solving the Kohn-Sham equations for the material under consideration. The Kohn-Sham potential we have in mind is, in the simplest implementation, the one obtained within a standard DFT-LDA (or GGA) electronic structure calculation of the local density. In a more sophisticated implementation, one may also correct the local density by correlation effects and use the associated Kohn-Sham potential (i.e modify the self-consistency cycle over the local density in comparison to standard LDA, see below). A many-body hamiltonian is then constructed as follows:

$$H = H_{KS} - H_{DC} + H_U \tag{117}$$

In this expression, H_U are many-body terms acting in the subset of correlated orbitals only. They correspond to matrix elements of the Coulomb interaction, and will in general involve arbitrary 2-particle terms $U_{abcd} c^\dagger_a c^\dagger_b c_d c_c$. In practice however, one often makes a further simplification and keep only density-density interactions (for technical reasons, this is always done when using QMC as a solver). To simplify notations, we shall limit ourselves here to this case, and use:

$$H_U = \frac{1}{2} \sum_{\mathbf{R}} \sum_{ab\sigma} U_{ab} \hat{n}_{\mathbf{R}a} \hat{n}_{\mathbf{R}b} \tag{118}$$

with:

$$U^{\uparrow\downarrow}_{mm'} = U_{mm'} \;,\; U^{\uparrow\uparrow}_{m\neq m'} = U^{\downarrow\downarrow}_{m\neq m'} = U_{mm'} - J_{mm'} \tag{119}$$

In this expression, $J_{mm'}$ is the Hund's coupling. For a more detailed discussion of the choice of the matrix of interaction parameters, see e.g Ref. [136].

The "double-counting" term H_{DC} needs to be introduced, since the contribution of interactions between the correlated orbitals to the total energy is already partially included in the exchange-correlation potential. Unfortunately, it is not possible to derive this term explicitly, since the energy within DFT is a functional of the total electron density, which combines all orbitals in a non-linear manner. In practice, the most commonly used form of the double-counting term is (for other choices, see e.g [137]):

$$H_{DC} = \sum_{\mathbf{R}\sigma ab} V^{DC}_{ab\sigma} c^{\dagger}_{a\sigma} c_{b\sigma}$$

$$V^{DC}_{ab\sigma} = \delta_{ab}\left[U(N-\frac{1}{2}) - J(N^{\sigma}-\frac{1}{2})\right] \tag{120}$$

The many-body hamiltonian (117) is then soved using the DMFT approximation. This means that a local self-energy matrix is assumed, which acts in the subset of correlated orbitals only:

$$\Sigma^{\mathbf{RR}'}_{LL'}(i\omega) = \delta_{\mathbf{R},\mathbf{R}'}\begin{pmatrix} 0 & 0 \\ 0 & \Sigma_{ab}(i\omega) \end{pmatrix} \tag{121}$$

In the DMFT framework, the local Green's function in the correlated subset:

$$G_{ab}(\tau-\tau') \equiv -\langle T c^{\dagger}_{a}(\tau) c_{b}(\tau')\rangle \tag{122}$$

is represented as the Green's function of the multi-orbital impurity model:

$$S = -\int_0^{\beta} d\tau \int_0^{\beta} d\tau' \sum_{ab} c^{\dagger}_{a}(\tau)[\mathcal{G}_0^{-1}]_{ab}(\tau-\tau')c_b(\tau') + \frac{1}{2}\sum_{ab} U_{ab}\int_0^{\beta} d\tau\, n_a(\tau)n_b(\tau) \tag{123}$$

The Weiss function (or alternatively the dynamical mean-field, or effective hybridisation function $\Delta_{ab} = (i\omega_n+\mu)\delta_{ab} - [\mathcal{G}_0^{-1}]_{ab})$ is determined, as before, from the self-consistency condition requesting that the on-site Green's function in the solid coincides with the impurity model Green's function. The components of the Green's function of the solid in the chosen basis set read:

$$[G^{-1}]_{LL'}(\mathbf{k}, i\omega_n) = (i\omega_n+\mu)\delta_{LL'} - h^{KS}_{LL'} + V^{DC}_{LL'} - \Sigma_{LL'}(i\omega_n) \tag{124}$$

In this expression, the self-energy matrix $\Sigma_{LL'}$ is constructed by using the components of the impurity self-energy $\Sigma_{ab} \equiv [\mathcal{G}_0^{-1}]_{ab} - [G^{-1}_{imp}]_{ab}$ into (121). The self-consistency condition relating implicitly $\mathcal{G}_0$ and G_{imp} finally reads:

$$G(i\omega_n)_{ab} = \sum_{\mathbf{k}}\left[(i\omega_n+\mu)\delta_{LL'} - h^{KS}_{LL'} + V^{DC}_{LL'} - \Sigma_{LL'}(i\omega_n)\right]^{-1}_{ab} \tag{125}$$

Note that this involves a matrix inversion at each **k**-point, as well as a **k**-summation over the Brillouin zone (which does not, in general, reduces to an integration over the band density of states, in contrast to the single-band case). Also let us emphasize that, even though the self-energy matrix has only components in the subspace of correlated orbitals, the components of the Green's function corresponding to all valence orbitals

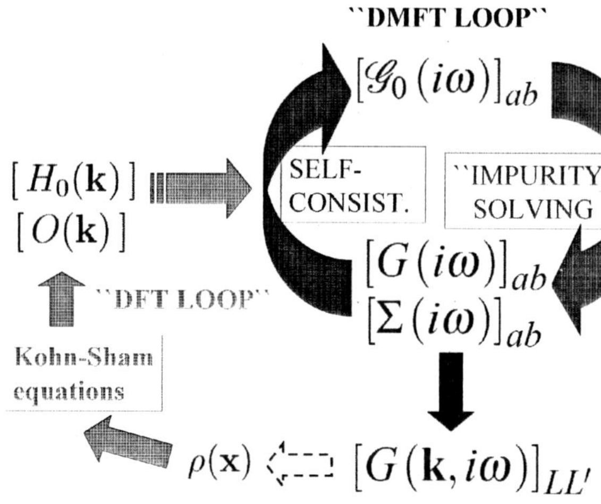

FIGURE 22. DMFT combined with electronic structure calculations. Starting from a local electronic density $\rho(\mathbf{r})$, the associated Kohn-Sham potential is calculated and the Kohn-Sham equations are solved. The Kohn-Sham hamiltonian $H^{KS}_{LL'}(\mathbf{k})$ is expressed in a localised basis set (e.g LMTOs). A double-counting term is substracted to obtain the one-electron hamiltonian $H_0 \equiv H^{KS} - H^{DC}$. The local self-energy matrix for the subset of correlated orbitals is obtained through the iteration of the DMFT loop: a multi-orbital impurity model for the correlated subset is solved (red arrow), containing as an input the dynamical mean-field (or Weiss field $\mathcal{G}_0$). The self-energy Σ_{ab} is combined with H_0 into the self-consistency condition Eq. (125) in order to update the Weiss field (blue arrow). At the end of the DMFT loop, the components of the full, $\mathbf{k}$-dependent, Green's function in the local basis set can be calculated and thus also an updated local density $\rho(\mathbf{r})$. This is used (dashed arrow) as a new starting density for the Kohn-Sham calculation until a converged local density is also reached . Alternatively, in a simplified implementation of this full scheme, the DFT-LDA calculation can be converged first and the corresponding H_0 injected into the DMFT loop without attempting to update $\rho(\mathbf{r})$.

$(s,p,d,\cdots)$ are modified due to the matrix inversion. Correlation effects encoded in the self-energy affect the local electronic density, which can be calculated from the full Green's function as:

$$\rho(\mathbf{r}) = \sum_{\mathbf{k}} \chi_{L\mathbf{k}}(\mathbf{r})\, G_{LL'}(\mathbf{k},\tau=0^-)\, \chi^*_{L'\mathbf{k}}(\mathbf{r}) \tag{126}$$

In a complete implementation, self-consistency over the local density should also be reached [73, 138]. The general structure of the combination of DMFT with electronic structure calculations, as well as the iterative procedure used in practice to solve the DMFT equations, is summarised on Fig. 22.

5.3. An application to d^1 oxides

On Fig. 23, I show the spectral functions recently obtained in Ref. [95] for $SrVO_3$, $CaVO_3$, $LaTiO_3$ and $YTiO_3$. These oxides have the same formal valence of the d-shell (d^1). The single electron sits in the t_{2g} multiplet, and the (empty) e_g doublet is well separated in energy. They have a perovskite structure with perfect cubic symmetry for the first one (Fig. 2) and increasing degree of structural distortion for the three others (corresponding mainly to the $GdFeO_3$-like tilting of oxygen octahedra). These calculations were performed in a downfolded (NMTO) basis set, including the off-diagonal components of the self-energy matrix. The latter are important for the compounds with the largest structural distortions. For comparison, the LDA density of states are shown on the same plot. For an independent DMFT calculation of the Ca/$SrVO_3$ compounds, see Ref. [94, 93] and Ref. [121, 139] for early calculations of the doped system $La_{1-x}Sr_xTiO_3$. The spectra in Fig. 23 have features which should be familiar to the reader at this point, namely:

- $SrVO_3$ and $CaVO_3$ are correlated metals with lower (~ -1.5 eV) and upper (~ 2.5 eV) Hubbard bands, as well as a relatively moderate narrowing of the quasiparticle bandwith. The calculated spectra compare favorably to the recent photoemission experiments of Fig. 11 (see [93] for a comparison).
- $LaTiO_3$ and $YTiO_3$ are Mott insulators, with quite different values of the Mott gap (~ 0.3 eV and ~ 1 eV, respectively) as observed experimentally. It was emphasized in [95] that the main reason for this difference is that the orbital degeneracy of the t_{2g} multiplet is lifted to a greater degree in $YTiO_3$ than in $LaTiO_3$ due to the larger structural distortion. Indeed, reducing orbital degeneracy is known to increase the effect of correlations (for comparable interaction strength) [140, 141, 142, 143]. It was also found in Ref. [95] that both compounds develop a very pronounced orbital polarization, of a quite different nature in each compound (see [144] for a discussion of orbital ordering in these materials and [145] for a recent experimental investigation).

This example, as well as several other recent studies, demonstrate that the embedding of DMFT within electronic structure calculations yields a powerful quantitative tool for understanding the rich interplay between correlation effects and material-specific aspects.

5.4. Functionals and total- energy calculations

In order to discuss total energy calculations in the LDA+DMFT framework[18], it is best to use a formulation of this scheme in terms of a (free-) energy functional. Kotliar and Savrasov [138, 147] have introduced for this purpose a ("spectral-density-

---

[18] I acknowledge a collaboration with B. Amadon and S. Biermann [146] on the topic of this section.

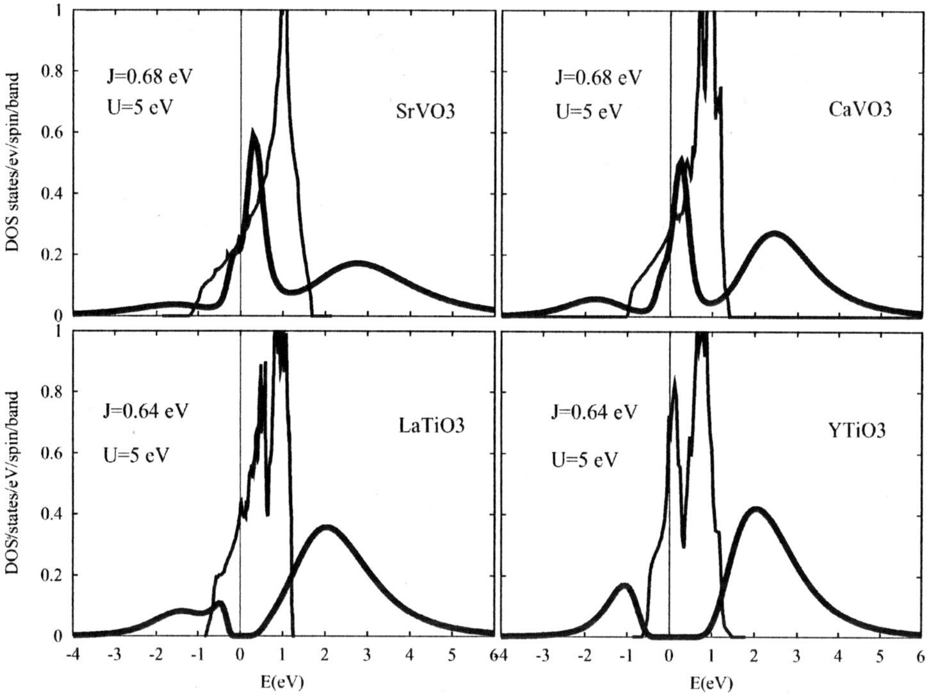

FIGURE 23. LDA+DMFT spectral densities of the transition-metal oxides discussed in the text, from Ref. [95]. The (QMC) calculations were made at $T = 770$K. For comparison, the LDA d.o.s are also displayed (thin lines).

") functional of both the total local electron density $\rho(\mathbf{r})$ and the on-site Green's function in the correlated subset: $G_{ab}^{\mathbf{RR}}$ (denoted G_{ab} for simplicity in the following). Let us emphasize that these quantities are independent, since G_{ab} is restricted to local components and to a subset of orbitals so that $\rho(\mathbf{r})$ cannot be reconstructed from it. The functional is constructed by introducing (see Section. 3) source terms $\lambda(\mathbf{r}) = v_{KS}(\mathbf{r}) - v_c(\mathbf{r})$ and $\Delta\Sigma_{ab}(i\omega_n)$ coupling to the operators $\psi^\dagger(\mathbf{r})\psi(\mathbf{r})$ and to $\sum_{\mathbf{R}} \chi_a^*(\mathbf{r}-\mathbf{R})\psi(\mathbf{r},\tau)\psi^\dagger(\mathbf{r}',\tau')\chi_b(\mathbf{r}'-\mathbf{R}) = c_{a\mathbf{R}}(\tau)c_{b\mathbf{R}}^\dagger(\tau')$, respectively. Furthermore, the Luttinger-Ward part of the functional is approximated by that of the on-site local many-body hamiltonian $H_U - H_{DC}$ introduced above. This yields:

$$\Omega[\rho(\mathbf{r}), G_{ab}; v_{KS}(\mathbf{r}), \Delta\Sigma_{ab}]_{LDA+DMFT} =$$
$$-\mathrm{tr}\ \ln[i\omega_n + \mu + \tfrac{1}{2}\nabla^2 - v_{KS}(\mathbf{r}) - \chi^*.\Delta\Sigma.\chi] - \int d\mathbf{r}\,(v_{KS} - v_c)\rho(\mathbf{r}) - \mathrm{tr}\,[G.\Delta\Sigma] +$$
$$+\tfrac{1}{2}\int d\mathbf{r}\,d\mathbf{r}'\rho(\mathbf{r})U(\mathbf{r}-\mathbf{r}')\rho(\mathbf{r}') + E_{xc}[\rho(\mathbf{r})] + \sum_{\mathbf{R}}\left(\Phi_{imp}[G_{ab}^{\mathbf{RR}}] - \Phi_{DC}[G_{ab}^{\mathbf{RR}}]\right)$$

In this expression, $\chi^*.\Delta\Sigma.\chi$ denotes the "upfolding" of the local quantity $\Delta\Sigma$ to the whole solid: $\chi^*.\Delta\Sigma.\chi = \sum_{\mathbf{R}}\sum_{ab}\chi_a^*(\mathbf{r}-\mathbf{R})\Sigma_{ab}(i\omega_n)\chi_b(\mathbf{r}'-\mathbf{R})$. Variations of this functional with respect to the sources $\delta\Omega/\delta v_{KS} = 0$ and $\delta\Omega/\delta\Sigma_{ab} = 0$ yield the standard expression of the local density and local Green's function in terms of the full Green's function in the solid:

$$\rho(\mathbf{r}) = \langle \mathbf{r}|\hat{G}|\mathbf{r}\rangle \ , \ G_{ab}(i\omega_n) = \langle \chi_{a\mathbf{R}}|\hat{G}|\chi_{b\mathbf{R}}\rangle \tag{127}$$

with:

$$\hat{G} = \left[i\omega_n + \mu + \frac{1}{2}\nabla^2 - v_{KS}(\mathbf{r}) - \chi^*.\Delta\Sigma.\chi\right]^{-1} \tag{128}$$

or, in the local basis set (see (124)):

$$\hat{G} = \sum_{\mathbf{k},LL'} |\chi_{L\mathbf{k}}\rangle \left[(i\omega_n + \mu).1 - \hat{h}^{KS}(\mathbf{k}) - \Delta\hat{\Sigma}(i\omega_n)\right]^{-1}_{LL'} \langle \chi_{L'\mathbf{k}}| \tag{129}$$

From these relations, the Legendre multiplier functions v_{KS} and $\Delta\Sigma$ could be eliminated in terms of ρ and G_{ab}, so that a functional of the local observables only is obtained:

$$\Gamma_{LDA+DMFT}[\rho, G_{ab}] = \Omega_{LDA+DMFT}\left[\rho(\mathbf{r}), G_{ab}; \lambda[\rho, G], \Delta\Sigma[\rho, G]\right] \tag{130}$$

Extremalisation of this functional with respect to ρ ($\delta\Gamma/\delta\rho = 0$) and G_{ab} ($\delta\Gamma/\delta G_{ab} = 0$) yields the expression of the Kohn-Sham potential and self-energy correction at self-consistency:

$$v_{KS}(\mathbf{r}) = v_c(\mathbf{r}) + \int d\mathbf{r}' U(\mathbf{r}-\mathbf{r}')\rho(\mathbf{r}') + \frac{\delta E_{xc}}{\delta\rho(\mathbf{r})} \tag{131}$$

$$\Delta\Sigma_{ab} = \frac{\delta\Phi_{imp}}{\delta G_{ab}} - \frac{\delta\Phi_{DC}}{\delta G_{ab}} \equiv \Sigma^{imp}_{ab} - V^{DC}_{ab} \tag{132}$$

Hence, one recovers from this functional the defining equations of the LDA+DMFT combined scheme, including self-consistency over the local density (127). Using (66) and (61), one notes that the free-energy can be written as:

$$\begin{aligned}\Omega_{LDA+DMFT} = \Omega_{DFT} &+ \mathrm{tr}\ln G_{KS}(\mathbf{k}, i\omega_n)^{-1} - \mathrm{tr}\ln G(\mathbf{k}, i\omega_n)^{-1} - \mathrm{tr}\left[G_{imp}\Sigma^{imp}\right] + \Sigma_{\mathbf{R}}\Phi_{imp} + \\ &+ \mathrm{tr}\left[G_{imp}V^{DC}\right] - \Sigma_{\mathbf{R}}\Phi_{DC}\end{aligned} \tag{133}$$

In this expression, Ω_{DFT} is the usual density-functional theory expression (66), while G_{KS} is the Green's function corresponding to the Kohn-Sham hamiltonian, i.e without the self-energy correction:

$$G^{-1}_{KS} \equiv i\omega_n + \mu - \hat{h}_{KS}(\mathbf{k}) \tag{134}$$

A careful examination of the zero-temperature limit of (133) leads to the following expression of the total energy [146]:

$$E_{LDA+DMFT} = E_{DFT} - \Sigma'_{\lambda}\, \varepsilon^{KS}_{\lambda} + \langle H_{KS}\rangle + \langle H_U\rangle - E_{DC} \tag{135}$$

$$= E_{DFT} + \Sigma_{\mathbf{k},LL'}\, h^{KS}_{LL'}\left[\langle c^{\dagger}_{L\mathbf{k}} c_{L'\mathbf{k}}\rangle_{DMFT} - \langle c^{\dagger}_{L\mathbf{k}} c_{L'\mathbf{k}}\rangle_{KS}\right] + \langle H_U\rangle - E_{DC} \tag{136}$$

The first term, E_{DFT} is the energy found within DFT(LDA), using of course the local density obtained at the end of the LDA+DMFT convergence cycle, namely:

$$E_{DFT} = \sum_{\lambda}{}' \varepsilon^{KS}_{\lambda} + \int d\mathbf{r}[v_c(\mathbf{r}) - v_{KS}(\mathbf{r})]\rho(\mathbf{r}) + \frac{1}{2}\int d\mathbf{r}d\mathbf{r}'\rho(\mathbf{r})u(\mathbf{r}-\mathbf{r}')\rho(\mathbf{r}') + E_{xc}[\rho] \tag{137}$$

Hence the total energy within LDA+DMFT is made of several terms. Importantly, it does *not* simply reduce to the expectation value $\langle H\rangle$ of the many-body hamiltonian (117) introduced in the previous section. Furthermore, $\langle H_{KS}\rangle = \mathrm{tr}\,[H_{KS}\hat{G}]$ must be evaluated with the full Green's function including the self-energy correction. Therefore, this quantity does not coincide with the sum of the (occupied) Kohn-Sham eigenvalues $\sum'_\lambda \varepsilon_\lambda^{KS} = \mathrm{tr}\, H_{KS}G_{KS}$. Eq. (135) expresses that the latter has to be removed from E_{DFT}, in order to correctly take into account the change of energy coming from the Kohn-Sham orbitals. This change can also be written $\langle H_{KS}\rangle_{DMFT} - \langle H_{KS}\rangle_{KS} = \mathrm{tr}\,[(G-G_{KS})H_{KS}]$. This is used in the second expression for the energy, which emphasizes the modification of the density matrix $\langle c^\dagger_{L\mathbf{k}} c_{L'\mathbf{k}}\rangle$ by correlations. Finally, the double-counting correction to the energy is the zero-temperature limit of $-\langle H_{DC}\rangle + \mathrm{tr}\,[G\Sigma^{DC}] - \Phi_{DC}$. The simplest form of double-counting correction (neglecting J for simplicity) corresponds to: $\Phi_{DC}[G_{ab}] = UN(N-1)/2$ with $N = \sum_a n_a = \sum_a \mathrm{tr}\, G_{aa}$. Hence $V^{DC}_{ab} = \delta\Phi_{DC}/\delta G_{ab} = U(N-1/2)n_a\delta_{ab}$, and $\langle H_{DC}\rangle = \mathrm{tr}\,[G\Sigma^{DC}] = UN(N-1/2)$ so that, finally: $E_{DC} = UN(N-1)/2$.

Another formula for the total energy within LDA+DMFT has been used by Held *et al.* in their investigation of the volume collapse transition of Cerium [148, 149].

Total energy calculations within LDA+DMFT, with full self-consistency on the local density have been performed by Savrasov, Kotliar and Abrahams [73, 138, 147] for metallic plutonium with fcc structure, corresponding to the δ-phase. The results are reproduced in Fig. 24, in which the total energy is plotted as a function of the unit-cell volume (normalised by the experimental value), for different values of the parameter U. It is seen that the GGA calculation underestimates the volume by more than 30%. As U increases, the minimum is pushed to higher volumes, and good agreement with experiments is reached for U in the range $3.8-4\,\mathrm{eV}$. Interestingly, in the presence of correlations, the energy curve develops a metastable shallow minimum at a lower volume, which can be interpreted as a manifestation of the α-phase (which has a more complicated crystal structure however). For the corresponding spectra, see [147]. In these DMFT calculations, the δ-phase of plutonium is described as a paramagnetic metal, in agreement with experiments. In contrast, a static LDA+U treatment[72, 150] also corrects the equilibrium volume, but at the expense of introducing an unphysical spin polarization[19].

5.5. A life without U: towards *ab-initio* DMFT

The combination of DMFT with electronic structure methods described in the previous section introduces a matrix U of local interaction parameters acting in the subset of correlated orbitals, as in the LDA+U scheme. Some of these parameters can be determined from constrained LDA calculations, or instead they can be viewed as adjustable.

---

[19] For an alternative description of the δ-phase of plutonium, in which a subset of the f-electrons are viewed as localised while the others are itinerant, see [151]

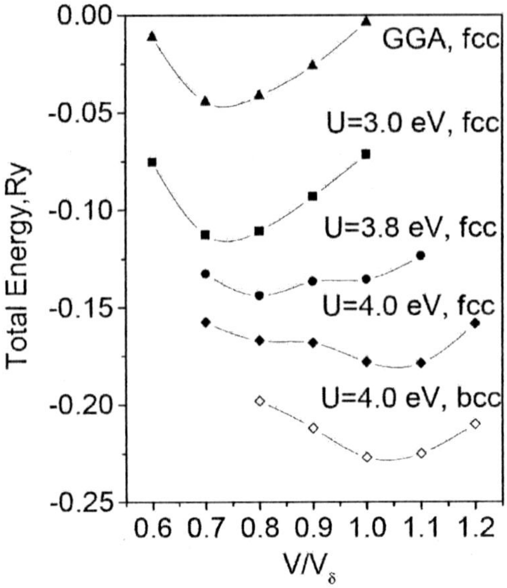

FIGURE 24. Total energy of fcc plutonium as a function of unit-cell volume (normalised by the experimental volume of the δ-phase), reproduced from Ref. [73, 147]. The upper curve is the GGA result. Other curves are from LDA+DMFT with different values of U. The lower curve is for the bcc structure.

Furthermore, introducing these interactions implies the need for a "double-counting" correction in order to remove the contribution to the total energy already taken into account in the (orbital-independent) exchange correlation potential. As such, this theory has great practical virtues. However, going beyond this framework and being able to treat the electron-electron interaction entirely from first- principles is a tempting and challenging project. Work in this direction have appeared recently [152, 125, 153, 154, 155, 156, 6].

Physically, the Hubbard interaction is associated with the screened Coulomb interaction as seen by a given atom in the solid. *Screening* is essential for estimating the order of magnitude of this parameter correctly. The naive view that U is simply the on-site matrix element of the Coulomb potential in the local basis- set would lead to values on the scale of tens of electron-volts, while the appropriate value in the solid is a few eV's ! This immediately points towards a key notion: that, in fact, the Hubbard U is a concept which *depends on the energy-scale*. At high energies (say, above the plasma frequency in a metal), it has a very large value associated with the bare, unscreened, matrix element, while at low energy screening takes place and it is considerably reduced. For first-principle RPA studies of the frequency dependence of the screened local interaction, see [5, 6].

In fact, the screened effective interaction in a solid can be related, quite generally, to the density-density correlation function. Let us start from the first-principles hamiltonian:

$$H = -\sum_i \frac{1}{2}\nabla_i^2 + \sum_i v(\mathbf{r}_i) + \frac{1}{2}\sum_{i\neq j} u(\mathbf{r}_i - \mathbf{r}_j)$$

$$= -\frac{1}{2}\int d\mathbf{r}\psi^\dagger\nabla^2\psi + \int d\mathbf{r}\, v(\mathbf{r})\hat{n}(\mathbf{r}) + \frac{1}{2}\int d\mathbf{r}d\mathbf{r}'\, u(\mathbf{r}-\mathbf{r}') : \hat{n}(\mathbf{r})\hat{n}(\mathbf{r}') : \quad (138)$$

in which $u(\mathbf{r}-\mathbf{r}') = e^2/|\mathbf{r}-\mathbf{r}'|$ is the bare Coulomb interaction, $\hat{n} \equiv \psi^\dagger(\mathbf{r})\psi(\mathbf{r})$ and $:():$ denotes normal ordering. The (connected) density-density correlation function is defined as:

$$\chi(\mathbf{r},\mathbf{r}';\tau-\tau') = \langle T\,(\hat{n}(\mathbf{r},\tau)-\rho(\mathbf{r}))\,(\hat{n}(\mathbf{r}',\tau')-\rho(\mathbf{r}'))\rangle \quad (139)$$

with $\rho(\mathbf{r}) = \langle\hat{n}(\mathbf{r})\rangle$ the local density. The screened effective interaction reads:

$$W(\mathbf{r},\mathbf{r}',i\omega) = u(\mathbf{r}-\mathbf{r}') - \int d\mathbf{r}_1\, d\mathbf{r}_2\, u(\mathbf{r}-\mathbf{r}_1)\chi(\mathbf{r}_1-\mathbf{r}_2;i\omega)u(\mathbf{r}_2-\mathbf{r}') \quad (140)$$

This can also be expressed in terms of the polarization $P \equiv -\chi.[1-u.\chi]^{-1}$ as $W = u.[1-P.u]^{-1}$ (the dot is an abbreviation for spatial convolutions). We emphasize that in this expression, P is the exact polarization operator, not its RPA approximation. The screened interaction W can be interpreted as the correlation function of the local scalar potential field conjugate to $\hat{n}(\mathbf{r})$, as can be shown from a Hubbard-Stratonovich transformation.

Armed with this precise formal definition of the screened interaction in the solid (and, naturally, also of the full Green's function $G(\mathbf{r},\mathbf{r}';\tau-\tau') \equiv -\langle T\psi(\mathbf{r},\tau)\psi^\dagger(\mathbf{r}',\tau')\rangle$), we would like to adopt now a local picture in which we focus on a given atom. This is done,as before, by specifying a complete basis set of functions $\chi_{L\mathbf{R}}(\mathbf{r})$ localised around the atomic positions $\mathbf{R}$. There is of course some arbitrariness in this choice, as already discussed. Adopting a local point of view, we focus on the matrix elements of the Green's function and of the screened effective interaction *on a given atomic site*:

$$G_{ab}(i\omega) = \langle\chi_{a\mathbf{R}}|G|\chi_{b\mathbf{R}}\rangle\ ,\ W_{a_1a_2a_3a_4}(i\omega) = \langle\chi_{a_1\mathbf{R}}\chi_{a_2\mathbf{R}}|W|\chi_{a_3\mathbf{R}}\chi_{a_4\mathbf{R}}\rangle \quad (141)$$

In this expression, the indices $a,b,\cdots$ can run over the full set of valence orbitals, or alternatively over a subset corresponding to the more strongly correlated ones. This is a matter of choice of the local quantities we decide to focus on. Following the point of view developed in the third section of these lectures, the key idea is again to introduce an *exact representation* of these local quantities as the solution of an atomic problem coupled to an effective bath. Because we want to represent the local components of both G and W, this effective problem now involves two Weiss functions, both in the one-particle and two-particle sectors. This is an extended form of dynamical mean-field theory (EDMFT). The action of the local problem reads:

$$S = \int d\tau d\tau' \left[-\sum c_a^\dagger(\tau)\mathcal{G}_{ab}^{-1}(\tau-\tau')c_b(\tau') + \right.$$
$$\left. +\frac{1}{2}\sum : c_{a_1}^+(\tau)c_{a_2}(\tau) : \mathcal{U}_{a_1a_2a_3a_4}(\tau-\tau') : c_{a_3}^+(\tau')c_{a_4}(\tau') : \right] \quad (142)$$

The local screened interaction is calculated from this effective action as: $W_{imp} = \mathcal{U} - \mathcal{U}\chi_{imp}\mathcal{U}$ with χ_{imp} the 2-particle impurity correlation funcition. The two Weiss fields $\mathcal{G}$ and $\mathcal{U}$ are adjusted in such a way that $G_{imp} = G_{ab}$ and $W_{imp} = W_{abcd}$, the local quantities in the solid. The impurity model (142) can be viewed as an atom hybridised with an effective bath of non-interacting fermions and also coupled to a bath of fluctuating electric scalar potentials.

This construction provides an unambiguous definition of the Hubbard interactions $\mathscr{U}_{abcd}(i\omega)$ in the solid (as well as of the usual dynamical mean-field $\mathscr{G}$), assuming of course that the local components of the screened interaction W and of the Green's function G are known. Frequency- dependence of $\mathscr{U}$ is essential in a proper definition of these Hubbard interactions, at least when a wide range of energy scale is considered. Naturally, one degree of arbitrariness remains, associated with the choice of the basis set: $\mathscr{U}$ will change when a different basis set is considered, keeping the same form of the effective interaction $W(\mathbf{r},\mathbf{r}';i\omega)$ in the full solid.

To proceed from these formal considerations to a practical scheme, we need to decide how W and G will actually be calculated, and this of course will involve approximations. Again, a free-energy functional is an excellent guidance and indeed such a functional of the full $G(\mathbf{r},\mathbf{r}';i\omega)$ and $W(\mathbf{r},\mathbf{r}';i\omega)$ has been introduced by Almbladh et al.[157], generalizing the Baym-Kadanoff construction (see also [47] for independent work). The functional reads:

$$\Gamma(G,W) = Tr\ln G - Tr[(G_H^{-1} - G^{-1})G] - \frac{1}{2}Tr\ln W + \frac{1}{2}Tr[(u^{-1} - W^{-1})W] + \Psi[G,W] \tag{143}$$

$G_H^{-1} = i\omega_n + \mu + \nabla^2/2 - v_H$ corresponds to the Hartree Green's function with v_H being the Hartree potential. For a derivation of (143) using a Hubbard-Stratonovich transformation and a Legendre transformation with respect to both G and W, see [47]. The functional $\Psi[G,W]$ is a generalization of the Luttinger-Ward functional $\Phi[G]$, whose derivative with respect to G gives the self-energy. Here we have, similarly (from $\delta\Gamma/\delta G = \delta\Gamma/\delta W = 0$):

$$G^{-1} = G_H^{-1} - \Sigma^{xc}\ ,\ \Sigma^{xc} = \frac{\delta\Psi}{\delta G}\ ;\ W^{-1} = u^{-1} - P\ ,\ P = -2\frac{\delta\Psi}{\delta W} \tag{144}$$

A well established electronic structure calculation method, which offers in part an alternative to DFT-LDA, is the so-called GW approach [158] (see [159] for a review). This corresponds to the following approximation to the Ψ-functional:

$$\Psi_{GWA} = -\frac{1}{2}\int d\mathbf{r}d\mathbf{r}' \int d\tau d\tau'\, G(\mathbf{r},\mathbf{r}',\tau-\tau')W(\mathbf{r},\mathbf{r}',\tau-\tau')G(\mathbf{r}',\mathbf{r},\tau'-\tau) \tag{145}$$

which yields the RPA-like approximation to the polarisation and exchange-correlation self-energy: $P = G \star G$ and $\Sigma^{xc} = -G \star W$. The GW approximation to the Ψ-functional is easily written in terms of the components of G and W in the chosen basis set:

$$\Psi_{GWA} = -\frac{1}{2}\int d\tau \sum_{L_1\cdots L_2'} \sum_{\mathbf{RR}'} G^{\mathbf{RR}'}_{L_1L_1'}(\tau) W^{\mathbf{RR}'}_{L_1L_2L_1'L_2'}(\tau) G^{\mathbf{R}'\mathbf{R}}_{L_2'L_2}(-\tau) \tag{146}$$

This can be separated into a contribution $\Psi^{non-loc}_{GWA}$ from non-local components (corresponding to the terms with $\mathbf{R} \neq \mathbf{R}'$ in (146)) and a contribution $\Psi^{loc}_{GWA}[G^{\mathbf{RR}}, W^{\mathbf{RR}}]$ from local components only ($\mathbf{R} = \mathbf{R}'$).

The GW approximation does treat the screened Coulomb interaction from first-principles, but does not treat successfully strong correlation effects. Recently, it has been suggested to improve on the GWA for the local contributions by using the DMFT framework [152, 153] (see also [125, 154, 155, 156]). One can think of different approximations to the Ψ-functional in this context, depending on whether the DMFT approach is used for all the valence orbitals $L = s, p, d, \cdots$, or for a subset (corresponding to the index $a, b, ...$) of correlated orbitals only. The corresponding Ψ-functional reads:

$$\Psi_{GW+DMFT}[G^{\mathbf{RR}'}_{L_1L'_1}, W^{\mathbf{RR}'}_{L_1L_2L'_1L'_2}] = \Psi^{non-loc}_{GWA} + [\Psi^{loc}_{GWA} - \Delta\Psi] + \sum_{\mathbf{R}} \Psi_{imp}[G^{\mathbf{RR}}_{ab}, W^{\mathbf{RR}}_{abcd}] \tag{147}$$

In this expression, Ψ_{imp} is the Ψ-functional corresponding to the local effective model (142), while $\Delta\Psi$ removes the components from Ψ^{loc}_{GWA} which will be taken into account in Ψ_{imp}, namely:

$$\Delta\Psi = -\frac{1}{2}\sum_{\mathbf{R}} \int d\tau \sum_{abcd} G^{\mathbf{R}}_{ab}(\tau) W^{\mathbf{RR}}_{abcd}(\tau) G^{\mathbf{RR}}_{DC}(-\tau) \tag{148}$$

If all valence orbitals are included in the DMFT treatment, the second term in the r.h.s of (147) is absent altogether. If only a correlated subset is treated with DMFT, $\Delta\Psi$ can be thought of as a term preventing double-counting of interactions in the correlated subset. In this context however, in contrast to LDA+DMFT, the form of this double-counting correction is known explicitly.

Taking derivatives of this functional with respect to the components of G and W, one sees that, in the GW+DMFT approach, the non-local components of the self-energy and of the polarization operator keep the same form as in the GWA, while the local components are replaced by the ones from the effective impurity model (possibly in the correlated subset only). The GW+DMFT theoretical framework is fully defined by (147) and the form of the impurity model (142). As before, an interative self-consistent process must be followed in order to obtain the self-energy and screened effective interaction, as well as the dynamical mean-field $\mathcal{G}$ and effective Hubbard interactions $\mathcal{U}$. This is described in more details in Refs. [152, 155, 156]. Concrete implementations of this scheme to electronic structure (and to model hamiltonians as well) is currently being pursued by several groups. For early results, see [152, 153, 154, 155, 156, 6].

6. CONCLUSION AND PERSPECTIVES

In these lectures notes, I have tried to give an introduction to some aspects of the physics of strong electron correlations in solids. Naturally, only a limited number of topics could be covered. The field is characterized by a fascinating diversity of material-dependent properties. It is, to a large extent, experimentally driven, and new discoveries are undoubtedly yet to come. Also, new territories outside the traditional boundaries of solid-state physics are currently being explored, such as correlation effects in nano-electronic devices or the condensed matter physics of cold atoms in optical lattices.

On the theory side, these lectures are influenced by the author's prejudice that (i) physics on intermediate energy scale matters and may be a key to the unusual behaviour of many strongly correlated materials and that (ii) quantitative theoretical techniques are essential to the development of the field, in combination with phenomenological considerations and experimental investigations.

Dynamical mean-field theory is a method of choice for treating these intermediate energy scales. The basic principles of this approach have been reviewed in these lectures. On the formal side, analogies with classical mean-field theory and density-functional theory have been emphasized, through the construction of free-energy functionals of local observables. A distinctive aspect of DMFT is that it treats quasi-particle excitations and higher energy incoherent excitations, on equal footing. As a result, it is able to describe transfers of spectra weight between quasiparticle and incoherent features as temperature, coupling strength, or some other external parameter (doping, pressure,...) is varied. I have emphasized that the *quasiparticle coherence scale* plays a key role in the physics of a strongly correlated metal. Above this scale, which can be dramatically reduced by correlations, unusual (non-Drude) transport and spectroscopic properties are observed, corresponding to an incoherent metallic regime. This is the case, in particular, for metals which are close to a Mott insulating phase. I have briefly reviewed the DMFT description of these effects in these lectures, in comparison to experiments, as well as the detailed theory of the Mott transition which has been one of the early successes of this approach. I have also provided an (admittedly quite succinct) introduction to the recent combination of DMFT with electronic structure calculations. These developments have been made possible by researchers from two communities joining forces towards a common goal. It provides us with a powerful quantitative tool for investigating material-dependent aspects of strong electron correlations.

Despite these successes, some key open questions in the physics of strongly correlated electron systems remain out of reach of the simplest version of DMFT. Indeed, in materials like cuprates, short-range spatial correlations play a key role (in particular magnetic correlations due to superexchange, leading to a strong tendency towards the formation of singlet bonds, as well as pair correlations). These correlations deeply affect the nature of quasiparticles. There is compelling experimental evidence that the quasiparticle coherence scale has thus a strong variation as the momentum **k** is varied along the Fermi surface, leading to the formation of "cold spots" and "hot regions". Extending the DMFT framework in order to take these effects into account may well be the most important frontier in the field.

ACKNOWLEDGMENTS

The content of these lecture notes has been greatly influenced by all the colleagues with whom I recently collaborated in this field, both theorists and experimentalists: B. Amadon, O.K. Andersen, F. Aryasetiawan, S. Biermann, S. Burdin, T.A. Costi, L. de Medici, S. Florens, T. Giamarchi, M. Grioni, S.R. Hassan, M. Imada, D. Jérome, G. Kotliar, H.R. Krishnamurthy, F. Lechermann, P. Limelette, A. Lichtenstein, S. Pankov, O. Parcollet, C. Pasquier, E. Pavarini, L. Perfetti, A. Poteryaev, M. Rozenberg, S. Sachdev, R. Siddharthan, P. Wzietek, as well as by all other colleagues with whom I have had profitable discussions over the last few years. I would like to thank particularly the members of the Ecole Polytechnique group for the friendly and stimulating atmosphere, and for the lively daily discussions. I am grateful to F. Mila for the invitation to lecture on this subject in the "Troisième cycle de la Suisse Romande" in may, 2002, to C. Berthier, G. Collin, C. Simon and the other organizers of the school on "Oxydes à propriétés remarquables" (GDR 2069) at Aussois in june, 2002 [160], to W.Temmermann and D.Szotek for organizing lectures in Daresbury in june, 2003, to F.Mancini and A. Avella for the organisation of the training course at Vietri in october, 2003, and to C. Ortiz and A. Ványolos for help with the notes of my lectures at Vietri. Hospitality of the KITP (Santa Barbara) and of ICTP (Trieste) is acknowledged. Support for research has been provided by CNRS, Ecole Polytechnique, the European Union (through the Marie Curie and RTN programs), and the Indo-French program of IFCPAR.

REFERENCES

1. Mott, N. F., *Proc. Phys. Soc. A*, **62**, 416 (1949).
2. Mott, N. F., *Metal-insulator transitions*, Taylor and Francis, London, 1990.
3. Georges, A., Kotliar, G., Krauth, W., and Rozenberg, M. J., *Reviews of Modern Physics*, **68**, 13–125 (1996).
4. T.Pruschke, M.Jarrell, and J.Freericks, *Adv. Phys.*, **42**, 187 (1995).
5. Springer, M., and Aryasetiawan, F., *Phys. Rev. B*, **57**, 4364–4368 (1998).
6. Aryasetiawan, F., Imada, M., Georges, A., Kotliar, G., Biermann, S., and Lichtenstein, A. I., *ArXiv Condensed Matter e-prints* (2004), cond-mat/0401620.
7. Hubbard, J., *Proc. Roy. Soc. (London)*, **A 276**, 238 (1963).
8. Hubbard, J., *Proc. Roy. Soc. (London)*, **A 277**, 237 (1964).
9. Hubbard, J., *Proc. Roy. Soc. (London)*, **A281**, 401 (1964).
10. Varma, C. M., and Giamarchi, T., *Model for oxide metals and superconductors*, Elsevier, 1991, les Houches Summer School.
11. Imada, M., Fujimori, A., and Tokura, Y., *Rev. Mod. Phys.*, **70**, 1039 (1998).
12. Tsuda, N., Nasu, K., Fujimori, A., and Siratori, K., *Electronic Conduction in Oxides*, Springer Series in Solid-State Sciences 94, Springer, Berlin, 2000, 2nd edn., ISBN 3-540-66956-6.
13. Harrison, W. A., *Electronic structure and the properties of solids*, Dover Pub., New York, 1989.
14. Friedel, J., *The physics of metals*, Cambridge University Press, New York, 1969, pp. 494–525.
15. McMahan, A. K., Huscroft, C., Scalettar, R. T., and Pollock, E. L., *J. Comput.-Aided Mater. Des.*, **5**, 131 (1998), cond-mat/9805064.
16. Wills, J., and Eriksson, O., *Los Alamos Science*, **26**, 128 (2000).
17. Anisimov, V. I., and Gunnarsson, O., *Phys. Rev. B*, **43**, 7570–7574 (1991).
18. Maiti, K., Ph.D. thesis, IISC, Bangalore (1997).
19. Fujimori, A., and Minami, F., *Phys. Rev. B*, **30**, 957 (1984).
20. Zaanen, J., Sawatzky, G. A., and Allen, J. W., *Phys. Rev. Lett.*, **55**, 418 (1985).
21. Kotliar, G., and Savrasov, S. Y., *International Journal of Modern Physics B*, **17**, 5101–5109 (2003).
22. Coleman, P., *Lectures on the Physics of Highly Correlated Electron Systems*, American Institute of Physics, New York, 2002, vol. VI, chap. Local moment physics in heavy electron systems, pp. 79–160, cond-mat/0206003.
23. Hewson, A. C., *The Kondo problem to heavy fermions*, Cambridge University Press, 1993.
24. Burdin, S., Georges, A., and Grempel, D. R., *Physical Review Letters*, **85**, 1048–1051 (2000).
25. Metzner, W., and Vollhardt, D., *Phys. Rev. Lett.*, **62**, 324 (1989).
26. Georges, A., and Kotliar, G., *Phys. Rev. B*, **45**, 6479–6483 (1992).
27. Jarrell, M., *Phys. Rev. Lett.*, **69**, 168–171 (1992).
28. Georges, A., "Exact functionals, effective actions and dynamical mean-field theories: some remarks.," in *Strongly Correlated Fermions and Bosons in Low-Dimensional Disordered Systems*, edited by I. et al., Kluwer Acad., 2002, vol. 72 of *NATO Science Series-II: Mathematics, Physics and Chemistry*.
29. Rozenberg, M. J., Zhang, X. Y., and Kotliar, G., *Phys. Rev. Lett.*, **69**, 1236 (1992).

30. Georges, A., and Krauth, W., *Phys. Rev. Lett.*, **69**, 1240–1243 (1992).
31. Hirsch, J. E., and Fye, R. M., *Phys. Rev. Lett.*, **25**, 2521 (1986).
32. Bulla, R., Costi, T. A., and Vollhardt, D., *Phys. Rev. B*, **64**, 45103 (2001).
33. Kajueter, H., and Kotliar, G., *Phys. Rev. Lett.*, **77** (1996).
34. Florens, S., and Georges, A., *Phys. Rev. B*, **66**, 165111–+ (2002).
35. Oudovenko, V., Haule, K., Savrasov, S. Y., Villani, D., and Kotliar, G., *ArXiv Condensed Matter e-prints* (2004), cond-mat/0401539.
36. Schiller, A., and Ingersent, K., *Phys. Rev. Lett.*, **75**, 113 (1996).
37. Hettler, M. H., Tahvildar-Zadeh, A. N., Jarrell, M., Pruschke, T., and Krishnamurthy, H. R., *Phys. Rev. B*, **58**, 7475– (1998).
38. Kotliar, G., Savrasov, S., Palsson, G., and Biroli, G., *Phys. Rev. Lett.*, **87**, 186401 (2001).
39. Biermann, S., Georges, A., Lichtenstein, A., and Giamarchi, T., *Physical Review Letters*, **87**, 276405 (2001).
40. Biroli, G., Parcollet, O., and Kotliar, G. (2003), cond-mat/0307587.
41. Onoda, S., and Imada, M. (2003), cond-mat/0304580.
42. Si, Q., and Smith, J. L., *Phys. Rev. Lett.*, **77**, 3391 (1996).
43. Kajueter, H., *PhD thesis, Rutgers University* (1996).
44. Sengupta, A. M., and Georges, A., *Phys. Rev. B*, **52**, 10295–10302 (1995).
45. Smith, J. L., and Si, Q., *Phys. Rev. B*, **61**, 5184 (2000).
46. Fukuda, R., Kotani, T., and Yokojima, S., *Prog. Theor. Phys. Suppl.*, **121**, 1 (1996).
47. Chitra, R., and Kotliar, G., *Phys. Rev. B*, **63**, 115110 (2001).
48. Plefka, T., *J. Phys. A*, **15**, 1971 (1982).
49. Georges, A., and Yedidia, J. S., *Journal of Physics A Mathematical General*, **24**, 2173–2192 (1991).
50. Polchinsky, J., *Nucl. Phys. B*, **231**, 269 (1984).
51. Schehr, G., and Doussal, P. L., *preprint cond-mat/030486* (2003).
52. Chauve, P., and Doussal, P. L., *Phys. Rev. E*, **64**, 051102 (2001).
53. Thouless, D., Anderson, P., and Palmer, R., *Phil. Mag.*, **35**, 593 (1977).
54. Mermin, N., *Phys. Rev.*, **137**, A1441 (1965).
55. Hohenberg, P., and Kohn, W., *Phys. Rev.*, **136**, B864 (1964).
56. Kohn, W., and Sham, L., *Phys. Rev.*, **140**, A1133 (1965).
57. Fukuda, R., Kotani, T., Suzuki, Y., and Yokojima, S., *Prog. Theor. Phys.*, **92**, 833 (1994).
58. Valiev, M., and Fernando, G., *Phys. Lett. A*, **227**, 265 (1997).
59. Argaman, N., and Makov, G., *Am. J. Phys.*, **68**, 69 (2000), preprint physics/9806013.
60. Dreizler, R., and Gross, E., *Density Functional Theory*, Springer-Verlag, 1990.
61. Jones, R., and Gunnarsson, O., *Rev. Mod. Phys.*, **61**, 689 (1989).
62. Chayes, J., and Chayes, L., *J. Stat. Phys.*, **36**, 471 (1984).
63. Chayes, J., Chayes, L., and Ruskai, M. B., *J. Stat. Phys.*, **38**, 497 (1985).
64. Harris, J., *Phys. Rev. A*, **29**, 1648 (1984).
65. Khodel, V. A., Shaginyan, V. R., and Khodel, V. V., *Physics Reports*, **249**, 1 (1994).
66. Amusia, M. Y., Msezane, A. Z., and Shaginyan, V. R. (2003), arXiv cond-mat/0312162.
67. Chitra, R., and Kotliar, G., *Phys. Rev. B*, **62**, 12715 (2000).
68. Kotliar, G., *Eur. J. Phys. B*, **27**, 11 (1999).
69. Johansson, B., *Phil. Mag.*, **30**, 469 (1974).
70. Skriver, H. L., Andersen, O. K., and Johansson, B., *Physical Review Letters*, **41**, 42–45 (1978).
71. Skriver, H. L., Andersen, O. K., and Johansson, B., *Physical Review Letters*, **44**, 1230–1233 (1980).
72. Savrasov, S. Y., and Kotliar, G., *Physical Review Letters*, **84**, 3670–3673 (2000).
73. Savrasov, S. Y., Kotliar, G., and Abrahams, E., *Nature*, **410**, 793 (2001).
74. Fujimori, A., *J. Phys. Chem. Solids*, **53**, 1595 (1992).
75. McWhan, D. B., Menth, A., Remeika, J. P., Brinckman, W. F., and Rice, T. M., *Phys. Rev. B*, **7**, 1920 (1973).
76. Lefebvre, S., Wzietek, P., Brown, S., Bourbonnais, C., Jérome, D., Mèziére, C., Fourmigué, M., and Batail, P., *Phys. Rev. Lett.*, **85**, 5420 (2000).
77. Bao, W., Broholm, C., Aeppli, G., Dai, P., Honig, J. M., and Metcalf, P., *Phys. Rev. Lett.*, **78**, 507 (1997).
78. Kotliar, G., *Physica B*, **259-261**, 711 (1999).

79. Misguich, G., and Lhuillier, C., *Frustrated spin systems*, World Scientific, Singapore, 2003, chap. Two-dimensional quantum antiferromagnets, cond-mat/0310405.
80. S.Sachdev (2004), cond-mat/0401041.
81. Shimizu, Y., Miyagawa, K., Kanoda, K., Maesato, M., and Saito, G. (2003), cond-mat/0307483.
82. Imada, M., Mizusaki, T., and Watanabe, S., *cond-mat/0307022* (2003).
83. Georges, A., and Krauth, W., *Phys. Rev. B*, **48**, 7167–7182 (1993).
84. Rozenberg, M. J., Kotliar, G., and Zhang, X. Y., *Phys. Rev. B*, **49**, 10181 (1994).
85. Laloux, L., Georges, A., and Krauth, W., *Phys. Rev. B*, **50**, 3092–3102 (1994).
86. Florens, S., *Cohérence et localisation dans les systèmes d'électrons fortement corrélés*, Ph.D. thesis, Université Paris 6 and Ecole Normale Supérieure, Paris (2003).
87. Georges, A., Florens, S., and Costi, T. A., *ArXiv Condensed Matter e-prints* (2003), cond-mat/0311520.
88. Fujimori, A., Hase, I., Namatame, H., Fujishima, Y., Tokura, Y., Eisaki, H., Uchida, S., Takegahara, K., and de Groot, F. M. F., *Phys. Rev. Lett.*, **69**, 1796 (1992).
89. Inoue, I. H., Hase, I., Aiura, Y., Fujimori, A., Haruyama, Y., Maruyama, T., and Nishihara, Y., *Physical Review Letters*, **74**, 2539–2542 (1995).
90. Rozenberg, M. J., Inoue, I. H., Makino, H., Iga, F., and Nishihara, Y., *Physical Review Letters*, **76**, 4781–4784 (1996).
91. Maiti, K., Sarma, D. D., Rozenberg, M., Inoue, I., Makino, H., Goto, O., Pedio, M., and Cimino, R., *Europhys. Lett.*, **55**, 246 (2001).
92. Sekiyama, A., Fujiwara, H., Imada, S., Eisaki, H., Uchida, S. I., Takegahara, K., Harima, H., Saitoh, Y., and Suga, S., *ArXiv Condensed Matter e-prints* (2002), cond-mat/0206471.
93. Sekiyama, A., Fujiwara, H., Imada, S., Suga, S., Eisaki, H., Uchida, S. I., Takegahara, K., Harima, H., Saitoh, Y., Nekrasov, I. A., Keller, G., Kondakov, D. E., Kozhevnikov, A. V., Pruschke, T., Held, K., Vollhardt, D., and Anisimov, V. I., *ArXiv Condensed Matter e-prints* (2003), cond-mat/0312429.
94. Nekrasov, I., Keller, G., Kondakov, D., Kozhevnikov, A., Pruschke, T., Held, K., Vollhardt, D., and Anisimov, V. (2002), cond-mat/0211508.
95. Pavarini, E., Biermann, S., Poteryaev, A., Lichtenstein, A. I., Georges, A., and Andersen, O. K., *ArXiv Condensed Matter e-prints* (2003), cond-mat/0309102.
96. Matsuura, A. Y., Watanabe, H., Kim, C., Doniach, S., Shen, Z. X., Thio, T., and Bennett, J. W., *Phys. Rev. B*, **58**, 3690 (1998).
97. Mo, S. K., Denlinger, J. D., Kim, H. D., Park, J. H., Allen, J. W., Sekiyama, A., Yamasaki, A., Kadono, K., Suga, S., Saitoh, Y., Muro, T., Metcalf, P., Keller, G., Held, K., Eyert, V., Anisimov, V. I., and Vollhardt, D., *Phys. Rev. Lett.*, **90**, 186403 (2003).
98. Liu, L., Allen, J., Gunnarsson, O., Christansen, N., and Andersen, O., *Phys. Rev. B*, **45**, 8934 (1992).
99. Rozenberg, M. J., Kotliar, G., Kajueter, H., Thomas, G. A., Rapkine, D. H., Honig, J. M., and Metcalf, P., *Phys. Rev. Lett.*, **75**, 105 (1995).
100. Eldridge, J., Kornelsen, K., Wang, H., Williams, J., Crouch, A., and Watkins, D., *Sol. State. Comm.*, **79**, 583 (1991).
101. Merino, J., and McKenzie, R. H., *Phys. Rev. B*, **61**, 7996 (2000).
102. Majumdar, P., and Krishnamurthy, H., *Phys. Rev. B*, **52** (1995).
103. Limelette, P., Wzietek, P., Florens, S., Georges, A., Costi, T. A., Pasquier, C., Jérome, D., Mézière, C., and Batail, P., *Physical Review Letters*, **91**, 016401 (2003).
104. Khurana, A., *Phys. Rev. Lett.*, **64**, 1990 (1990).
105. Moeller, G., Si, Q., Kotliar, G., Rozenberg, M., and Fisher, D. S., *Phys. Rev. Lett.*, **74**, 2082 (1995).
106. Pálsson, G., and Kotliar, G., *Phys. Rev. Lett.*, **80**, 4775–4778 (1998).
107. Valla, T., Johnson, P. D., Yusof, Z., Wells, B. O., Loureiro, Li, Q., M., S., Cava, R. J., Mikami, M., Mori, Y., Yoshimura, M., and Sasaki, T., *Nature*, **417**, 627 (2002).
108. Castellani, C., DiCastro, C., Feinberg, D., and Ranninger, J., *Phys. Rev. Lett.*, **43**, 1957 (1979).
109. Jayaraman, A., McWhan, D. B., Remeika, J. P., and Dernier, P. D., *Phys. Rev. B*, **2**, 3751 (1970).
110. Kotliar, G., Lange, E., and Rozenberg, M. J., *Phys. Rev. Lett.*, **84**, 5180 (2000).
111. Rozenberg, M. J., Chitra, R., and Kotliar, G., *Phys. Rev. Lett.*, **83**, 3498 (1999).
112. Limelette, P., Georges, A., Jérome, D., Wzietek, P., Metcalf, P., and Honig, J. M., *Science*, **302**, 89–92 (2003).
113. Kagawa, F., Itou, T., Miyagawa, K., and Kanoda, K. (2003), preprint cond-mat/0307304.
114. Majumdar, P., and Krishnamurthy, H., *Phys. Rev. Lett.*, **73** (1994).

115. Cyrot, M., and Lacour-Gayet, P., *Sol. State Comm.*, **11**, 1767 (1972).
116. Merino, J., and McKenzie, R. H., *Phys. Rev. B*, **62**, 16442–16445 (2000).
117. Hassan, S. R., Georges, A., and Krishnamurthy, H. R. (2004), preprint.
118. Fournier, D., Poirier, M., Castonguay, M., and Truong, K., *Phys. Rev. Lett.*, **90**, 127002 (2003).
119. Parcollet, O., Biroli, G., and Kotliar, G. (2003), cond-mat/0308577.
120. Giamarchi, T., Biermann, S., Georges, A., and Lichtenstein, A., *ArXiv Condensed Matter e-prints* (2004), cond-mat/0401268.
121. Anisimov, V. I., Poteryaev, A. I., Korotin, M. A., Anokhin, A. O., and Kotliar, G., *J. Phys. Cond. Matter*, **9**, 7359–7367 (1997).
122. Lichtenstein, A. I., and Katsnelson, M. I., *Phys. Rev. B*, **57**, 6884–6895 (1998).
123. Held, K., Nekrasov, I. A., Blümer, N., Anisimov, V. I., and Vollhardt, D., *Int. J. Mod. Phys. B*, **15**, 2611 (2001), cond-mat/0010395.
124. Held, K., Nekrasov, I. A., Keller, G., Eyert, V., Blümer, N., McMahan, A., Scalettar, R. T., Pruschke, T., Anisimov, V. I., and Vollhardt, D., *The LDA+DMFT Approach to Materials with Strong Electronic Correlations*, J. Grotendorst, D. Marks, and A. Muramatsu (ed.), NIC Series Volume 10, p. 175-209 (2002), 2001, proceedings of the Winter School on "Quantum Simulations of Complex Many-Body Systems: From Theory to Algorithms", February 25 - March 1, 2002, Rolduc/Kerkrade (NL); cond-mat/0112079.
125. Kotliar, G., and Savrasov, S. Y., *Dynamical Mean Field Theory, Model Hamiltonians and First Principles Electronic Structure Calculations*, In "New Theoretical Approaches to Strongly Correlated Systems, A.M. Tsvelik Ed., Kluwer Academic Publishers, 2001, proc. of the Nato Advanced Study Institute on New Theoretical Approaches to Strongly Correlated Systems, Cambridge, UK, 1999; preprint cond-mat/0208241.
126. Lichtenstein, A. I., Katsnelson, M. I., and Kotliar, G., "Spectral density functional approach to electronic correlations and magnetism in crystals," in *Electron Correlations and Materials Properties 2*, edited by A. Gonis, Kluwer, New York, 2002, cond-mat/0211076.
127. *Workshop on Realistic Theories of Correlated Electron Materials (online material)*, Kavli Institute for Theoretical Physics, UCSB, Santa Barbara, USA, 2002, http://online.itp.ucsb.edu/online/cem02/.
128. *Conference on Realistic Theories of Correlated Electron Materials (online material)*, Kavli Institute for Theoretical Physics, UCSB, Santa Barbara, USA, 2002, http://online.itp.ucsb.edu/online/cem02/si-conf-schedule.html.
129. *Euroconference on Ab-initio Many-Body Theory for Correlated Electron Systems (online material)*, The Abdus Salam International Centre for Theoretical Physics, Trieste, Italy, 2003, http://www.ictp.trieste.it/ smr1512/contributionspage.html.
130. Andersen, O. K., *Phys. Rev. B*, **12**, 3060–3083 (1975).
131. Skriver, H. L., *The LMTO method*, Springer, Berlin, 1984.
132. Andersen, O., Dasgupta, T. S., Ezhov, S., Tsetseris, L., Jepsen, O., Tank, R., Arcangeli, C., and Krier, G., "Third generation MTOs," 2000, http://psi-k.dl.ac.uk/newsletters/News45/ (online material).
133. Andersen, O. K., Saha-Dasgupta, T., Tank, R. W., Arcangeli, C., Jepsen, O., and Krier, G., "Developing the MTO Formalism," in *Electronic Structure and Physical Properties of Solids. The Use of the LMTO Method, Lectures of a Workshop Held at Mont Saint Odile, France, October 2-5, 1998. Edited by H. Dreyssé, Lecture Notes in Physics, vol. 535, p.3*, 2000, p. 3.
134. Marzari, N., and Vanderbilt, D., *Phys. Rev. B*, **56**, 12847–12865 (1997).
135. Anisimov, V. I., Zaanen, J., and Andersen, O. K., *Phys. Rev. B*, **44**, 943–954 (1991).
136. Anisimov, V. I., Aryasetiawan, F., and Lichtenstein, A. I., *J. Phys. Condensed Matter*, **9**, 767–808 (1997).
137. Lichtenstein, A. I., Katsnelson, M. I., and Kotliar, G., *Phys. Rev. Lett.*, **87**, 067205 (2001).
138. Kotliar, G., and Savrasov, S. Y. (2001), cond-mat/0106308.
139. Nekrasov, I. A., Held, K., Blümer, N., Poteryaev, A. I., Anisimov, V. I., and Vollhardt, D., *European Physical Journal B*, **18**, 55–61 (2000).
140. Gunnarsson, O., Koch, E., and Martin, R. M., *Phys. Rev. B*, **56**, 1146–1152 (1997).
141. Koch, E., Gunnarsson, O., and Martin, R. M., *Phys. Rev. B*, **60**, 15714–15720 (1999).
142. Florens, S., Georges, A., Kotliar, G., and Parcollet, O., *Phys. Rev. B*, **66**, 205102–+ (2002).
143. Manini, N., Santoro, G. E., dal Corso, A., and Tosatti, E., *Phys. Rev. B*, **66**, 115107 (2002).

144. Mochizuki, M., and Imada, M., *Phys. Rev. Lett.*, **91**, 167203 (2003).
145. Cwik, M., Lorenz, T., Baier, J., Müller, R., André, G., Bourée, F., Lichtenberg, F., Freimuth, A., Schmitz, R., Müller-Hartmann, E., and Braden, M., *Phys. Rev. B*, **68**, 060401 (2003).
146. Amadon, A., Biermann, S., and Georges, A. (2003), unpublished, and in preparation.
147. Savrasov, S. Y., and Kotliar, G. (2003), preprint cond-mat/0308053.
148. Held, K., McMahan, A. K., and Scalettar, R. T., *Phys. Rev. Lett.*, **87**, A266404+ (2001).
149. McMahan, A. K., Held, K., and Scalettar, R. T., *Phys. Rev. B*, **67**, 075108 (2003).
150. Bouchet, J., Siberchicot, B., Jollet, F., and Pasturel, A., *J. Phys. Condensed Matter*, **12**, 1723–1733 (2000).
151. Eriksson, O., Becker, J. D., Balatsky, A. V., and Wills, J. M., *J. Alloys and Comp.*, **287**, 1 (1999).
152. Biermann, S., Aryasetiawan, F., and Georges, A., *Physical Review Letters*, **90**, 086402 (2003).
153. Sun, P., and Kotliar, G., *Phys. Rev. B*, **66**, 085120 (2002).
154. Sun, P., and Kotliar, G., *ArXiv Condensed Matter e-prints* (2003), cond-mat/0312303.
155. Biermann, S., Aryasetiawan, F., and Georges, A., "Electronic Structure of Strongly Correlated Materials: towards a First Principles Scheme," in *Physics of Spin in Solids: Materials, Methods, and Applications*, NATO Science Series II, Kluwer Acad. Pub., 2004, arXiv cond-mat/0401653.
156. Aryasetiawan, F., Biermann, S., and Georges, A., "A First Principles Scheme for Calculating the Electronic Structure of Strongly Correlated Materials: GW+DMFT," in *Coincidence Studies of Surfaces,Thin Films and Nanostructures*, Wiley, 2003, arXiv cond-mat/0401626.
157. Almbladh, C. O., von Barth, U., and van Leeuwen, R., *Int. J. Mod. Phys. B*, **13**, 535 (1999).
158. Hedin, L., *Phys. Rev.*, **139** (1965).
159. Aryasetiawan, F., and Gunnarsson, O., *Rep. Prog. Phys.*, **61**, 237 (1998).
160. *School on Oxides with remarkable properties (online material)*, GDR 2069 (CNRS), Aussois, France, 2002, http://www-lsp.ujf-grenoble.fr/vie_scientifique/gdr/GDROX/.

Electron-Phonon Interaction and Strong Correlations in High-Temperature Superconductors: One can not avoid the unavoidable

Miodrag L. Kulić

Johann Wolfgang Goethe-University,
Institute for Theoretical Physics,
P.O.Box 111932, 60054 Frankfurt/Main, Germany

Abstract. The important role of the electron-phonon interaction (EPI) in explaining the properties of the normal state and pairing mechanism in high-T_c superconductors (HTSC) is discussed. A number of experimental results are analyzed such as: dynamical conductivity, Raman scattering, neutron scattering, ARPES, tunnelling measurements, isotope effect and etc. They give convincing evidence that the EPI is strong and dominantly contributes to pairing in HTSC oxides. It is argued that strong electronic correlations in conjunction with the pronounced (in relatively weakly screened materials) EPI are unavoidable ingredients for the microscopic theory of pairing in HTSC oxides. I present the well defined and controllable theory of strong correlations and the EPI. It is shown that strong correlations give rise to the pronounced *forward scattering peak* in the EPI - the FSP theory. The FSP theory explains in a consistent way several (crucial) puzzles such as much smaller transport coupling constant than the pairing one ($\lambda_{tr} \ll \lambda_{ph}$), which are present if one interprets the results in HTSC oxides by the old Migdal-Eliashberg theory for the EPI. The ARPES non-shift puzzle - where the nodal kink at 70 meV is unshifted in the superconducting state while the anti-nodal one at 40 meV is shifted, can be explained at present only by the FSP theory. It predicts also: (1) a knee-like shape of the imaginary part of the self-energy at $\omega < \omega_{ph}^{(70)}$ what has been recently confirmed in ARPES measurements; (2) that the Coulomb scattering gives very small coupling constant $\lambda_C << \lambda_{ph}$, which is also confirmed in ARPES spectra where $\lambda_C < 0.4$ and $\lambda_{ph} > 1$. A number of other interesting predictions of the FSP theory are also discussed.

1. INTRODUCTION

1.1. Importance of strong electronic correlations and EPI

Seventeen years after the discovery of the high-T_c superconductors (HTSC) [1] there is still no consensus about the pairing mechanism in these materials. At present two possible theories are in the focus, the first one based on the electron-phonon interaction (EPI) and the second one based on spin fluctuation interaction (SFI). In the meantime it was well established that metallic compounds of HTSC oxides are obtained from insulating parent compounds by doping with small number of carriers - usually called holes. It turns out that the parent insulating state is far from being conventional band insulator where usually an even number of electrons (holes) per lattice site fill Bloch bands completely. By counting the electron number one comes (naively) to the conclusion that the

CP715, *Lectures on the Physics of Highly Correlated Electron Systems VIII: Eighth Training Course*, edited by A. Avella and F. Mancini

parent compounds of copper oxides (for instance La_2CuO_4 and $YBa_2Cu_3O_6$) should be metallic, because in the unit cell there is odd (nine) number of d-electrons per Cu^{2+} ion. The way out from this controversy is in the presence of strong electronic correlations. They are due to the localized d-orbital on the Cu^{2+} ion giving rise to the strong Coulomb repulsion U of two $3d_{x^2-y_2}$ electrons (or holes) at a given lattice site with opposite spins. This repulsion keeps electrons apart making them to be localized on the lattice, but with localized spins ($S = 1/2$). This type of insulating state is called the *Mott-Hubbard insulator*. Speaking in language of electronic bands, for large on-site repulsion $U \gg W$ and for one electron per lattice site the original conduction band (with the width W) is split into the lower Hubbard band with localized spins and the empty upper band separated by U from the lower one - see more in [2] and Section 4.

The relevance of strong correlations is well documented experimentally: (**i**) The electron-energy-loss spectroscopy [24] shows a transfer of intensity (which is a measure of the number of states) from higher to lower energies by doping. Such a property is characteristic for the class of Hubbard models where the number of states in the upper Hubbard band decreases by increasing the hole doping. For comparison, in typical semiconductors the number of states in the valence band is determined by the number of atoms, i.e. it is fixed and doping independent. (**ii**) The self-consistent band-structure calculations and the photoemission experiments gave that the effective Hubbard interaction (U) for the Cu ions is of the order $U \approx 6-10\ eV$ [25], which is much larger than the observed band width W ($\sim 2\ eV$) [26]. (**iii**) A rather direct evidence for strong correlations comes from the doping dependence of the dynamic conductivity $\sigma(\omega)$ in $La_{2-x}Sr_xCuO_4$ and $Nd_{2-x}Ce_xCuO_{4-y}$, particularly from the observed shift of the spectral weight from high to low energies with doping [27]. Besides the development of the Drude peak around $\omega = 0$ in the underdoped systems the so called mid-infrared (*MIR*) peak is also developed around $0.4\ eV$.

Regarding the EPI one can put an "old fashioned" question: Does the EPI makes (contributes to) the superconducting pairing in HTSC oxides? Surprisingly, most of researchers in the field believe that the EPI is irrelevant and that the pairing mechanism is due to spin fluctuations and strong correlations alone- see [29]. This belief is mainly based on an incorrect stability criterion (which, if true, would strongly limit T_c in the EPI mechanism), and also on a number of experimental results which give evidence for strong anisotropic ($d-wave$ like) pairing with gapless regions on the Fermi surface [6], etc. Moreover, the phase sensitive *SQUID* measurements of the Josephson effect [30], [31] in the orthorhombic material $YBa_2Cu_3O_{6+x}$ are strongly in favor of an "orthorhombic" $d-wave$ superconducting order parameter, for instance $\Delta(\mathbf{k}) = \Delta_s + \Delta_d(\cos k_x - \cos k_y)$. As experiments of Tsuei et al. [30], [31] show one has $\Delta_s < 0.1\Delta_d$ in optimally doped $YBa_2Cu_3O_{6+x}$, which means that zeros of $\Delta(\mathbf{k})$ are near intersections of the Fermi surface and the lines $k_x \approx \pm k_y$. Recent experiments on the single-layer crystals $Tl_2Ba_2CuO_{6+x}$ and on $Bi_2Sr_2CaCu_2O_{8+x}$ ($Bi2212$) done by Tsuei group [32], [33], [34], prove the existence of pure $d-wave$ pairing in underdoped, optimally and overdoped systems. The recent interference experiments on $Nd_{2-x}Ce_xCuO_{4-y}$ point also to d-wave pairing in this compound [35]. In that respect, we point out that there is also an widespread (and unfounded) belief that $d-wave$ is incompatible with the EPI pairing mechanism.

Another argument used against the EPI as an origin of superconductivity in HTSC oxides is based on the small value of the oxygen isotope effect α_O ($\alpha = \alpha_O + \alpha_{Cu} + \alpha_Y + \alpha_{Ba}$) in optimally doped materials, such as $YBCO$ with highest critical temperature $T_c \approx 92K$ where $\alpha_O \approx 0.05$ [104], instead of the canonical value $\alpha = 1/2$ which would be in the case of the EPI pairing mechanism alone and in the presence of O-vibrations only.

On the other hand, there is good experimental evidence that the EPI is sufficiently large in order to produce superconductivity in HTSC oxides, i.e. $\lambda > 1$. Let us quote some of them: (**1**) The superconductivity induced *phonon renormalization* [3], [36], [4], [5] is much larger in HTSC oxides than in $LTSC$ superconductors. This is partially due to the larger value of Δ/E_F in HTSC than in $LTSC$; (**2**) the *line-shape* in the phonon Raman scattering is very asymmetric (Fano line), which points to a substantial interaction of the lattice with some quasiparticle (electronic liquid) continuum. For instance, the recent phonon Raman measurements [4] on $HgBa_2Ca_3Cu_4O_{10+x}$ at $T < T_c$ give very large softening (self-energy effects) of the A_{1g} phonons with frequencies 240 and 390 cm^{-1} by 6 % and 18 %, respectively. At the same time there is a dramatic increase of the line-width immediately below T$_c$, while above T$_c$ the line-shape is strongly asymmetric. A substantial phonon renormalization was obtained in $(Cu,C)Ba_2Ca_3Cu_4O_{10+x}$ [5]; (**3**) the *large isotope coefficients* ($\alpha_O > 0.4$) in $YBCO$ away from the optimal doping [104] and $\alpha_O \approx 0.15 - 0.2$ in the optimally doped $La_{1.85}Sr_{0.15}CuO_4$. At the same time one has $\alpha_O \approx \alpha_{Cu}$ making $\alpha \approx 0.25 - 0.3$. This result tell us that other, besides O, ions participate in pairing; (**4**) the most important evidence that the EPI plays an important role in pairing comes from *tunnelling spectra* in HTSC oxides, where the phonon-related features have been clearly seen in the $I - V$ characteristics [37], [38], [39], [40], [41]; (4) the *penetration depth* in the a-b plane of YBCO is increased significantly after the substitution $O^{16} \rightarrow O^{18}$, i.e. $(\Delta\lambda_{ab}/\ \lambda_{ab}) = (^{18}\lambda_{ab} - ^{16}\lambda_{ab})/^{16}\lambda_{ab} = 2.8$ % at 4 K [42]. Since $\lambda_{ab} \sim m^*$ the latter result, if confirmed, could be due to the nonadiabatic increase of the effective mass m^*.

Recent ARPES measurements on HTSC oxides [43], [108] show a *kink* in the quasiparticle spectrum at characteristic (oxygen) phonon frequencies in the normal and superconducting state. This is clear evidence that the EPI is strong and involved in pairing.

On the *theoretical side* there are self-consistent LDA band-structure calculations which (in spite of their shortcomings) give a rather large bare EPI coupling constant $\lambda \sim 1.5$ in $La_{1.85}Sr_{0.15}CuO_4$ [51], [53]. The *nonadiabatic effects* due to poor metallic screening along the c-axis may increase λ additionally [52], [53]. All these facts are in favor of the substantial EPI in HTSC oxides. However, if the properties of the normal and superconducting state in HTSC oxides are interpreted in terms of the standard EPI theory, which holds in $LTSC$ systems, some puzzles arise. One of them is related to the normal-state conductivity (resistivity) - in optimally doped systems the width of the Drude peak in $\sigma(\omega)$ and the temperature dependence of the resistivity $\rho(T)$ are not incompatible with the strong-coupling theory with $\lambda \sim 3$ and $\lambda_{tr} \sim 1$ (if $\omega_{pl} \sim 3\ eV$), where λ_{tr} is the transport EPI coupling constant [50]. On the other side the combined resistivity and low frequency conductivity (Drude part) measurements give $\lambda_{tr} \approx 0.3$ if the plasma frequency takes the value $\omega_{pl} \sim 1\ eV$ - see more below. If one assumes that $\lambda_{tr} \approx \lambda$, which is the case in most low temperature superconductors ($LTSC$), such a small λ can not give large $T_c (\approx 100\ K)$.

In the past there were doubts on the ability of the EPI to explain the linear temperature dependence of the resistivity in the underdoped system [58] $Bi_{2+x}Sr_{2-y}CuO_{6\pm\delta}$, which starts at low $T > 10-20\ K$. Because the asymptotic T^5 behavior of $\rho(T)$ (for $T \ll \Theta_D$) is absent in this sample, then it seems that this experiment is questioning seriously the contribution of the EPI to the resistivity. However, there are other measurements [59] on $Bi_{2+x}Sr_{2-y}CuO_{6\pm\delta}$ where the linear behavior starts at higher temperature, i.e. at $T > 50\ K$. Additionally, the resistivity measurements [14] on Bi_2SrCuO_x samples with low $T_c \simeq 3\ K$ show saturation to finite value at $T = 0\ K$. After subtraction of this constant part one obtains the Bloch-Grüneisen behavior between $T_c \simeq 3\ K$ and $300\ K$, which is due to the EPI.

Concerning the EPI, the above results imply the following possibilities: (**a**) $\lambda_{tr} \ll 1 < \lambda$ and the pairing is due to the *EPI*, or (**b**) $\lambda_{tr} \simeq \lambda \approx 0.4-0.6$ and the EPI is ineffective (although present) in pairing; (**c**) $\lambda_{tr} \simeq \lambda$ but the EPI is responsible for pairing on the expense of some peculiarities of equations describing superconductivity. In Section 5. we present a theory of the EPI renormalized by strong electronic correlations, which is in favor of the case (**a**). It is interesting that the similar puzzling situation ($\lambda_{tr} \ll \lambda$) is realized in $Ba_xK_{1-x}BiO_3$ compound (with $T_c \simeq 30\ K$), where optical measurements give $\lambda_{tr} \approx 0.1-0.3$ [15], while tunnelling measurements [16] give $\lambda \sim 1$. Note, in $Ba_xK_{1-x}BiO_3$ there are no magnetic fluctuations (or magnetic order) and no signs of strong electronic correlations. Therefore, the EPI is favored as the pairing mechanism in $Ba_xK_{1-x}BiO_3$. It seems that in this compound *long-range forces*, in conjunction with some nesting effects, may be responsible for this discrepancy?

One can summarize, that the EPI theory, which pretends to explain the normal metallic state and superconductivity in HTSC oxides, is confronted with the problem of explaining why the EPI coupling is present in self-energy effects (governed by the coupling constant $\lambda > 1$) but it is suppressed in transport properties (which depend on $\lambda_{tr} < 1$), i.e. why λ_{tr} is (much) smaller than λ. One of the possibilities is that strong electronic correlations, as well as the long-range Madelung forces, affect the EPI significantly. This will be discussed in forthcoming sections. In light of the above discussion it is also important to know the role of the EPI in the formation of $d-wave$ superconducting state in HTSC oxides, i.e. why it is compatible with d-wave pairing?

In this review we discuss theoretical and experimental results in HTSC oxides and mostly those which are related to: (i) strong quasiparticle scattering in the normal state, (ii) the pairing mechanism [17], [18], [19], [20].

The paper is organized as follows. In Section 2. we review important physical properties of HTSC oxides in the normal and superconducting state, whose understanding is a basis for the microscopic theory of superconductivity. Only those experiments (and theoretical interpretations) are discussed here which are in our opinion most important in getting information on the pairing mechanism in HTSC oxides. In Section 3. we discuss the general theory of the EPI and its low-energy version. The theory of strong electronic correlations is studied in Section 4., where much space is devoted to a systematic, recently elaborated, method for strongly correlated electrons [17], [18], [19], [20] - the X-method. The latter considers strongly interacting quasiparticles as *composite objects*, contrary to the slave-boson method which at some stage assumes spin and charge separation [28]. A systematic theory of the renormalization of the EPI coupling by strong electronic correlations [17], [18] is exposed in Section 5. It is shown there, that the *for-*

ward scattering peak develops in the EPI by lowering doping, while the coupling at large transfer momenta (the backward scattering) is suppressed.

In Section 6. we summarize the basic predictions of the theory based on the existence of the forward scattering peak in the EPI, impurity and Coulomb scattering, and possible relation between the forward scattering peak in the EPI and pseudogap. The comparison between the EPI and SFI prediction is given in Section 7. The (im)possibility of superconductivity in the Hubbard and t-J mode is studied in Section 8., while the obtained results are summarized in Section 9.

1.2. Prejudices on the EPI

In spite of the reach experimental evidence in favor of the strong EPI in HTSC oxides there was a disproportion in the research (especially theoretical) activity, since the investigation of the spin fluctuations mechanism of pairing prevailed in the literature. This was partly due to a theoretically unfounded statement - given in [10], on the upper limit of T_c in the phonon mechanism of pairing. It is well known that in an electron-ion system besides the EPI there is also the repulsive Coulomb interaction and these are not independent. In the case of an isotropic and homogeneous system with a weak (quasi)particle interaction the effective potential $V_{eff}(\mathbf{k},\omega)$ in the leading approximation looks like as for two external charges (e) embedded in the medium with the *total macroscopic longitudinal dielectric function* $\varepsilon_{tot}(\mathbf{k},\omega)$ ($\mathbf{k}$ is the momentum and ω is the frequency) [12], i.e.

$$V_{eff}(\mathbf{k},\omega)=\frac{V_{ext}(\mathbf{k})}{\varepsilon_{tot}(\mathbf{k},\omega)}=\frac{4\pi e^2}{k^2\varepsilon_{tot}(\mathbf{k},\omega)}. \tag{1}$$

In the case when the interaction between quasiparticles is strong, the state of embedded quasiparticles changes substantially due to the interaction with other quasiparticles, giving rise to $V_{eff}(\mathbf{k},\omega)\neq 4\pi e^2/k^2\varepsilon_{tot}(\mathbf{k},\omega)$. In that case V_{eff} depends on other (than $\varepsilon_{tot}(\mathbf{k},\omega)$) response functions. However, in the case when Eq.(1) holds the weak-coupling limit is realized where T_c is given by $T_c=\bar{\omega}\exp(-1/(\lambda-\mu^*)$ [9], [12]). Here, λ is the EPI coupling constant, $\bar{\omega}$ is the average phonon frequency and μ^* is the Coulomb pseudo-potential, $\mu^*=\mu/(1+\mu\ln E_F/\bar{\omega})$ (E_F is the Fermi energy). λ and μ are expressed by $\varepsilon_{tot}(\mathbf{k},\omega=0)$

$$\langle N(0)V_{eff}(\mathbf{k},\omega=0)\rangle\equiv\mu-\lambda=N(0)\int_0^{2k_F}\frac{kdk}{2k_F^2}\frac{4\pi e^2}{k^2\varepsilon_{tot}(\mathbf{k},\omega=0)}, \tag{2}$$

where $N(0)$ is the density of states at the Fermi surface and k_F is the Fermi momentum - see more in [11]. In [10] it was claimed that the lattice stability of the system with respect to the charge density wave formation implies that the condition $\varepsilon_{tot}(\mathbf{k},\omega=0)>1$ must be fulfilled for all $\mathbf{k}$. If this were correct then from Eq.(2) follows that $\mu>\lambda$, which limits the maximal value of T_c to the value $T_c^{\max}\approx E_F\exp(-4-3/\lambda)$. In typical metals $E_F<(1-10)\ eV$ and if one accepts this (unfounded) statement that $\lambda\leq\mu\leq 0.5$ one obtains $T_c\sim(1-10)\ K$. The latter result, of course if it would be true, means mean that the EPI is ineffective in producing high-T_c superconductivity, let say not higher than 20 K? However, this result is apparently in conflict with a number of experimental results in low-T_c superconductors (LTS), where $\mu \leq \lambda$ and $\lambda > 1$. For instance, $\lambda \approx 2.5$ is realized in $PbBi$ alloy, which is definitely much higher than $\mu(< 1)$ thus contradicting the statement made in Ref.[10].

The statement in [10] that $\varepsilon_{tot}(\mathbf{k}, \omega = 0) > 1$ must be fulfilled for all $\mathbf{k}$ is in an apparent conflict with the basic theory [12], which tells us that $\varepsilon_{tot}(\mathbf{k}, \omega)$ is not the response function. If a small external potential $\delta V_{ext}(\mathbf{k}, \omega)$ is applied to the system it induces screening by charges of the medium and the total potential is given by $\delta V_{tot}(\mathbf{k}, \omega) = \delta V_{ext}(\mathbf{k}, \omega)/\varepsilon_{tot}(\mathbf{k}, \omega)$ which means that $1/\varepsilon_{tot}(\mathbf{k}, \omega)$ is the response function. The latter obeys the Kramers-Kronig dispersion relation which implies the following stability condition: $1/\varepsilon_{tot}(\mathbf{k}, \omega = 0) < 1$ for $\mathbf{k} \neq 0$, i.e. either $\varepsilon_{tot}(\mathbf{k} \neq 0, \omega = 0) > 1$ or $\varepsilon_{tot}(\mathbf{k} \neq 0, \omega = 0) < 0$. This important theorem has been first proved in the seminal article by David Abramovich Kirzhnits [12] and it invalidates the formula for $T_c^{\max}$ by setting aside the above restriction on the maximal value of T_c.

Is $\varepsilon_{tot}(\mathbf{k} \neq 0, \omega = 0) < 0$ realized in real systems? This question was thoroughly studied in Ref. [13] and in the context of HTSC in [11], while here we enumerate the main results. In the inhomogeneous system, such as a crystal, the total longitudinal dielectric function is matrix in the space of reciprocal lattice vectors ($\mathbf{Q}$), i.e. $\hat{\varepsilon}_{tot}(\mathbf{k}+\mathbf{Q}, \mathbf{k}+\mathbf{Q}', \omega)$, and $\varepsilon_{tot}(\mathbf{k}, \omega)$ is defined by $\varepsilon_{tot}^{-1}(\mathbf{k}, \omega) = \hat{\varepsilon}_{tot}^{-1}(\mathbf{k}+\mathbf{0}, \mathbf{k}+\mathbf{0}, \omega)$. For instance in the dense metallic systems with one ion per cell (such as the metallic hydrogen) and with the electronic dielectric function $\varepsilon_{el}(\mathbf{k}, 0)$ one has [13]

$$\varepsilon_{tot}(\mathbf{k}, 0) = \frac{\varepsilon_{el}(\mathbf{k}, 0)}{1 - \frac{1}{\varepsilon_{el}(\mathbf{k},0) G_{EP}(\mathbf{k})}}. \tag{3}$$

At the same time the frequency of the longitudinal phonon $\omega_l(\mathbf{k})$ is given by

$$\omega_l^2(\mathbf{k}) = \frac{\Omega_{pl}^2}{\varepsilon_{el}(\mathbf{k}, 0)} [1 - \varepsilon_{el}(\mathbf{k}, 0) G_{EP}(\mathbf{k})], \tag{4}$$

where G_{EP} is the local field correction G_{EP} - see Ref. [13]. The right condition for the lattice stability requires that the phonon frequency must be positive, $\omega_l^2(\mathbf{k}) > 0$, which implies that for $\varepsilon_{el}(\mathbf{k}, 0) > 0$ one has $\varepsilon_{el}(\mathbf{k}, 0) G_{EP}(\mathbf{k}) < 1$. The latter gives $\varepsilon_{tot}(\mathbf{k}, 0) < 0$. The calculations [13] show that in the metallic hydrogen crystal $\varepsilon_{tot}(\mathbf{k}, 0) < 0$ for all $\mathbf{k} \neq \mathbf{0}$. The sign of $\varepsilon_{tot}(\mathbf{k}, 0)$ for a number of crystals with more ions per unit cell is thoroughly analyzed in [13], where it is shown that $\varepsilon_{tot}(\mathbf{k} \neq \mathbf{0}, 0) < 0$ is *more a rule than an exception*. The physical reason for $\varepsilon_{tot}(\mathbf{k} \neq \mathbf{0}, 0) < 0$ is due to local field effects described by $G_{EP}(\mathbf{k})$. Whenever the local electric field $\mathbf{E}_{loc}$ acting on electrons (and ions) is different from the average electric field $\mathbf{E}$, i.e. $\mathbf{E}_{loc} \neq \mathbf{E}$ there are corrections to $\varepsilon_{tot}(\mathbf{k}, 0)$ (or in the case of the electronic subsystem to $\varepsilon_e(\mathbf{k}, 0)$) which may lead to $\varepsilon_{tot}(\mathbf{k}, 0) < 0$.

The above analysis tells us that in real crystals $\varepsilon_{tot}(\mathbf{k}, 0)$ can be negative in the large portion of the Brillouin zone giving rise to $\lambda - \mu > 0$, due to local field effects. This means that the dielectric function ε_{tot} *does not limit T_c in the phonon mechanism of pairing*. The latter does not mean that there is no limit on T_c at all. We mention in advance that the local field effects play important role in HTSC oxides, due to their layered structure with ionic-metallic binding, thus giving rise to large *EPI* - see more subsequent sections.

In concluding we point out, that there are no theoretical and experimental arguments for ignoring the EPI in HTSC oxides. However, it is necessary to answer several important questions which are also related to experimental findings in HTSC oxides: (**1**) If the EPI is responsible for pairing in HTSC oxides and if superconductivity is of $d-wave$ type, how these two facts are compatible? (**2**) Why is the transport EPI coupling constant λ_{tr} (entering the resistivity formula) much smaller than the pairing EPI coupling constant $\lambda(>1)$ (entering the formula for T_c), i.e. why one has $\lambda_{tr}(\approx 0.4-0.9) \ll \lambda(\sim 2)$? (**3**) Is the high T_c value possible for a moderate EPI coupling constant, let say for $\lambda \leq 1$? (**4**) Finally, if the EPI interaction is ineffective for pairing in HTSC oxides why it is so?

2. EXPERIMENTS RELATED TO PAIRING MECHANISM

A much more extensive discussion (than here) of the experimental situation in HTSC oxides is given in a number of papers - see reviews [2], [11]. In the following we discuss briefly experimental results, by including the most recent ones, which can give us a clue for the pairing mechanism in the HTSC oxides.

2.1. Magnetic neutron scattering

2.1.1. Normal state

The cross-section for the *inelastic neutron magnetic scattering* is expressed via the Fourier transform of the spin-correlation function (the spin structure factor) $S^{\alpha\alpha}(\mathbf{k},\omega)$ which is proportional to the imaginary part of the susceptibility $Im\chi(\mathbf{k},k_z,\omega)$. In the (normal) metallic state of doped HTSC oxides without magnetic order the inelastic scattering (in absence of the AF magnetic order) is of interest and in most systems $Im\chi(\mathbf{k},k_z,\omega)$ is peaked around the *AF* wave-vector $\mathbf{Q}=(\pi,\pi)$. The pronounced magnetic fluctuations in the underdoped metallic state is contrary to usual metals (described by the Landau-Fermi liquid) where the magnetic fluctuations are much weaker. In HTSC oxides $Im\chi(\mathbf{k},\omega)$ depends on hole doping, and for instance, in the bilayer (two layers per the unit cell) compound $YBa_2Cu_3O_{6+x}$ the low energy spectra is peaked at $\mathbf{Q}$, whose width δ_m broadens by increasing doping concentration - see review [66]. Around the optimal doping the magnetic correlation length $\xi_m=(2/\delta_m)\sim(1-2)a$ is almost temperature independent. This fact contradicts the assumption of the theory of spin fluctuation mechanism by the Pines group [29], where ξ_m is strongly T-dependent. We stress that in the SFI theory $Im\chi(\mathbf{k},\omega)$ is important quantity since the effective pairing potential $V_{eff}(\mathbf{k},\omega)$ and the self-energy $\Sigma_{sf}(\mathbf{k},\omega)$ are approximately given by (on the real

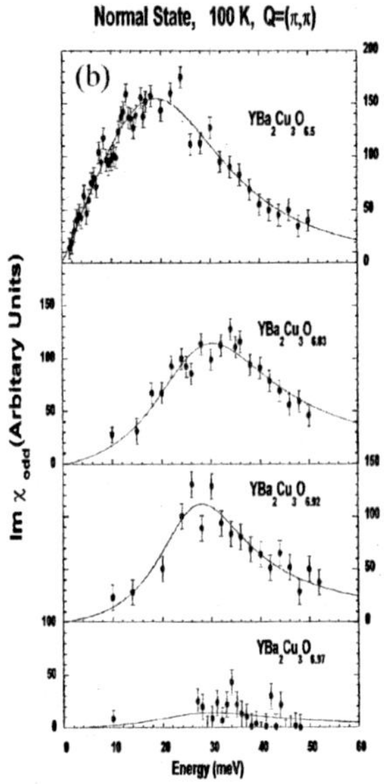

FIGURE 1. Magnetic spectral function $Im\chi^{(-)}(\mathbf{k},\omega)$: (a) $I_Q(T_c)$ values at $T = 200\ K$ for various HTSC oxides: $LSCO$ - $La_{2x}Sr_xCuO_4$; $TBCO$ - $Tl_2Ba_2CuO_{6+x}$ and $Tl_2Ba_2CaCu_2O_8$; $YBCO$ - and $YBa_2Cu_4O_8$ - from [67]; (b) for $YBa_2Cu_3O_{6+x}$ in the normal state at $T = 100\ K$ and at $Q = (\pi,\pi)$. 100 counts in the vertical scale corresponds to $\chi_{max}^{(-)} \approx 350\mu_B^2/eV$ - from [66]; (c) for $YBa_2Cu_3O_{6+x}$ in the superconducting state at $T = 5\ K$ and at $Q = (\pi,\pi)$ - from [66].

frequency axis - see [2])

$$\Sigma_{sf}(\mathbf{k},\omega) \approx \sum_{\mathbf{q}} \int \frac{d\Omega}{\pi} G(\mathbf{q},\Omega) V_{eff}(\mathbf{k}+\mathbf{q},\omega+\Omega),$$

$$V_{SF}(\mathbf{q},\omega+i0^+) = g_{SF}^2 \int_{-\infty}^{\infty} \frac{d\Omega}{\pi} \frac{Im\chi(\mathbf{q},\Omega+i0^+)}{\Omega-\omega} \qquad (5)$$

where $G(\mathbf{q},\Omega)$ is the electron Green's function. This approach can be theoretically justified in the weak coupling limit ($U \ll W$) only. Although the HTSC oxides are far from this limit this expression is frequently used in the SFI theories of pairing, where larger $Im\chi(\mathbf{k},\omega)$ should give larger T_c.

What is the experimental situation? The antibonding (odd) spectral function $Im\chi^{(odd)}(\mathbf{k},\omega)$ for $YBa_2Cu_3O_{6+x}$ is strongly doping dependent as it is seen in Fig. 1. By comparing the magnetic neutron scattering (normal state) spectra in $YBa_2Cu_3O_{6.92}$ and $YBa_2Cu_3O_{6.97}$ in Fig. 1a the difference is reflected in their spectral functions $Im\chi^{(odd)}(\mathbf{k},\omega)$. Namely, in the frequency interval which is important for superconducting pairing $Im\chi^{(odd)}(\mathbf{Q},\omega)$ of

$YBa_2Cu_3O_{6.92}$ is much larger than that in $YBa_2Cu_3O_{6.97}$ although the differences in their critical temperatures T_c is very small, i.e. $T_c = 91\ K$ for $YBa_2Cu_3O_{6.92}$ and

$T_c = 92.5\ K$ for $YBa_2Cu_3O_{6.97}$. This result, in conjunction with the anti-correlation between the NMR spectral function $I_{\mathbf{Q}} = \lim_{\omega \to 0} Im\chi(\mathbf{Q},\omega)/\omega$ and T_c - shown in Fig. 1a, is apparently against the SFI theoretical models for pairing mechanism [29], [65].

2.1.2. Superconducting state

In the superconducting state the magnetic fluctuations are drastically changed, what is in fact expected for the singlet pairing state which induces spin gap in the magnetic excitation spectrum of s-wave superconductors. However, the spectrum in the superconducting state of HTSC oxides is more complex due to d-wave pairing and specificity of the band structure. For instance, at $T < T_c$ the sharp peak in $Im\chi^{(odd)}(\mathbf{k},\omega)$ is seen at $\omega_{reson} = 41\ meV$ and at $\mathbf{k}_{2D} = (\pi/a, \pi/a)$ of the fully oxygenated (optimally doped) $YBa_2Cu_3O_{6+x}$ ($x \sim 1$, $T_c \approx 92$ K) [71], [72]. The doping dependence of the peak position and its width [66] is shown in Fig. 1c, where it is seen that by increasing doping the peak in the superconducting state becomes sharper and moves to higher frequencies (scaling with T_c), while its height is decreasing. This can be qualitatively explained by using the *RPA* susceptibility

$$\chi(\mathbf{k},\omega) = \frac{\chi_0(\mathbf{k},\omega)}{1 - U_{eff}(\mathbf{q})\chi_0(\mathbf{k},\omega)}, \tag{6}$$

where the bare susceptibility $\chi_0(\mathbf{k},\omega)$ contains the coherence factor $[1-(\xi_{\mathbf{k}+\mathbf{q}}\xi_{\mathbf{q}} + \Delta_{\mathbf{k}+\mathbf{q}}\Delta_{\mathbf{q}})/E_{\mathbf{k}+\mathbf{q}}E_{\mathbf{q}}]$ - see [2]. This (type II) coherence factor reflects the (well known) fact that the magnetic scattering is not the time reversal symmetry. In the case when $\mathbf{k}$ and $\mathbf{k}+\mathbf{q}$ are near the Fermi surface and when $\Delta_{\mathbf{k}+\mathbf{q}} \approx -\Delta_{\mathbf{q}}$ at $\mathbf{k} = \mathbf{Q} = (\pi/a, \pi/a)$ the coherence factor is of the order of one at or near the Fermi surface (note $\xi_{\mathbf{k}+\mathbf{q}}\xi_{\mathbf{q}} \leq 0$) and therefore contributes significantly to $\chi_0(\mathbf{k} = \mathbf{Q},\omega)$. The case $\Delta_{\mathbf{q}}\Delta_{\mathbf{k}+\mathbf{q}} < 0$ is realized when the $d-wave$ order parameter, for instance $\Delta_{\mathbf{k}} = (\Delta_0/2)[\cos k_x - \cos k_y]$. So, the mechanism of the peak formation (below T_c) is the consequence of the electron-pair creation with an electron in the $(+)$ lobe and a hole in the $(-)$ lobe of the superconducting order parameter. Note, that the $(\pm)$ lobes of $\Delta_{\mathbf{k}}$ are separated approximately by the wave-vector $\mathbf{Q} = (\pi/a, \pi/a)$. Due to the large density of states near the lobes a large peak in $Im\chi(\mathbf{k} = \mathbf{Q}, k_z, \omega)$ is expected to be realized, i.e. $\omega_{reson} \geq 2\Delta_0$. Of course the better (than RPA) calculations of $\chi(\mathbf{k} = \mathbf{Q}, k_z, \omega)$ is needed for a full quantitative analysis, where a possible resonance in $\chi(\mathbf{k},\omega)$ with $\omega_{reson} \leq 2\Delta_0$ can also contribute. It is important to stress, that the magnetic resonance in the superconducting state is *consequence of superconductivity* but not its cause as it was stated in some papers. It can not be the cause for superconductivity simply because its intensity at T around T_c is vanishing small and not affecting T_c at all. If the magnetic resonance would be the origin for superconductivity (and high T_c) the phase transition at T_c must be first order, contrary to experiments where it is second order.

The next very serious argument against the SFI pairing mechanism is the *smallness of the coupling constant* g_{sf}. Namely, the real spin-fluctuation coupling constant is rather small $g_{sf} \leq 0.2\ eV$, what is in contrast to the large value ($g_{sf}^{(MMP)} \sim 0.6\ eV$) assumed in the SFI theory by the Pines group - see the MMP model in Section 7.1. The upper limit of $g_{sf}(\leq 0.2\ eV)$ is extracted from : (**i**) the width of the resonance peak [68], and (**ii**) the small magnetic moment ($\mu < 0.1\ \mu_B$) in the antiferromagnetic state of LASCO and YBCO [69]. Note, that the pairing coupling in the SFI theory is $\lambda_{sf} \sim g_{sf}^2$, and for the realistic value of $g_{sf} \leq 0.2\ eV$ it would produce $\lambda_{sf} \sim 0.2$ and very small $T_c \sim 1\ K$. The *SFI* model roots on its basic $t-J$ Hamiltonian. However, recently it was shown in [70] that *there is no superconductivity in the t-J model* at temperatures characteristic for HTSC oxides - see Fig. 27 below. If it exists T_c must be very low.

In conclusion, the inelastic magnetic neutron scattering give evidence that the spin fluctuations interaction (SFI), although pronounced in underdoped systems, is *ineffective* in the pairing mechanism of HTSC oxide. However, the SFI in conjunction with the residual Coulomb repulsion *triggers* superconductivity from s-wave to d-wave, whose strength is predominantly due to the EPI - see discussion in Sections 5.-7..

2.2. Dynamical conductivity and resistivity $\rho(T)$

Since $\sigma(\omega)$ and $\rho(T)$ give important information on the dominant scattering mechanism, in the following we analyze their properties in more details.

2.2.1. Dynamical conductivity $\sigma(\omega)$

$\sigma(\omega)$ is in fact *derived quantity* since it is extracted from the *measured* optic reflectivity $R(\omega)$ and absorption $A(\omega)$. By measuring the normal-incident (of light) reflectivity $R(\omega)$ in the whole frequency region ($0 \leq \omega < \infty$) one can determine the phase $\phi(\omega)$ of the complex reflectivity

$$r(\omega) = \sqrt{R(\omega)}e^{i\phi(\omega)} = \frac{\sqrt{\varepsilon(\omega)}-1}{\sqrt{\varepsilon(\omega)}+1} \tag{7}$$

by the Kramers-Kronig relation, and accordingly to determine in principle the complex dielectric function

$$\varepsilon(\omega) = \varepsilon_\infty + \varepsilon_{latt}(\omega) + \frac{4\pi i\sigma(\omega)}{\omega}, \tag{8}$$

where ε_∞ and $\sigma(\omega)$ are electronic contributions and ε_{latt} is the lattice contribution. However, $R(\omega)$ is usually measured in a finite ω region and extrapolations is needed, especially at very low frequencies. This extrapolation of $R(\omega)$ also contains some model assumptions on the scattering processes in the system (on $\sigma(\omega)$), i.e. $1-R(\omega) \sim \sqrt{\omega}$ - the Hagen-Rubens relation for the standard (with elastic scattering only) Drude metal, or $1-R(\omega) \sim \omega$ for strong EPI (or for marginal Fermi liquid). So, one should be always cautious not to overinterpret the meaning of $\sigma(\omega)$ obtained in such a way.

In HTSC oxides $R(\omega)$, $A(\omega)$ are usually measured in a broad frequency region - up to several eV. At such high frequencies the interband transitions take place and in order to calculate $\sigma(\omega)$ the knowledge of the band structure is needed. This problem was analyzed in the framework of the LDA band structure calculations [73] by taking into account the interband transitions, where a rather good agreement with experiments for $\omega > 1\ eV$ was found. This is surprising since the *LDA*-method does not contain the Hubbard bands, and according to [73] there is no sign of transitions between Hubbard sub-bands in the high energy region of $\sigma(\omega)$. This very interesting result deserves to be further analyzed since it contradicts the physics of the Hubbard models.

Here we discuss briefly the normal state $\sigma(\omega)$ in the low frequency region $\omega < 1\ eV$ where the *intraband* effects dominate the quasiparticle scattering. In the low ω regime the processing of the data in the metallic state of HTSC oxides is usually done by using the generalized Drude formula for the inplane conductivity $\sigma(\omega) = \sigma_1 + i\sigma_2$ [85], [86], [87]

$$\sigma_{ii}(\omega) = \frac{\omega_{p,ii}^2}{4\pi} \frac{1}{\Gamma_{tr}(\omega, T) - i\omega m_{tr}(\omega)/m_\infty}. \tag{9}$$

$i = a, b$ enumerates the plane axis, $\Gamma_{tr}(\omega, T)$ and $m_{tr}(\omega)$ are the transport scattering rate and optic mass, respectively. Sometimes in the analysis of experimental data the effective transport scattering rate $\Gamma_{tr}^*(\omega, T)$ and the effective plasma frequency $\omega_p^*(\omega)$ are used, which are defined by

$$\Gamma_{tr}^*(\omega, T) = \frac{m}{m_{tr}(\omega)} \Gamma_{tr}(\omega, T) = \frac{\omega \sigma_1(\omega)}{\sigma_2(\omega)}, \tag{10}$$

and

$$\omega_p^{*2}(\omega) = \frac{m}{m_{tr}(\omega)} \omega_p^2. \tag{11}$$

For best optimally doped HTSC systems the best fit for $\Gamma_{tr}^*(\omega, T)$ is given by $\Gamma_{tr}^*(\omega, T) \approx \max\{\alpha T, \beta \omega\}$ in the temperature and frequency range from very low ($\sim 100\ K$) up to $2000\ K$, where α, β are of the order one - see Fig. 2. These results tell us that the quasiparticle liquid, which is responsible for transport properties in HTSC, is not a simple (weakly interacting) Fermi liquid. We remind the reader that in the usual (canonical) normal Fermi liquid with the Coulomb interaction on has $\Gamma_{tr}(\omega, T) \sim \Gamma_{tr}^*(\omega, T) \sim \Gamma(\omega, T) \sim \max\{T^2, \omega^2\}$ at low T and ω, which means that quasiparticles are well defined objects near (and at) the Fermi surface since $\omega \gg \Gamma(\omega, T)$. In case of HTSC oxides with $\Gamma(\omega, T) \sim \max(\omega, T)$ in the broad regions and the quasiparticles decay rapidly and therefore are not well defined objects. At these temperatures and frequencies the simple canonical Landau quasiparticle concept fails. The latter behavior can be due to the strong electron-electron inelastic scattering, or due to the quasiparticle scattering on phonons (or on other bosonic excitations). It is important to stress, that quasiparticles

interacting with phonons at finite T are not described with the standard Fermi liquid, in particular at $T > \Theta_D/5$, since the scattering rate is larger than the quasiparticle energy, i.e. one has $\Gamma \sim \max(\omega, T)$. Such a system is well described by the *Migdal-Eliashberg theory* whenever $\omega_D \ll E_F$ is fulfilled, which in fact treats quasiparticles beyond the original Landau quasiparticle concept. Note, that even when the original Landau quasiparticle concept fails the transport properties may be described by the Boltzmann equation, which is a wider definition of the Landau-Fermi liquid.

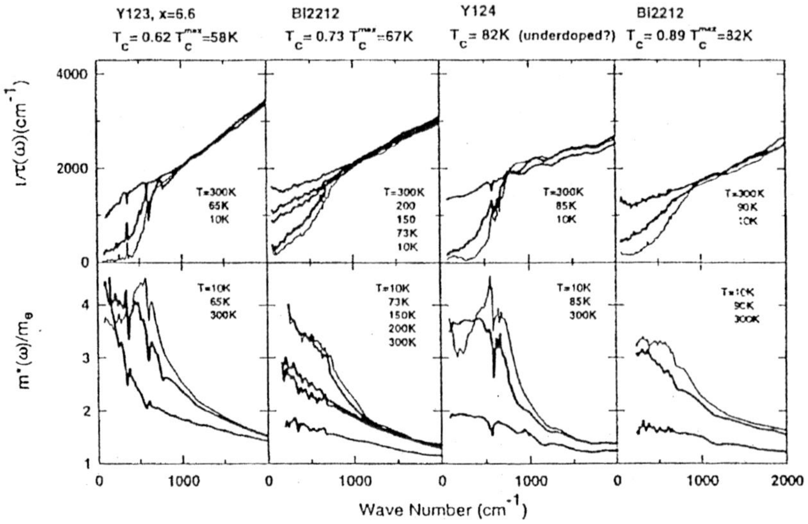

FIGURE 2. The transport scattering rate $1/\tau(\omega)$ (in the text $\Gamma_{tr}(\omega)$) and the transport effective mass $m^*(\omega)/m_e$ (in the text $m_{tr}(\omega)/m_\infty$) for series of underdoped HTSC oxides. $\Gamma_{tr}(\omega)$ is temperature independent above 1000 cm^{-1} but it is depressed at low T and low ω- from [62] © 1996 IOP Publishing Ltd.

We point out, that in a number of articles it was *incorrectly* assumed that $\Gamma(\omega,T) \approx \Gamma_{tr}(\omega,T) \approx \Gamma^*_{tr}(\omega,T)$ holds in HTSC oxides. The above discussed experiments (see Fig. 2) give that $\Gamma^*_{tr}(\omega,T)$ is linear in the broad region of ω and T up to 2500 K - see in Fig. 2. However, if $\Gamma(\omega,T)$ is due to the EPI it saturates at the maximum phonon frequencies $\omega^{ph}_{\max}(\leq 1000\ K)$. By assuming also that $\Gamma^{EPI}(\omega,T) \approx \Gamma^{EPI}_{tr}(\omega,T)$ holds for all ω, in a number of papers it was concluded, that the EPI does not contribute to the inelastic scattering of quasiparticles and to the Cooper pairing in HTSC oxides. Does it hold $\Gamma^{EPI}(\omega,T) \approx \Gamma^{EPI}_{tr}(\omega,T)$ in HTSC oxides? The answer is NO.

$\sigma(\omega)$ of HTSC oxides was theoretically analyzed [86], [87] in terms of the EPI, where it was found that $\Gamma_{tr}(\omega,T)$ and $m_{tr}(\omega,T)$ depend on the transport spectral function $\alpha^2_{tr,EP}F(\omega)$ - see more in [2]. Their analysis is based on: (i) the assumption that $\alpha^2_{tr,EP}F(\omega) \approx \alpha^2_{EP}F(\omega)$ - the Eliashberg spectral function: (ii) the shape of $\alpha^2_{EP}F(\omega)$ is extracted from various tunnelling conductivity measurements, [38], [39], [40], [41], which makes a rather large EPI coupling constant and the critical temperature $\lambda = 2\int_0^\infty d\omega\alpha^2(\omega)F(\omega)/\omega \approx 2$ and $T_c \approx 90\ K$, respectively; (iii) the plasma frequency is taken to be $\omega_{pl} = 3\ eV$. It was obtained that $\Gamma^{EP}_{tr}(\omega,T) \sim \omega$ in a very broad ω-interval (up to 250 meV), which is much larger than the maximum phonon frequency $\omega^{ph}_{\max} \approx 80$ *mev*. This is illustrated in Fig. 3. Moreover, $\Gamma^{EP}_{tr}(\omega,T)$ *differs significantly* from the quasiparticle scattering rate $\Gamma^{EP}(\omega,T) = -2Im\Sigma(\omega)$ [86], [87]. We see from Fig. 3 that $\Gamma^{EP}(\omega,T)$ is much steeper function than $\Gamma^{EP}_{tr}(\omega,T)$ and the former saturates at much lower frequency - of the order of the maximum phonon frequency $\omega^{ph}_{\max}$.

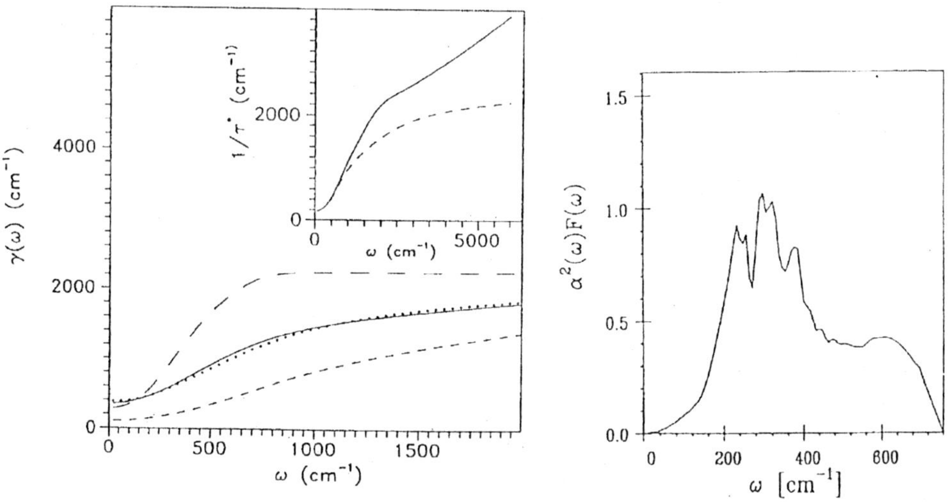

FIGURE 3. The theoretical predictions for the frequency dependence of the various relaxation rates $\gamma(=\Gamma)$ with $\alpha^2F(\omega)$ - right: the generalized Drude fit for $\Gamma_{tr}(\omega)$ - solid line; $\Gamma^*_{tr}(\omega)$ - short-dashed line; $\Gamma(\omega)$ - long dashed line; $\Gamma_{tr}(\omega)$ calculated - dotted line. In the inset the calculated $\Gamma^*_{tr}(\omega)(=1/\tau^*(\omega))$ with (solid line) and without (dashed line) the interband contributions with $\alpha^2F(\omega)$ from right and at $T=100\ K$ - from [87] © Elsevier 1989, reprinted with permission.

Note, that $\Gamma^{*,EP}_{tr}(\omega,T)=(m/m_{tr}(\omega))\Gamma^{EP}_{tr}(\omega,T)$ is also quasi-linear function in a very broad region $150\ K<\omega<3000\ K$ - see Fig. 3. The slope of $\Gamma^{*,EP}_{tr}(\omega,T)$ is of the order of one, in accordance with experiments results [86], [87], and it (and $\Gamma^{EP}_{tr}(\omega,T)$) saturates at $\omega_{sat}\simeq -Im\Sigma_{tr}(\omega_{sat})\gg\omega^{ph}_{\max}$ only. The transport spectral function $\alpha^2_{tr}(\omega)F(\omega)$ can be also extracted from the transport scattering rate $\Gamma_{tr}(\omega,T=0)$ - see [2], [11], since the theory gives that

$$\Gamma_{tr}(\omega,T=0)=\frac{2\pi}{\omega}\int_0^{\omega}d\Omega(\omega-\Omega)\alpha^2_{tr}(\Omega)F(\Omega). \tag{12}$$

However, real measurements are performed at finite$T(>T_c)$ where $\alpha^2_{tr}(\omega)F(\omega)$ is the solution of the Fredholm integral equation (of the first kind). Such an inverse problem at finite temperatures in HTSC oxides is studied first in [86] (see also [87]), where the smeared structure of $\alpha^2_{tr}(\omega)F(\omega)$ in $YBa_2Cu_3O_{7-x}$ was obtained, which is in qualitative agreement with the shape of the phonon density of states $F(\omega)$. At finite T the problem is more complex because the fine structure of $\alpha^2_{tr}(\omega)F(\omega)$ gets blurred as the calculations in [74] show. The latter gave that $\alpha^2_{tr}(\omega)F(\omega)$ ends up at $\omega_{\max}\approx 70-80\ meV$, which is the maximal phonon frequency in HTSC oxides. This result indicates strongly that the EPI in HTSC oxides is dominant in the IR optics. We point out, that if $R(\omega)$ (and $\sigma(\omega)$) are due to some bosonic process with large frequency cutoff ω_c in the spectrum, as it is the case with the spin-fluctuation (SFI) scattering where $\omega_c\approx 400\ meV$, the extracted $\alpha^2_{tr}(\omega)F(\omega)$ should end up at this high ω_c. The latter is *not seen* in optic measurements at $T>T_c$, which tells us that the SFI scattering, with $\alpha^2_{tr}(\omega)F(\omega)\sim g^2_{sf}Im\chi_s(\omega)$ and with the cutoff $\omega_c \geq 400\ meV$, is rather weak and ineffective in optics of HTSC oxides.

We stress that the extraction of Γ_{tr} from $R(\omega)$ is subtle procedure, because it depends also on the assumed value of ε_∞. For instance, if one takes $\varepsilon_\infty = 1$ then Γ_{tr}^{EP} is linear up to very high ω, while for $\varepsilon_\infty > 1$ the linearity of Γ_{tr}^{EP} saturates at lower ω. Since $\Gamma_{tr}^{EP}(\omega, T)$, extracted in [62], and recently also in [63], is linear up to very high ω it may be that the ion background and interband transitions (contained in ε_∞) are not properly taken into account in these papers. As a curiosity in a number of papers, even in the very cited ones such as [62], [63], there is no information which value for ε_∞ they take. We stress again, that the behavior of $\Gamma_{tr}(\omega)$ is linear up to much higher frequencies for $\varepsilon_\infty = 1$ than for $\varepsilon_\infty \approx 4-5$ - the characteristic value for HTSC, giving a lot of room for inadequate interpretations of results. In that respect, the recent elipsometric optic measurements on YBCO [75] confirm the results of the previous ones [76] that $\varepsilon_\infty \geq 4$ and that Γ_{tr}^{EP} saturates at lower frequency than it was the case in Ref. [62]. We stress again that the reliable estimation of the value and ω, T dependence of $\Gamma_{tr}(\omega)$ and $m(\omega)$ can be done, not from the reflectivity measurements [62], [63], but from elipsometric ones only [76], [75].

In concluding this part we stress two facts: (**1**) The large difference in the ω, T behavior of $\Gamma_{tr}^{EP}(\omega, T)$ and $\Gamma^{EP}(\omega, T)$ is not a specificity of HTSC oxides but it is realized also in a number of *LTSC* materials. In fact this is a common behavior even in simple metals, such as Al, Pb, as shown in [118], where $\Gamma^{EP}(\omega, T)$ saturates at much lower (Debay) frequency than $\Gamma_{tr}^{EP}(\omega, T)$ and $\Gamma_{tr}^{*,EP}(\omega, T)$ do. In that respect the difference between simple metals and HTSC oxides is in the scale of phonon frequencies, i.e. $\omega_{\max}^{ph} \sim 100\ K$ in simple metals, while $\omega_{\max}^{ph} \sim 1000\ K$ in HTSC oxides. Having in mind these well established and well understood facts, it is very surprising that even nowadays, 18 years after the discovery of HTSC oxides, the principal and quantitative difference between Γ and Γ_{tr} is neglected in the analysis of experimental data. For instance, by neglecting the pronounced (qualitative and quantitative) difference between $\Gamma_{tr}(\omega, T)$ and $\Gamma(\omega, T)$, in the recent papers [63], [64] were made far reaching, but unjustified, conclusions that the magnetic pairing mechanism prevails; (**2**) It is worth of mentioning, that quite similar (to HTSC oxides) properties, of $\sigma(\omega)$, $R(\omega)$ and $\rho(T)$ were observed in experiments [91] on isotropic metallic oxides $La_{0.5}Sr_{0.5}CoO_3$ and $Ca_{0.5}Sr_{0.5}RuO_3$ - see Fig. 4. We stress that in these compounds there are no signs of antiferromagnetic fluctuations (which are present in HTSC oxides) and the peculiar behavior is probably due to the EPI.

It is worth of mentioning that after the discovery of HTSC in 1986 a number of controversial results related to $\sigma(\omega)$ were published, followed by a broad spectrum of results and interpretations, from standard approaches up to highly exotic ones. For example, the reported experimental values for ω_{pl} were in the surprisingly large range $(0.06 - 25)\ eV$, causing a number of exotic (and confusing) theoretical models for electronic dynamics - see more in [76]. (The similar situation was with ARPES measurements - see below.) So, one should be very cautious in interpreting experimental and theoretical results. In that respect, recent experiments related to the *optical sum-rule* is an additional example for controversies in this field coming from inadequate interpretations of results. This is the reason why we devote more space to the problem of "violation" of partial sum-rule.

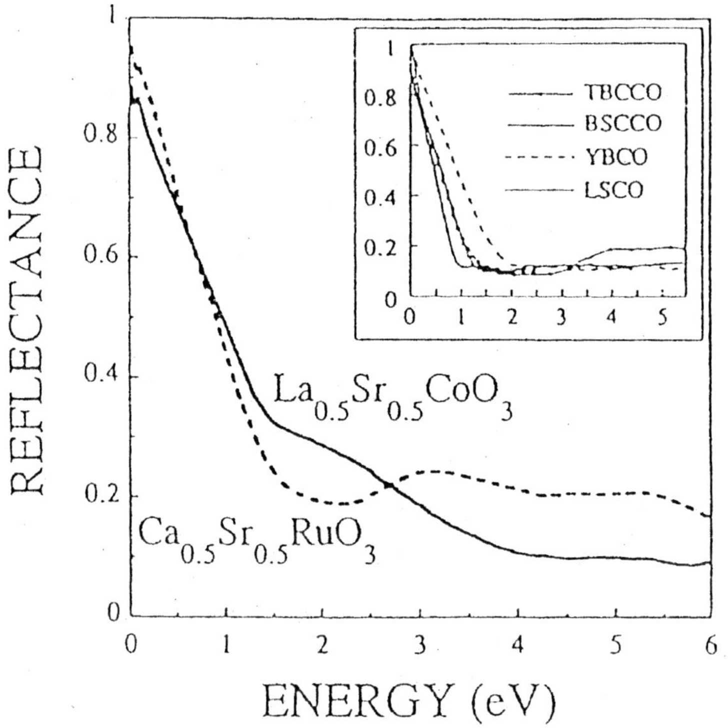

FIGURE 4. Broad range specular reflectance spectra of $Ca_{0.5}Sr_{0.5}RuO_3$ (broken line) and $La_{0.5}Sr_{0.5}CoO_3$ (solid line). Inset spectra of $Tl_2Ba_2Ca_2Cu_3O_{10}$, $Bi_2Sr_2CaCu_2O_8$, $YBa_2Cu_3O_7$ and $La_{1.85}Sr_{0.15}CuO_4$. From [91].

There are two kinds of sum rules which are used in interpreting results on $\sigma(\omega)$. The first one is the *total sum rule* and in the normal state it reads

$$\int_0^{\infty} \sigma_1^N(\omega)d\omega = \frac{\omega_{pl}^2}{8} = \frac{\pi ne^2}{2m}, \qquad (13)$$

while in the *superconducting state* [77] it is given by the Tinkham-Ferrell-Glover (TFG) sum-rule

$$\frac{c^2}{8\lambda_L^2} + \int_{+0}^{\infty} \sigma_1^S(\omega)d\omega = \frac{\omega_{pl}^2}{8}. \qquad (14)$$

Here, n - is the total electron density, e - the electron charge, m - the bare electron mass, λ_L - the London penetration depth. The first term $c^2/8\lambda_L^2$ is due to the appearance of the superconducting condensate (ideal conductivity) which contributes $\sigma_{1,cond}^S(\omega) = (c^2/4\lambda_L^2)\delta(\omega)$. The total sum rule represents the fundamental property of matter - the conservation of the electron number. To calculate it one should use the total Hamiltonian

$\hat{H}_{tot} = \hat{T}_e + \hat{H}_{int}$ by taking into account all electrons, bands and their interactions $\hat{H}_{int}$ (Coulomb, EPI, with impurities,etc.). Here T_e is the kinetic energy of bare electrons

$$\hat{T}_e = \sum_{\sigma} \int d^3x \psi_\sigma^\dagger(x) \frac{\hat{\mathbf{p}}^2}{2m} \psi_\sigma(x) = \sum_{\mathbf{p},\sigma} \frac{\mathbf{p}^2}{2m_e} c_{\mathbf{p}\sigma}^\dagger c_{\mathbf{p}\sigma}. \tag{15}$$

The *partial sum rule* is related to the energetics in the conduction (valence) band. Usually it is derived by using the Hamiltonian of the valence electrons

$$\hat{H}_v = \hat{T}_v + \hat{V}_{v,Coul} = \sum_{\mathbf{p},\sigma} \varepsilon_{\mathbf{p}} c_{v,\mathbf{p}\sigma}^\dagger c_{v,\mathbf{p}\sigma} + \hat{V}_{v,Coul}, \tag{16}$$

which contains the band-energy (with dispersion $\varepsilon_{\mathbf{p}}$) and the Coulomb interaction of valence electrons $\hat{V}_{v,Coul}$. In the normal state the *partial sum-rule* reads [78] (for general form of $\varepsilon_{\mathbf{p}}$)

$$\int_0^\infty \sigma_{1,v}^N(\omega) d\omega = \frac{\pi e^2}{2V} \sum_{\mathbf{p}} \frac{\langle n_{v,\mathbf{p}} \rangle_{H_v}}{m_{\mathbf{p}}} \equiv \frac{\omega_{pl,v}^2(T)}{8} \tag{17}$$

where $n_{v,\mathbf{p}} = c_{\mathbf{p}\sigma}^\dagger c_{\mathbf{p}\sigma}$ and the reciprocal mass is given by $1/m_{\mathbf{p}} = \partial^2 \varepsilon_{\mathbf{p}} / \partial p_x^2$. To simplify further discussion we assume for $\varepsilon_{\mathbf{p}} = -2t(\cos p_x a + \cos p_y a)$ the *tight-binding model with nearest neighbors* (n.n.) where $1/m_{\mathbf{p}} = -2ta^2 \cos p_x a$. In practice measurements are performed up to finite ω and the integration over ω goes up to some cutoff frequency ω_c (of the order of band plasma frequency). It is straightforward to show that one has (for the n.n. tight-binding model)

$$\int_0^{\omega_c} \sigma_{1,v}^N(\omega) d\omega = \frac{\pi e^2 a^2}{2} \langle -T_v \rangle \equiv \frac{\omega_{pl,v}^2(T)}{8} \tag{18}$$

where $\langle -T_v \rangle_{H_v} = -\sum_{\mathbf{p}} \varepsilon_{\mathbf{p}} \langle n_v \rangle_{H_v}$ and by $\omega_{pl,v}^2$ is defined the band plasma frequency.

In that case the *partial sum-rule in the superconducting state* reads

$$\frac{c^2}{8\lambda_L^2} + \int_{+0}^{\omega_c} \sigma_{1,v}^S(\omega) d\omega = \frac{\pi e^2 a^2}{2} \langle -T_v \rangle. \tag{19}$$

The sum-rule was studied intensively in optimally and underdoped $Bi_2Sr_2CaCu_2O_{8-x}$ and in $YBa_2Cu_3O_{7-x}$ for the *intraplane conductivity*, where the whole frequency region is separated into the low ("intraband")- and high ("interband")-frequency parts A_L and $(A_H + A_{VH})$, respectively

$$\bar{A}_L(0,\omega_c) + A_H(\omega_c, \alpha\omega_c) + A_{VH}(\alpha\omega_c, \infty) = \omega_{pl}^2/8 \tag{20}$$

with $\bar{A}_L(0,\omega_c) = A_L(0,\omega_c) + \delta_{SN}\omega_{pl,S}^2$ where $\delta_{SN} = 1$ in superconducting state and $A(\omega_1,\omega_2) = \int_{\omega_1}^{\omega_2} \sigma_1(\omega) d\omega$. The temperature dependence of A_L and A_H in the above HTSC oxides was studied in [79] and [75], by assuming that "intraband" effects are exhausted for $\omega_c \approx 1.25\ eV$ and the main temperature dependence of the high-frequency

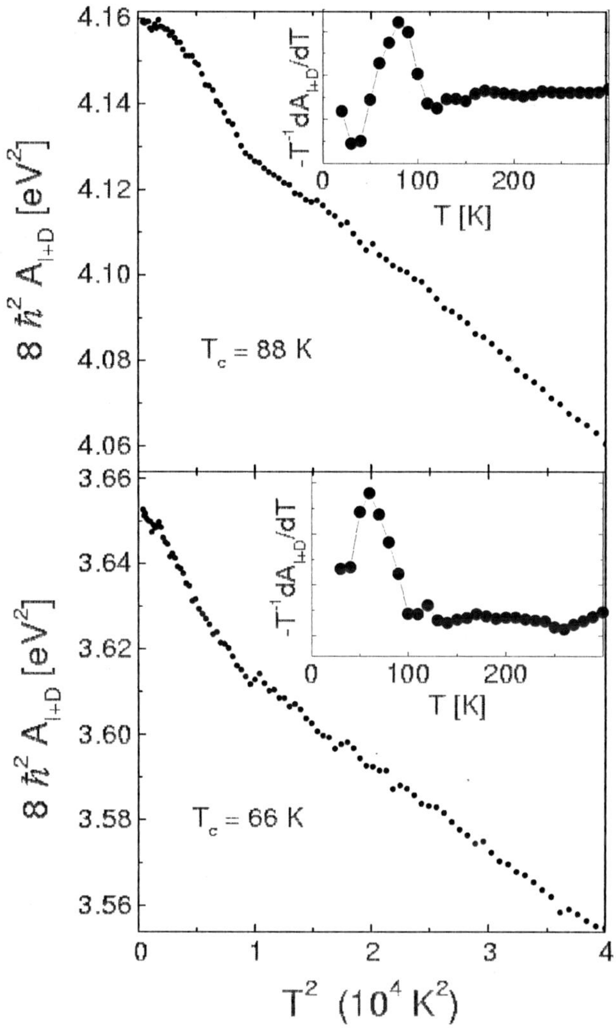

FIGURE 5. Measured T-dependence of $\bar{A}_L(0,\omega_c)$ and $\bar{A}_H(\omega_c,2\omega_c)$ for $\omega_c \approx 1.25 eV$ of $Bi2212$ ($T_c = 88$ K). The data from [79]

region comes for $\alpha = 2$ in Eq.(22), i.e. for $1.25\ eV < \omega < 2.5\ eV$, while the temperature dependence of the very high energy part A_{VH} is negligible. It was found that $\bar{A}_L(0,\omega_c)$ grows (quadratically with T) about 6% between 300 and 4 K, while $A_H(\omega_c,\alpha\omega_c)$ decreases with decreasing T^2. In the superconducting state there is a small extra increase of $\bar{A}_L(0,\omega_c)$. The results are shown in Fig. 5.

In connection with this experiment let us stress that in the BCS superconductor the TFG sum-rule is practically satisfied if the integration goes up to $\omega_s \approx (4-6)\Delta$, where Δ is the superconducting gap. This means that in the BCS superconductors the spectral weight appearing in the condensate (at $\omega = 0$) is transferred from the region $0^+ - \omega_S$. However, the experiment in [79] shows a transfer of the spectral-weight from the high ($\omega > 1\ eV$) to low energies - see below. This fact was interpreted by some researchers [80], [81] as a "violation" of the TFG sum-rule, i.e. that there is more spectral weight in the condensate (at $\omega = 0$) than it is expected from the TFG sum-rule and effectively means the decrease of the kinetic energy in the superconducting state. This is in contrast to the increase of the kinetic energy in the BCS superconducting state. We are going to discuss this problem in details and to demonstrate that the analysis in terms of the kinetic energy only, is untenable.

What is the origin of the spectral-weight transfer, especially in the superconducting state of HTSC oxides? Here, we shall study the *inplane* $\sigma_{a-b}(\omega)$ only, since the origin of the quasiparticle dynamics along the c-axis is still unclear.

The *first theoretical interpretation* of the spectral-weight transfer was based on the partial sum-rule in which the temperature dependence is related to the temperature change of the kinetic energy $\langle -T_v \rangle$ (or for more realistic spectrum of $\omega_{pl,v}^2(T)$) - see Eq.(17-19). In this framework the extra increase of $\bar{A}_L(0, \omega_c)$ is related to the lowering of the band kinetic energy in the superconducting state [80], [81]. If this would be true, then the lowering of the band kinetic energy (per particle) is approximately $(\langle T_v \rangle_{N,T>T_c} - \langle T_v \rangle_{S,T<T_c})/N \sim 1\ meV$, what is approximately by factor ten larger than the superconducting condensation energy. Note, that in the weak coupling BCS theory of superconductivity the kinetic energy is increased in the superconducting state. So, the alleged large lowering of the kinetic energy in the superconducting state is interpreted as a result of some exotic pairing mechanism in which the kinetic energy (or $\omega_{pl,v}^2(T)$ for more general spectrum) is significantly lowered but the potential energy is increased in the superconducting state, contrary to the case of BCS approach. However, this interpretation misses a very important contribution to the partial sum rule, which is due to the large and strongly T, ω dependent transport scattering rate $\Gamma_{tr}(T, \omega)$.

Before discussing the partial sum rule more adequately, let us mention that the separation of the valence-band kinetic energy from the potential one in strongly correlated systems is *not well defined procedure*. For instance, in the Hubbard model with $U >> t$ and nearest neighbor hoping t one has (see below and also in [2]) the low-energy (valence) Hamiltonian H_v is given

$$H_v = -t \sum_{i,\delta,\sigma} X_i^{\sigma 0} X_{i+\delta}^{0\sigma}, \tag{21}$$

where the Hubbard operators X_i describe the motion of composite quasiparticles with excluded doubly occupancy - see more in Section 4. They have complicated non-canonical (anti)commutation rules, which means that Eq.(21) mixes the kinetic energy with the (kinematical) potential energy of band (valence) quasiparticles.

The *second theoretical approach is* proposed recently [82], which is in principle exact, is based on the fact that in HTSC oxides there is strong electron scattering - direct or via phonons, on impurities, etc. So, the presence of an inelastic (and elastic) scattering prevents the interpretation of the partial sum rule in terms of the band kinetic energy only. As an illustrative example for this assertion may serve the scattering of electrons on impurities, where the intraband contribution to $\sigma_{1,v}(\omega)$ is given by

$$\sigma_{1,v}(\omega) = \frac{\omega_{pl,v}^2}{4\pi} \frac{\Gamma_{i,tr}}{\omega^2 + \Gamma_{i,tr}^2}. \tag{22}$$

Here, $\Gamma_{i,tr}/2 = 1/\tau_{i,tr}$ is the quasiparticle transport relaxation rate due to impurities. In this case the partial sum-rule reads

$$\int_0^{\omega_c} \sigma_{1,v}^N(\omega)d\omega = \frac{\omega_{pl,v}^2}{8}(1 - \frac{2\Gamma_{i,tr}}{\omega_c}). \tag{23}$$

This result means that the intraband sum-rule can be satisfied in the presence of impurities only for $\omega_c \to \infty$. The similar conclusion holds in the case of inelastic scattering via phonons although in that case Γ_{tr} is ω- and T-dependent. The similar reasoning holds for interband transitions which in the presence of scattering have also the low-frequency tail. Since in HTSC oxides $\Gamma_{tr}(\omega)$ is dominantly due to the EPI and reaches values up to 100 meV, there is no other way to study the partial sum-rule (the value of $\int_0^{\omega_c} \sigma_{1,v}(\omega)d\omega$) than to calculate $\sigma_{1,v}(\omega)$ directly from a microscopic model. The latter must incorporate relevant scattering mechanisms and bands. Such calculations were done in [82] by taking into account the large EPI interaction. By using the EPI spectral function $\alpha^2(\omega)F(\omega)$ from tunnelling measurements and by assuming $\alpha_{tr}^2(\omega)F(\omega) \approx \alpha^2(\omega)F(\omega)$ the authors have calculated $A_H(\omega_c, 2\omega_c)$ (and $A_L(0, \omega_c)$) in the normal state and found a good agreement with experiments [79] - see Fig. 6. We stress that the recent elipsometric measurements of the dielectric function $\varepsilon(\omega)$ [75] confirms this theoretical prediction [82].

From the above analysis we conclude that: (i) the interpretation of the partial sum-rule in HTSC oxides only in terms of the kinetic energy (or $\omega_{pl,v}^2(T)$) is physically unjustified; (ii) *the EPI interaction is strong and dominating scattering mechanism in the optical properties of the normal state*. Reliable calculations of the partial sum-rule in the superconducting state are still missing, since in that case one should know much more details on the superconducting order $\Delta(\mathbf{k}, \omega)$ and $\Gamma_{tr}(\mathbf{k}, \omega)$, which are at present too ambitious task.

2.2.2. Resistivity $\rho(T)$

A lot of experimental and theoretical works were devoted to the temperature dependence of resistivity $\rho(T)$ in HTSC oxides. General properties of the resistivity in HTSC oxides are the following: (**1**) The resistivity is very anisotropic in single crystals where one has $r_c \equiv (\rho_c(T)/\rho_{a-b}(T)) \gg 1$ at T above T_c - see [89], i.e. $r_c \approx 300$ in $La_{1.85}Sr_{0.15}CuO_4$ and $Nd_{1.85}Ce_{0.15}CuO_4$, $r_c \approx 20 - 150$ in $YBa_2Cu_3O_{7-x}$, $r_c \approx 10^5$ in $Bi_2Sr_2CaCu_2O_8$ depending also on the sample preparation, temperature etc. The anisotropy of the in-plane resistivity is much less, i.e. $r_a \equiv (\rho_{aa}(T)/\rho_{bb}(T)) \sim 1-2$, depending also on the sample preparation, temperature etc.; (**2**) The in-plane resistivity $\rho_{a-b}(T)$ at room temperature is more than two orders of magnitude higher than that of the metallic Cu (where $\rho_{Cu}(T_{room}) \approx 1.5$ μΩcm), i.e. $\rho_{a-b}(T)$ of HTSC oxides lies more in the semiconductor range and $\rho_{a-b}(T) \gg \rho_{Cu}(T)$; (**3**) $\rho_{a-b}(T) \sim T$ for $T > T_c$, which

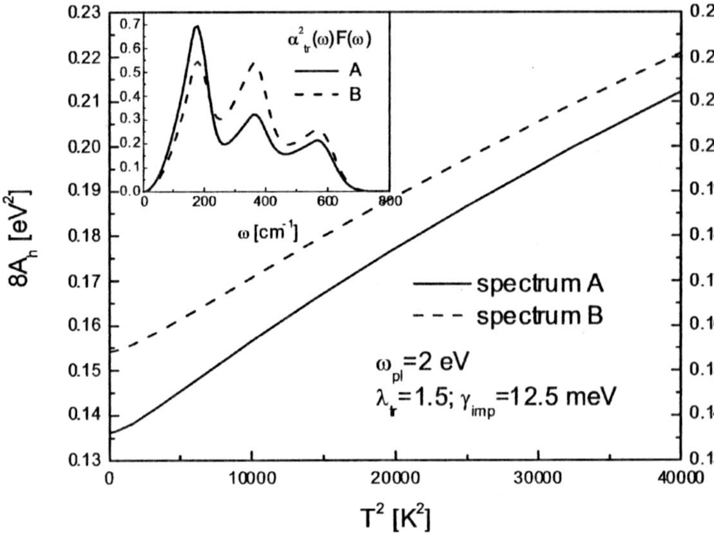

FIGURE 6. Calculated theoretical T-dependence of high-energy part of the sum rule $\bar{A}_H(\omega_c = 1.25eV, 2\omega_c = 2.5eV)$ by taking into account the electron-phonon interaction. The data from [82]

deviates at $T > (800 - 1000)\ K$ and saturates at even higher temperatures, depending on samples etc.; (**4**) ρ_{a-b} varies from $\rho_{a-b}(T) \sim T$ (with small residual resistivity) in optimally doped systems being $\rho_{a-b}(T) \sim T^{3/2}$ in overdoped systems, as experiments on $La_{2-x}Sr_xCuO_4$ show [88]; (**5**) In most samples of HTSC oxides the *c*-axis resistivity $\rho_c(T)$ shows a non-metallic behavior especially in samples with huge anisotropy along the *c*-axis, growing by decreasing temperature, i.e. $(d\rho_c(T)/dT) < 0$, being superconducting below T_c.

We discuss briefly the *in-plane resistivity* $\rho_{a-b}(T)$ only, because its temperature behavior is a direct consequence of the quasi-2*D* motion of quasiparticles and of the inelastic scattering which they suffer. At present there is no consensus on the origin of the linear temperature dependence of the inplane resistivity $\rho_{a-b}(T)$ in the normal state. As it is stressed several times many researchers are (erroneously) believing that such a behavior can not be due to the EPI? The inadequacy of this claim was already demonstrated by analyzing the dynamical conductivity $\sigma(\omega)$. The inplane resistivity in HTSC oxides is usually analyzed by the Kubo approach, or by the Boltzmann equation. In the latter case $\rho(T)$ is given by

$$\rho(T) = \frac{4\pi}{\omega_p^2}\Gamma_{tr}(T) \tag{24}$$

$$\Gamma_{tr}(T) = \frac{\pi}{T}\int_0^\infty d\omega \frac{\omega}{\sin^2(\omega/2T)}\alpha_{tr}^2(\omega)F(\omega), \tag{25}$$

where $\alpha_{tr}^2(\omega)F(\omega)$ is the EPI transport spectral function. It is well-known that at $T > \Theta_D/5$ and for the Debay spectrum one has

$$\rho(T) \simeq 8\pi^2\lambda_{tr}^{EP}\frac{k_B T}{\hbar\omega_p^2} = \rho' T. \qquad (26)$$

In HTSC oxides the reach and broad spectrum of $\alpha_{tr}^2(\omega)F(\omega)$ is favorable for such a linear behavior. The measured transport coupling constant λ_{tr} contains in principle all scattering mechanisms, although usually some of them dominate. For instance, the proponents of the spin-fluctuations mechanism assume that λ_{tr} is entirely due to the scattering on spin fluctuations. However, by taking into account specificities of HTSC oxides the experimental results for the inplane resistivity $\rho_{a-b}(T)$ can be satisfactory explained by the *EPI* mechanism. From tunnelling experiments [37], [38], [39], [40], [41] one obtains that $\lambda \approx 2-3$ and if one assumes that $\lambda_{tr} \approx \lambda$ and $\omega_{pl} \geq (3-4)$ eV (the value obtained from the band-structure calculations) then Eq.(26) describes the experimental situation rather well. The plasma frequency ω_{pl} which enters Eq.(26) can be extracted from optic measurements ($\omega_{pl,ex}$), i.e. from the width of the Drude peak at small frequencies. However, since $\lambda_{tr} \approx 0.25\omega_{pl}^2(eV)\rho'(\mu\Omega\text{cm}/K)$ there is an experimental constraint on λ_{tr}. The experiments [76] give that $\omega_{pl} \approx (2-2.5)$ eV and $\rho' \approx 0.6$ in oriented YBCO films, and $\rho' \approx 0.3$ in single crystals of BISCO. These results makes a limit on $\lambda_{tr} \approx 0.9-0.4$.

So, in order to explain $\rho(T)$ with small λ_{tr} and high T_c (which needs large λ) by the EPI it is necessary to have $\lambda_{tr} \leq (\lambda/3)$. This means that in HTSC oxides the *EPI* is reduced in transport properties where $\lambda_{tr} \ll \lambda$. This reduction of ω_p^2 and λ_{tr} means that they contain renormalization (with respect to the *LDA* results) due to various quasiparticle scattering processes and interactions, which do not enter in the *LDA* theory. In subsequent chapters we shall argue that the strong suppression of λ_{tr} may have its origin in strong electronic correlations [17], [18], [19].

In conclusion, optic and resistivity measurements in normal state of HTSC oxides are much more in favor of the EPI than against it. However, some intriguing questions still remains to be answered: **(i)** which are the values of λ_{tr} and ω_{pl}: (ii) why one has $\lambda_{tr} \ll \lambda$: (iii) what is the role of the Coulomb scattering in $\sigma(\omega)$ and $\rho(T)$. The ARPES measurements (see discussion below) give evidence for the appreciable Coulomb scattering at higher frequencies, where $\Gamma(\omega) \approx \Gamma_0 + \lambda_c\omega$ for $\omega > \omega_{\max}^{ph}$ with $\lambda_c \approx 0.4$. So, in spite of the fact that the EPI is suppressed in transport properties it is sufficiently strong in order to dominate in some temperature regime. It may happen that at higher temperatures the Coulomb scattering dominates in $\rho(T)$, which certainly does not disqualify the EPI as the pairing mechanism in HTSC oxides. For better understanding of $\rho(T)$ we need a controllable theory for the Coulomb scattering in strongly correlated systems, which is at present lacking.

2.3. Raman scattering in HTSC oxides

If the elementary excitation involved in the Raman scattering are electronic we deal with the *electronic Raman effect*, while if an optical phonon is involved we deal with the *phonon Raman effect*. The Raman scattering in the normal and superconducting state of HTSC oxides is an important spectroscopic tool which gives additional information on quasiparticle properties - the electronic Raman scattering, as well as on phonons and their renormalization by electrons - the phonon Raman scattering.

2.3.1. Electronic Raman scattering

The Raman measurements on various HTSC oxides show a remarkable correlation between the Raman cross-section $\tilde{S}_{\exp}(\omega)$ and the optical conductivity $\sigma_{a-b}(\omega)$, i.e.

$$\tilde{S}_{\exp}(\omega) \sim [1+n_B(\omega)]\langle | \gamma_{sc}(\mathbf{q}) |^2\rangle_F \omega\sigma_{a-b}(\omega), \tag{27}$$

where $n_B(\omega)$ is the Bose function and $\gamma_{sc}(\mathbf{q})$ screened Raman vertex - see more in [2]. Previously it was demonstrated that $\sigma_{a-b}(\omega)$ depends on the transport scattering rate $\Gamma_{tr}(\omega,T)$ where $\Gamma_{tr}(\omega,T) \sim T$ and $n_B(\omega) \sim T/\omega$ for $\omega < T$, thus giving $\tilde{S}(\mathbf{q},\omega) \approx Const_1$ in that range. For $\omega > T$ one has $\omega\sigma_{a-b}(\omega) \approx Const$ giving also $\tilde{S}(\mathbf{q},\omega) \approx Const_2$. We have also demonstrated that the *EPI* with the very broad spectral function $\alpha^2F(\omega)$ (see Fig. 11 below) explains in a natural way ω,T dependence of $\sigma_{a-b}(\omega)$ and $\Gamma_{tr}(\omega,T)$. So, the Raman spectra in HTSC oxides can be explained by the EPI in conjunction with strong correlations. This conclusion is supported by calculations of the Raman cross-section [90] which take into account the EPI with $\alpha^2F(\omega)$ extracted from the tunnelling measurements on $YBa_2Cu_3O_{6+x}$ and $Bi_2Sr_2CaCu_2O_{8+x}$ [37]. They are in a good qualitative agreement with experimental results - see more in [2].

We stress again, that quite similar (to HTSC oxides) properties of the electronic Raman scattering (besides $\sigma(\omega)$, $R(\omega)$ and $\rho(T)$) were observed in experiments [91] on isotropic metallic oxides $La_{0.5}Sr_{0.5}CoO_3$ and $Ca_{0.5}Sr_{0.5}RuO_3$ - see Fig. 7. To repeat again, in these compounds there are no signs of antiferromagnetic fluctuations (which are present in HTSC oxides) and the peculiar behavior is probably due to the EPI.

2.3.2. Phonon Raman scattering

Normal state. - The effect of the EPI on the Raman scattering is characterized by the *Fano asymmetry parameter* $q(\omega)$ - see more in [2]. If it is finite the line shape is *asymmetric* - the *Fano effect (resonance)*. By decreasing $q(\omega)$ the phonon line shape becomes more asymmetric, which means stronger EPI (in case when continuum states are due to conduction carriers). The Fano resonance is experimentally found in HTSC oxide $YBa_2Cu_3O_{7-\delta}$ [93], where the line asymmetry is clearly seen in optimally doped ($\delta \ll 1$) systems, while it is absent in the insulating state ($\delta = 1$). The existence of the Fano (asymmetric) line shape in HTSC oxides is a direct proof that the discrete phonon level interacts with continuum of states, which are conduction electrons in the metallic state - see Fig. 8.

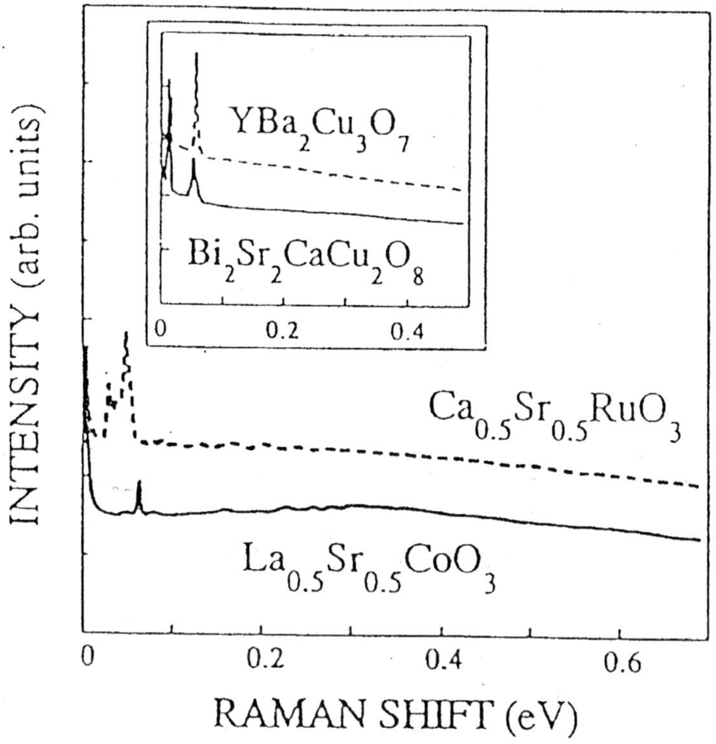

FIGURE 7. Broad range Raman scattering spectra of $Ca_{0.5}Sr_{0.5}RuO_3$ (broken line) and $La_{0.5}Sr_{0.5}CoO_3$ (solid line). Inset spectra of $Tl_2Ba_2Ca_2Cu_3O_{10}$, $Bi_2Sr_2CaCu_2O_8$, $YBa_2Cu_3O_7$ and $La_{1.85}Sr_{0.15}CuO_4$. From [91].

Superconducting state. - It is well known that the renormalization of phonon frequencies and their life-times by superconductivity in $LTSC$ materials is rather small - around one percent. The smallness of the effect is characterized by the parameter Δ/E_F which is very small in low temperature superconductors. However, Δ/E_F is much larger in HTSC oxides and already from that point of view one expects much stronger renormalization effects. At the very beginning several Raman active phonon modes, with frequencies $128, 153, 333, 437$ and 501 cm^{-1}, were detected in $YBa_2Cu_3O_7$ and these modes are totally symmetric modes (with respect to the orthorhombic point group D_{2h}). ($1\text{cm}^{-1} = 29.98GHz = 0.123985meV = 1.44K$) However, by using the approximate tetragonal symmetry (with the point group D_{4h}) the mode at $\omega_{B_{1g}} = 333$ cm^{-1} transforms according to the B_{1g} representation, while the other modes according to the A_{1g} one - see Fig. 9. The *Fano resonance* (asymmetric line shape) of the B_{1g} mode indicates an appreciable coupling of the lattice to the continuum, which in fact corresponds to the charge carries. It is interesting to note that the A_{1g} modes in $YBCO$ are weakly affected in the presence of superconductivity, while the B_{1g} mode *softens* by 9 cm^{-1} (by approximately 3 %) [92]. It is well established also that this softening is due to super-

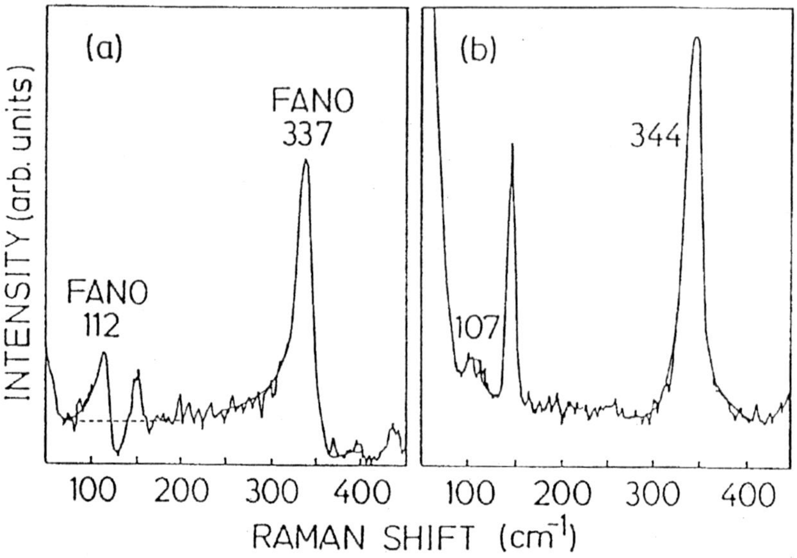

FIGURE 8. Fano resonance in $YBa_2Cu_3O_{7-x}$. The asymmetry is seen for 112 and 337 cm^{-1} phonons in the superconductor ($x = 0$). The semiconductor ($x = 1$) has Lorenzian line shapes. From [93] © Springer-Verlag GmbH 1991 with permission.

conductivity and not due to, for instance, structural changes, because it disappears in magnetic fields higher than H_{c2}.

The frequency shift $\delta\omega_\lambda$ and the phonon line width Γ_λ in the superconducting state have been studied numerically in [96] for the case of the isotropic $s-wave$ superconducting gap ($\Delta(\mathbf{k}) = \Delta = const$) and for strong coupling superconductivity. They have predicted the *phonon-softening* and line-width *narrowing* for $\omega_0 < 2\Delta$, while for $\omega_0 > 2\Delta$ there is a *phonon-hardening* and line-width broadening. These predictions are surprisingly in agreement with experiments [92], in spite of the assumed isotropic $s-wave$ pairing what is contrary to the experimentally well established $d-wave$ pairing in *YBCO*. Later calculations of the renormalization of the B_{1g} Raman phonon mode in the presence of the weak coupling $d-wave$ superconductivity [99] show that if one assumes that $\omega_{B_{1g}} < 2\Delta_{\max}$ there is phonon softening accompanied with the line broadening below T$_c$. The latter is possible because of the gapless character (on a part of the Fermi surface) of $d-wave$ pairing. In that respect the calculations of the phonon renormalization based on the strong coupling $d-wave$ superconductivity are of significant interest and still awaiting.

Recent report [4] on the superconductivity-induced strong phonon renormalization of the A_{1g} phonons at 240 and 390 cm^{-1} (by 6 and 18 % respectively) in $HgBa_2Ca_3Cu_4O_{10+x}$ ($T_c = 123\ K$) - the so called $Hg-1234$ compound, renders an additional evidence for the strong EPI in HTSC oxides. In [4] the EPI coupling constant is estimated to be rather large for the A_{1g} phonons ($\lambda_{A_{1g}} \approx 0.08$). Since there are 60 phonon modes in $HgBa_2Ca_3Cu_4O_{10+x}$ they are capable to produce large EPI coupling

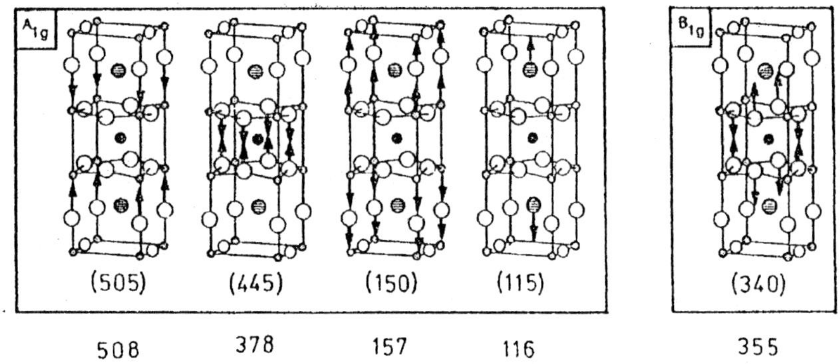

FIGURE 9. Assignation of A_g modes according to calculations in [94]. In brackets are experimental phonon frequencies in cm^{-1}. 115 cm^{-1} is the *Ba* mode, 150 cm^{-1} is the *Cu*2 mode,340 cm^{-1} (B_{1g} mode) and 445 cm^{-1} modes are due to vibration of $O(2,3)$ ions in the CuO_2 , while 505 cm^{-1} mode is due to *O*4 ions. From [95].

constant $\lambda = \sum_{\nu=1}^{60} \lambda_\nu > 1$ - see Fig. 10(a-b). A conservative estimation of the upper limit of λ_{max} gives $\lambda_{max} \approx 60 \times 0.08 = 4.8$ which is, of course, far from the realistic value of $\lambda \leq 2$. In any case this analysis confirm that the EPI of some Raman modes in HTSC oxides is strong. To this point, very recent Raman scattering measurements on the $(Cu,C) - 1234$ compound with $T_c = 117\ K$ reveal strong superconductivity induced phonon self-energy effects [5]. The A_{1g} phonons at 235 cm $^{-1}$ and 360 cm^{-1} (note $\omega_{pl} < 2\Delta_0$), which involve vibrations of the plane oxygen with some admixture of *Ca* displacements, exhibit pronounced Fano line shape (in the normal and superconducting state) with the following interesting properties in the superconducting state: *(i)* the phonon intensity is increased substantially; *(ii)* both phonons soften; *(iii)* the phonon line width (of both phonons) increases dramatically below T_c passing through a maximum slightly below T_c, and decreases again at low T but remaining broader than immediately below T_c. This line broadening is difficult to explain by $s - wave$ pairing, where the line narrowing is expected, but it can be explained by superconducting pairing with nodes in the quasiparticle spectrum, for instance by $d - wave$ pairing [97]. The large EPI coupling constants for these two modes are estimated from the asymmetric Fano line shape, which gives $\lambda_{235} = 0.05$ and $\lambda_{360} = 0.07$ (note in *YBCO* $\lambda_{A_{1g}} = 0.01$ for $\omega_{A_{1g}} = 440$ cm^{-1} and $\lambda_{B_{1g}} = 0.02$ for $\omega_{B_{1g}} = 340$ cm^{-1}, rather small values) giving the upper value for the total coupling constant $\lambda_{max} = 4$. This result gives additional important evidence for the strong *EPI* in HTSC oxides.

2.3.3. Electron-phonon coupling in Raman scattering

We would like to stress the importance of the (phonon) Raman scattering measurements for the theory of the EPI in HTSC oxides. The *covalent part* of the EPI is due to

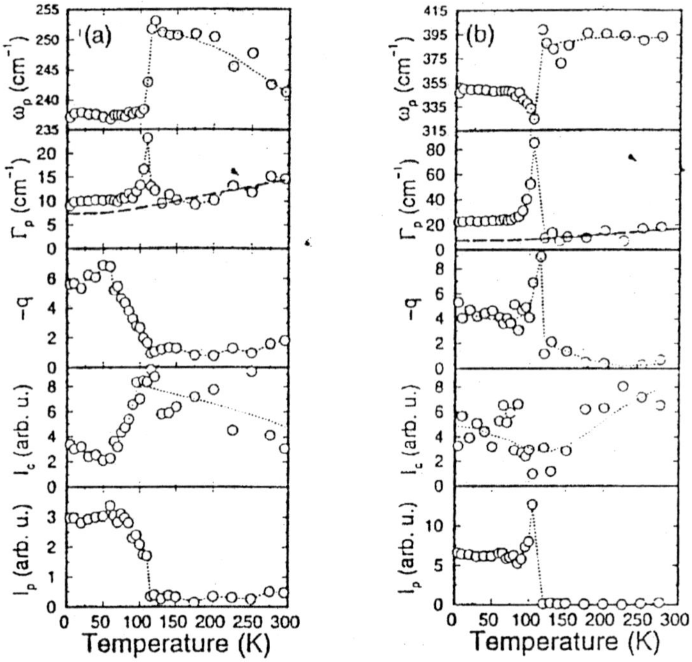

FIGURE 10. The fitted frequency ω_p, line-width Γ_p, asymmetry parameter q, and the phonon intensity I_p of the $Hg-1234$ Raman spectra in the A_{1g} mode measured in $xtxt$ polarization with 647.1 nm laser line: (a) at 240 cm^{-1}; (b) at 390 cm^{-1} - from [4].

the strong *covalency* of the Cu and O orbitals in the CuO_2 planes. In that case the EPI coupling constant is characterized by the parameter ("field") $E^{cov} \sim \partial t_{p-d}/\partial R \sim q_0 t_{p-d}$, where t_{p-d} is the hopping integral between $Cu(d_{x^2-y^2})$ and $O(p_{x,y})$ orbitals and the length q_0^{-1} characterizes the spacial exponential fall-off of the hopping integral t_{p-d}. The covalent EPI is unable to explain the strong phonon renormalization (the self-energy features) in the B_{1g} mode in $YBa_2Cu_3O_7$ by superconductivity, since in this mode the O-ions move along the $c-axis$ in opposite directions, while for this mode $\partial t_{p-d}/\partial R$ is zero in the first order in the phonon displacement. Therefore the EPI in this mode must be due to the *ionic contribution* to the EP interaction which comes from the change in the Madelung energy as it was first proposed in [47], [48]. Namely, the Madelung interaction creates an electric field perpendicular to the CuO_2 planes, which is due to the surrounding ions which form an asymmetric environment. In that case the site energies ε_i^0 contain the matrix element $\varepsilon^{ion} = \langle \psi_i \mid \phi(\mathbf{r}) \mid \psi_i \rangle$, where $\mid \psi_i \rangle$ is the atomic wave function at the i-th site, while the potential $\phi(\mathbf{r})$ steams from surrounding ions. In simple and transition metals the surrounding ions are well screened and therefore the change of ε^{ion} in the presence of phonons is negligible, contrary to HTSC oxides which are almost *ionic compounds* (along the c-axis) where the change of ε^{ion} is appreciable and characterized by the field strength $E^{ion} = V/a_n$. Here, V is the characteristic potential due to surrounding ions and a_n is the distance of the neighboring ions. Immediately after the discovery of HTSC oxides in many papers [114], [60], [98] it was (incorrectly) assumed that the covalent part dominates the EPI in these materials. The calculation of T_c by considering only covalent effects [114], [60] gave rather small T_c ($\sim 10-20\ K$ in $YBCO$, and $20-30\ K$ in $La_{1.85}Sr_{0.15}CuO_4$). It turns out that in HTSC oxides the opposite inequality $E^{ion} \gg E^{cov}$ is realized for most c-axis phonon modes, on which basis the renormalization of the Raman B_{1g} mode can be explained - see more in [2]. This is supported by detailed theoretical studies in for the $YBCO$ compound [48], [49], where it is calculated the change in the *ionic Madelung energy* due to the out of plane oxygen vibration in the B_{1g} mode. Similarly as in $YBCO$, the large superconductivity-induced phonon self-energy effects in $HgBa_2Ca_3Cu_4O_{10+x}$ and in $(Cu,C)Ba_2Ca_3Cu_4O_{10+x}$ for the A_{1g} modes are also due to the ionic (Madelung) coupling. In these modes oxygen ions move also along the $c-axis$ and the ionicity of the structure is involved in the EPI. This type of the (*long-range*) EPI is absent in usual isotropic metals ($LTSC$ superconductors), where the large Coulomb screening makes it to be local. Similar ideas are recently incorporated into the Eliashberg equations in [133], [132]. The weak screening along the c-axis, which is due to the very small hopping integral for carrier motion, is reflected in the very small plasma frequency $\omega_p^{(c)}$ along this axis. Since for some optical phonon modes one has $\omega_{ph} > \omega_p^{(c)}$ then nonadiabatic effects in the screening are important. The latter can give rise to much larger EPI coupling constant for this modes [52], [53].

In conclusion, the electron and phonon Raman scattering measurements in the normal and superconducting state of HTSC oxides give the following important results: (a) phonons interact strongly with the electronic continuum, i.e. the EPI is substantial; (b) the ionic contribution (the Madelung energy) to the EPI interaction for c-axis phonon modes gives substantional contribution to the (large) EPI coupling constant ($\lambda > 1$).

2.4. Tunnelling spectroscopy in HTSC oxides

Tunnelling methods are important tools in studying the electronic density of states $N(\omega)$ in superconductors and in the past they have played very important role in investigating of low T_c-superconductors. By measuring the current-voltage ($I-V$) characteristic in typical tunnelling junctions (with large tunnelling barrier) it was possible from the tunnelling conductance $G(V)(= dI/dV)$ to determine $N(\omega)$ and the superconducting gap as a function of temperature, magnetic field etc. Moreover, by measuring of $G(V)$ at voltages $eV > \Delta$ in the NIS (normal metal - isolator - superconductor) junctions it was possible to determine the Eliashberg spectral function $\alpha^2F(\omega)$ (which is due to some bosonic mechanism of quasiparticle scattering) and finally to confirm (definitely) the phonon mechanism of pairing in $LTSC$ materials, except maybe heavy fermions [100]. We shall discuss here only the results for $\alpha^2F(\omega)$ obtained from $I-V$ measurements, while a more extensive discussion of other aspects is given in [2].

2.4.1. $I-V$ characteristic and $\alpha^2F(\omega)$

If one considers a *NIS* contact where the left (L) and right (R) banks of the contact can be normal (N) metal or superconductor (S), respectively, with very small transparency then tunnelling effects are studied in the framework of the tunnelling Hamiltonian $\hat{H}_T = \sum_{\mathbf{k},\mathbf{p}}(T_{\mathbf{k},\mathbf{p}}c^{\dagger}_{\mathbf{k}L}c_{\mathbf{q}R} + h.c)$. In that case the single-particle tunnelling current is given by the formula [101]

$$I_{qp}(V) = 2e\sum_{\mathbf{k},\mathbf{p}} \mid T_{\mathbf{k},\mathbf{p}} \mid^2 \times$$

$$\times \int_{-\infty}^{\infty} d\omega A_N(\mathbf{k},\omega)A_S(\mathbf{p},\omega+eV)[n_F(\omega)-n_F(\omega+eV)]. \quad (28)$$

The single-particle spectral function $A_{N(S)}(\mathbf{k},\omega)$ is related to the imaginary part of the retarded single particle Green's function, i.e. $A(\mathbf{k},\omega) = -ImG^{ret}(\mathbf{k},\omega)/\pi$, while the tunnelling matrix element $\mid T_{\mathbf{k},\mathbf{p}} \mid^2$ is derived in the quantum-mechanical theory of tunnelling through the barrier - see [2]. Note, in the superconducting state $A(\mathbf{k},\omega)$ depends on the superconducting gap function $\Delta(\mathbf{k},\omega)$, which is on the other hand a functional of the spectral function $\alpha^2F(\omega)$. The fine structure in the second derivative d^2I/dV^2 at voltages above the superconducting gap is related to the spectral function $\alpha^2F(\omega)$. For instance, plenty of break-junctions made from $Bi-2212$ single crystals [37] show that negative peaks in d^2I/dV^2, although broadened, coincide with the peaks in the generalized phonon density of states $G_{ph}(\omega)$ measured by neutron scattering - see more in [2]. Note, the reported broadening of these peaks might be partly due to $d-wave$ pairing in HTSC oxides. The tunnelling density of states $N_T(V) \sim dI/dV$ shows a gap structure and it was found that $2\bar{\Delta}/T_c = 6.2-6.5$, where $T_c = 74-85\ K$ and $\bar{\Delta}$ is some average value of the gap. By assuming $s-wave$ superconductivity [37] and by solving the *MR* problem (inversion of Eliashberg equations), the spectral function $\alpha^2F(\omega)$ is obtained which gives $\lambda \approx 2.3$. Note, in extracting λ [37] the standard value of the effective Coulomb parameter $\mu^* \approx 0.1$ is assumed. Although this analysis [37] was done by assuming $s-wave$ pairing it is qualitatively valuable procedure also in the case of $d-wave$ pairing, because one expects that $d-wave$ pairing does not spoil significantly the global structure of d^2I/dV^2 at $eV > \Delta$, but introducing mainly a broadening of peaks. The latter effect can be partly due to an inhomogeneity of the gap. The results obtained in [37] were *reproducible* on more than 30 junctions, while in $Bi(2212)-GaAs$ and $Bi(2212)-Au$ planar tunnelling junctions similar results were. Several groups [39], [40], [41] have obtained similar results for the shape of the spectral function $\alpha^2F(\omega)$ from the $I-V$ measurements on various HTSC oxides as shown in Fig. 11. The results shown in Fig. 11 leave no much doubts on the effectiveness of the EPI in pairing mechanism of HTSC oxides. In that respect recent tunnelling measurements on $Bi_2Sr_2CaCu_2O_8$ [102] are impressive, since the Eliashberg spectral function $\alpha^2F(\omega)$ was extracted from the measurements of d^2I/dV^2. The obtained $\alpha^2F(\omega)$ has several peaks in the broad frequency region up to 80 meV - see Fig. 11 (curve Shimada et al.), which coincide rather well with the peaks in the phonon density of states $F(\omega)$. Moreover, the authors of [102] were able to extract the coupling constant for modes laying in (and around) these peaks and their contribution to T_c. They managed to extract the EPI coupling constant, which is unexpectedly very large, i.e. $\lambda(=2\int d\omega\alpha^2F(\omega)/\omega) = \sum\lambda_i \approx 3.5$. Since almost all

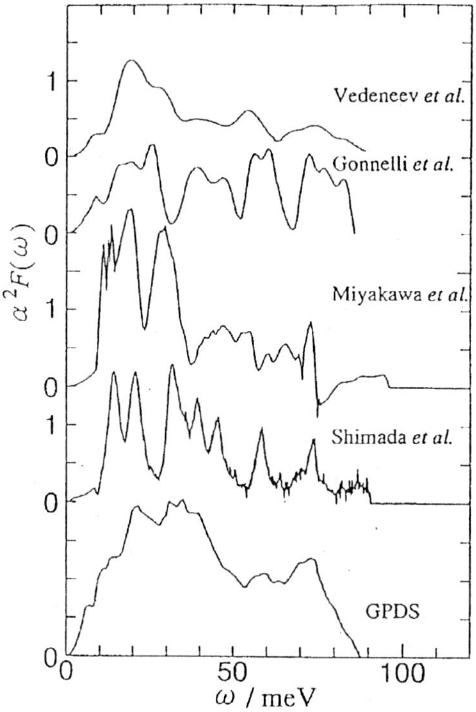

FIGURE 11. The spectral function $\alpha^2F(\omega)$ obtained from measurements of $G(V)$ by various groups on various junctions: Vedeneev et al. [37], Gonnelli et al. [41], Miyakawa et al. [39], Shimada et al.[38]. The generalized density of states GPDS for *Bi*2212 is plotted at the bottom - from [38].

phonon mode contributes to λ, this means that on the average each particular phonon mode is moderately coupled to electrons thus keeping the lattice stable. Additionally, they have found that some low-frequency phonon modes corresponding to Cu, Sr and Ca vibrations are rather strongly coupled to electrons, similarly as the high frequency oxygen vibrations along the c-axis do. These results confirm the importance of the axial modes in which the change of the Madelung energy is involved, thus supporting the idea conveyed through this article of the importance of the ionic Madelung energy in the EPI interaction of HTSC oxides.

In conclusion, the common results for all reliable tunnelling measurements in HTSC oxides, including $Ba_{1-x}K_xBiO_3$ too [103], [16], is that no particular mode can be singled out in the spectral function $\alpha^2F(\omega)$ as being the only one which dominates in pairing mechanism. This important result means that the high T_c is not attributable to a particular phonon mode in the EPI mechanism, since all phonon modes contribute to λ. Having in mind that the phonon spectrum in HTSC oxides is very broad (up to 80 meV), then the large EPI coupling constant ($\lambda \approx 2$) in HTSC oxides is not surprising at all. We stress, that compared to neutron scattering experiments the tunnelling experiments are superior in determining the EPI spectral function $\alpha^2F(\omega)$.

2.5. Isotope effect in HTSC oxides

The isotope effect has played an important role in elucidating the pairing mechanism in *LTSC* materials. Note, the standard BCS theory predicts that for the pure phonon-mediated mechanism of pairing the isotope coefficient $\alpha = -d\ln T_c/d\ln M$, where M is the ionic mass, takes its canonical value $\alpha = 1/2$. However, later on it was clear that α can take values less (even negative) then its canonical value in the phonon-mediated mechanism of pairing if there is pronounced Coulomb pseudopotential $\mu*$ - see more in [2].

2.5.1. Experiments on the isotope coefficient α

A lot of measurements of α_O and α_{Cu} were performed on various hole-doped and electron-doped HTSC oxides and we give a brief summary of the main results [104]: **(1)** The O isotope coefficient α_O strongly depends on the hole concentration in the hole-doped materials where in each group of HTSC oxides ($YBa_2Cu_3O_{7-x}$, or $La_{2-x}Sr_xCuO_4$ etc.) a small oxygen isotope effect is observed in the optimally doped (maximal T_c) samples. For instance $\alpha_O \approx 0.02-0.05$ in $YBa_2Cu_3O_7$ with $T_{c,\max} \approx 91\ K$, $\alpha_O \approx 0.1-0.2$ in $La_{1.85}Sr_{0.15}CuO_4$ with $T_{c,\max} \approx 35\ K$; $\alpha_O \approx 0.03-0.05$ in $Bi_2Sr_2CaCu_2O_8$ with $T_{c,\max} \approx 76\ K$; $\alpha_O \approx 0.03$ and even negative (-0.013) in $Bi_2Sr_2Ca_2Cu_2O_{10}$ with $T_{c,\max} \approx 110\ K$; the experiments on $Tl_2Ca_{n-1}BaCu_nO_{2n+4}$ $(n = 2,3)$ with $T_{c,\max} \approx 121\ K$ are still unreliable and α_O is unknown; $\alpha_O < 0.05$ in the electron-doped $(Nd_{1-x}Ce_x)_2CuO_4$ with $T_{c,\max} \approx 24\ K$. **(2)** For hole concentrations away from the optimal one, T_c decreases while α_O increases and in some cases reaches large value $\alpha_O \approx 0.5$ - see Fig. 12 for *La* compounds. This holds not only for parent compounds but also for systems with substitutions, like $(Y_{1-x-y}\mathrm{Pr}_x Ca_y)Ba_2Cu_3O_7$, $Y_{1-y}Ca_yBa_2Cu_4O_4$ and $Bi_2Sr_2Ca_{1-x}Y_xCu_2O_8$. Note, the decrease of T_c is not a prerequisite for the increase of α_O. This became clear from the *Cu* substituted experiments $YBa_2(Cu_{1-x}Zn_x)_3O_7$ where the decrease of T_c (by increasing of the *Zn* concentration) is followed by only small increase of α_O [105]. Only in the case of very low $T_c < 20\ K$ then α_O becomes large, i.e. $\alpha_O > 0.1$. **(3)** The largest α_O is obtained even in the optimally doped compounds like in systems with substitution, such as $La_{1.85}Sr_{0.15}Cu_{1-x}M_xO_4$, $M = Fe, Co$, where $\alpha_O \approx 1.3$ for $x \approx 0.4$ %. **(4)** In $La_{2-x}M_xCuO_4$ there is a *Cu* isotope effect which is of the order of the oxygen one, i.e. $\alpha_{Cu} \approx \alpha_O$ giving $\alpha_{Cu} + \alpha_O \approx 0.25-0.35$ for optimally doped systems $(x = 0.15)$. In the case when $x = 0.125$ with $T_c \ll T_{c,\max}$ one has $\alpha_{Cu} \approx 0.8-1$ with $\alpha_{Cu} + \alpha_O \approx 1.8$. The appreciate copper isotope effect in $La_{2-x}M_xCuO_4$ tells us that vibrations of other than oxygen ions could be important in giving high T_c. The latter property is more obvious from tunnelling measurements, which are discussed above. **(5)** There is *negative* Cu isotope effect in the oxygen-deficient system $YBa_2Cu_3O_{7-x}$ where α_{Cu} is between -0.14 and -0.34 if T_c lies in the 60 K plateau. **(6)** There are reports on *small negative* α_O in

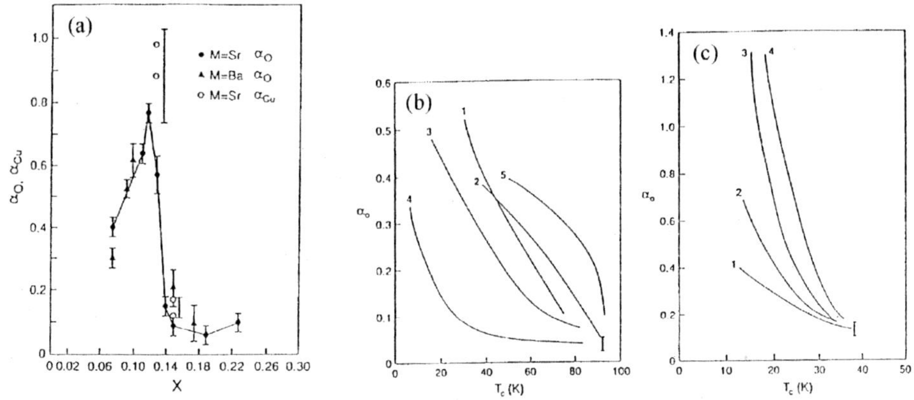

FIGURE 12. The oxygen isotope exponent α_O for: (a) $La_{2-x}Sr_xCuO_4$ as a function of Sr concentration - from [104]. The oxygen isotope exponent α_O as a function of T_c for: (b) $YBa_2Cu_3O_7$. 1: $(Y_{1-x}Pr_x)Ba_2Cu_3O_7$; 2: $YBa_{2-x}La_xCu_3O_7$; 3: $YBa_2(Cu_{1-x}Co_x)_3O_7$; 4: $YBa_2(Cu_{1-x}Zn_x)_3O_7$; 5: $YBa_2(Cu_{1-x}Fe_x)_3O_7$. (c) $La_{1.85}Sr_{0.15}CuO_4$. 1: $La_{1.85}Sr_{0.15}(Cu_{1-x}Ni_x)O_4$; 2: $La_{1.85}Sr_{0.15}(Cu_{1-x}Zn_x)O_4$; 3: $La_{1.85}Sr_{0.15}(Cu_{1-x}Co_x)O_4$; 4: $La_{1.85}Sr_{0.15}(Cu_{1-x}Fe_x)O_4$ - from [104].

some systems like $YSr_2Cu_3O_7$ with $\alpha_O \approx -0.02$ and in $BISCO-2223$ ($T_c = 110\ K$) where $\alpha_O \approx -0.013$ etc. However, the systems with negative α_O present considerable experimental difficulties, as it is pointed out in [104].

The above enumerated results, despite experimental difficulties, are more in favor than against of the hypothesis that the EPI interaction is strongly involved in the pairing mechanism of HTSC oxides. By assuming that the experimental results on the isotope effect reflect an intrinsic property of HTSC oxides one can rise a question: which theory can explain these results? Since at present there is no consensus on the pairing mechanism in HTSC materials there is also no definite theory for the isotope effect. Besides the calculation of the coupling constant λ any microscopic theory of pairing is confronted also with the following questions: (**a**) why is the isotope effect small in optimally doped systems and (**b**) why α increases rapidly by further under(over)doping of the system?

It should be stressed, that at present all theoretical approaches are semi-microscopic, but what is interesting most of them indicate that in order to explain the rather unusual isotope effect in HTSC materials one should invoke the *forward scattering peak* in the EPI [2].

In conclusion, experimental investigations of the isotope effect in HTSC oxides have shown the importance of the EPI interaction in the pairing mechanism.

2.6. ARPES experiments in HTSC oxides

2.6.1. *Spectral function* $A(\vec{k},\omega)$ *from ARPES*

The *angle-resolved photoemission spectroscopy* (ARPES) is nowadays a leading spectroscopy method in the solid state physics. The method consists in shining light (photons) with energies between $20-1000\ eV$ on the sample and by detecting momentum $(\mathbf{k})$- and energy(ω)-distribution of the outgoing electrons. The resolution of ARPES is drastically increased in the last decade with the energy resolution of $\Delta E \approx 2\ meV$ (for photon energies $\sim 20\ eV$) and angular resolution of $\Delta\theta \approx 0.2°$. The ARPES method is surface sensitive technique, since the average escape depth (l_{esc}) of the outgoing electrons is of the order of $l_{esc} \sim 10$ Å. Therefore, one needs very good surfaces in order that the results be representative for the bulk sample. In that respect the most reliable studies were done on the bilayer $Bi_2Sr_2CaCu_2O_8$ (Bi2212) and its single layer counterpart $Bi_2Sr_2CuO_6$ (Bi2201), since these materials contain weakly coupled BiO planes with the longest interplane separation in the HTSC oxides. This results in a *natural cleavage* plane making these materials superior to others in ARPES experiments. After a drastic improvement of sample quality in others families of HTSC materials, became the ARPES technique a central method in theoretical considerations. Potentially, it gives information on the quasiparticle Green's function, i.e. on the quasiparticle spectrum and life-time effects. The ARPES can indirectly give information on the momentum and energy dependence of the pairing potential. Furthermore, the electronic spectrum of the HTSC oxides is highly *quasi-2D* which allows an unambiguous determination of the momentum of the initial state from the measured final state momentum, since the component parallel to the surface is conserved in photoemission. In this case the ARPES probes (under some favorable conditions) directly the single particle spectral function $A(\mathbf{k},\omega)$.

In the following we discuss only those ARPES experiments which give evidence for the importance of the EPI in HTSC oxides - see detailed reviews in [106], [107].

The *photoemission* measures a nonlinear response function of the electron system, since the photo-electron current $\langle \mathbf{j}(1)\rangle$ at the detector is proportional to the incident photon flux (square of the vector potential $\mathbf{A}$), i. e. schematically one has

$$\langle \mathbf{j}(1)\rangle \sim \langle \mathbf{j}(\bar{2})\mathbf{j}(1)\mathbf{j}(\bar{3})\rangle \mathbf{A}(\bar{2})\mathbf{A}(\bar{3}), \tag{29}$$

and integration over bar $1 = (\mathbf{x},t)$ indices is understood. The correlation function $\langle \mathbf{j}(\bar{2})\mathbf{j}(1)\mathbf{j}(\bar{3})\rangle$ describes all processes related to electrons, such as photon absorption, electron removal and electron detection, are treated as a single coherent process. In this case the bulk, surface and evanescent states, as well as surface resonances should be taken into account - the so called *one-step model*.

Under some conditions the one-step model can be simplified by an approximative, but physically plausible, *three-step model*. In this model the photoemission intensity

$$I_{tot}(\mathbf{k},\omega) = I \cdot I_2 \cdot I_3 \tag{30}$$

is the product of three independent terms: **(1)** I - describes optical excitation of the electron in the bulk; **(2)** I_2 - the scattering probability of the travelling electrons; **(2)** I_3 -

the transmission probability through the surface potential barrier. The central quantity in the three-step model is $I(\mathbf{k},\omega)$. To calculate it one assumes the *sudden approximation*, i.e. that the outgoing electron is moving so fast that it has no time to interact with the photo-hole - see more in [106],[107]. It turns out that $I(\mathbf{k},\omega)$ can be written in the form [106], [107] (for $\mathbf{k}=\mathbf{k}_{\parallel}$)

$$I(\mathbf{k},\omega) \simeq I_0(\mathbf{k},\upsilon)f(\omega)A(\mathbf{k},\omega). \tag{31}$$

$I_0(\mathbf{k},\upsilon) \sim \mid \langle \psi_f \mid \mathbf{pA} \mid \psi_i \rangle \mid^2$ where $\langle \psi_f \mid \mathbf{pA} \mid \psi_i \rangle$ is the dipole matrix element and depends on $\mathbf{k}$, polarization and energy υ of the incoming photons. $f(\omega) = 1/(1+\exp\{\omega/T\})$ is the Fermi function and $A(\mathbf{k},\omega) = -\mathrm{Im}G(\mathbf{k},\omega)/\pi$ is the quasiparticle spectral function. In reality because of finite resolution of experiments, in $\mathbf{k}$ and ω, $I(\mathbf{k},\omega)$ should be convoluted by the ω-convolution function $R(\omega)$ and $\mathbf{k}$-convolution function $Q(\mathbf{k})$. It must be also added the extrinsic background B, which is due to secondary electrons (those which escape from the sample after having suffered inelastic scattering events coming out with reduced kinetic energy).

By measuring $A(\mathbf{k},\omega)$ one can determine $\Sigma(\mathbf{k},\omega) = \Sigma_1(\mathbf{k},\omega) + i\Sigma_2(\mathbf{k},\omega)$

$$A(\mathbf{k},\omega) = -\frac{1}{\pi}\frac{\Sigma_2(\mathbf{k},\omega)}{[\omega - \xi_0(\mathbf{k}) - \Sigma_1(\mathbf{k},\omega)]^2 + [\Sigma_2(\mathbf{k},\omega)]^2}. \tag{32}$$

$\xi_0(\mathbf{k}) = \varepsilon_{\mathbf{k}} - \mu$ is the bare quasiparticle energy. For instance in the case of the Landau-Fermi liquid $A(\mathbf{k},\omega)$ can be separated into the coherent and incoherent part

$$A(\mathbf{k},\omega) = Z_{\mathbf{k}}\frac{\Gamma_{\mathbf{k}}}{(\omega - \xi(\mathbf{k}))^2 + \Gamma_{\mathbf{k}}^2} + A_{inch}(\mathbf{k},\omega), \tag{33}$$

where $Z_{\mathbf{k}} = 1/(1 - \partial\Sigma_1/\partial\omega)$, $\xi(\mathbf{k}) = Z_{\mathbf{k}}(\xi_0(\mathbf{k}) + \Sigma_1)$ and $\Gamma_{\mathbf{k}} = Z_{\mathbf{k}} \mid \Sigma_2 \mid$ calculated at $\omega = \xi(\mathbf{k})$. For small ω one has $\xi(\mathbf{k}) >> \mid \Sigma_2 \mid$ and $\Gamma_{\mathbf{k}} \sim [(\pi T)^2 + \xi^2(\mathbf{k})]$.

In some period of the HTSC era there were a number of controversial ARPES results and interpretations, due to bad samples and to the euphoria with exotic theories. For instance, a number of (now well) established results were *questioned* in the first ARPES measurements, such as: the shape of the Fermi surface, which is correctly predicted by the LDA band-structure calculations; bilayer splitting in $Bi2212$, etc.

We summarize here the important ARPES results which were obtained recently, first in the *normal state* [106], [107]: **(N1)** There is well defined Fermi surface in the metallic state - with the topology predicted by the LDA; **(N2)** the spectral line are broad with $\mid \Sigma_2(\mathbf{k},\omega) \mid \sim \omega$ (or $\sim T$ for $T > \omega$); **(N3)** there is a bilayer band splitting in $Bi2212$ (at least in the overdoped state); **(N4)** at temperatures $T_c < T < T^*$ and in the underdoped HTSC oxides there is a d-wave like pseudogap $\Delta_{pg}(\mathbf{k}) \sim \Delta_{pg,0}(\cos k_x - \cos k_y)$ in the quasiparticle spectrum; **(N5)** the pseudogap $\Delta_{pg,0}$ increases by lowering doping; **(N6)** there is evidence for the strong EPI interaction with the *characteristic phonon* energy ω_{ph} - at $T > T_c$. The latter means, that in all HTSC oxides which show superconductivity there are *kinks* in the quasiparticle dispersion in the nodal direction (along the $(0,0)-(\pi,\pi)$ line) at around $\omega_{ph}^{(70)} \sim (60-70)\ meV$ [43] - see Fig. 13, and around the anti-nodal point $(\pi,0)$ at 40 meV [108] - see Fig. 14.

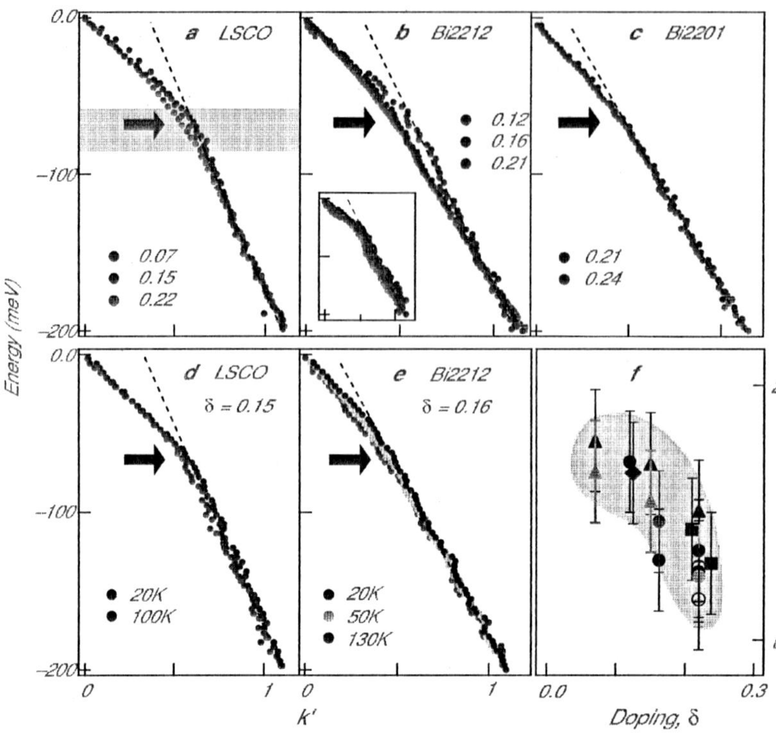

FIGURE 13. Quasiparticle dispersion of *Bi*2212, *Bi*2201 and *LSCO* along the nodal direction, plotted vs the momentum k for $(a)-(c)$ different doings, and $(d)-(e)$ different T; black arrows indicate the kink energy; the red arrow indicates the energy of the $q=(\pi,0)$ oxygen stretching phonon mode; inset of (e)- T-dependent Σ' for optimally doped *Bi*2212; (f) - doping dependence of λ' along $(0,0)-(\pi,\pi)$ for the different HTSC oxides. From [43], reprinted by permission from *Nature* (http://www.nature.com), © 2001.

In the *superconducting state* ARPES results are the following [106], [107]: **(S1)** there is an anisotropic superconducting gap in most HTSC compounds, predominately of d-wave like, $\Delta_{sc}(\mathbf{k}) \sim \Delta_0(\cos k_x - \cos k_y)$ with $2\Delta_0/T_c \approx 5-6$; **(S2)** there are dramatic changes in the spectral shapes near the point $(\pi,0)$, i.e. a *sharp quasiparticle peak* develops at the lowest binding energy followed by a dip and a broader hump, giving rise to the so called *peak-dip-hump structure*; **(S3)** the kink at $(60-70)$ meV is surprisingly *unshifted* in the superconducting state - Fig. 13 from [43]. To remind the reader, in the standard Eliashberg theory the kink should be shifted to $\omega_{ph}+\Delta_0$. **(S4)** the anti-nodal kink at $\omega_{ph}^{(40)} \sim 40$ meV is shifted in the superconducting state by Δ_0, i.e. $\omega_{ph}^{(40)} \rightarrow \omega_{ph}^{(40)} + \Delta_0 = (65-70)meV$ since $\delta = (25-30)meV$ - see Fig. 14 from [108].

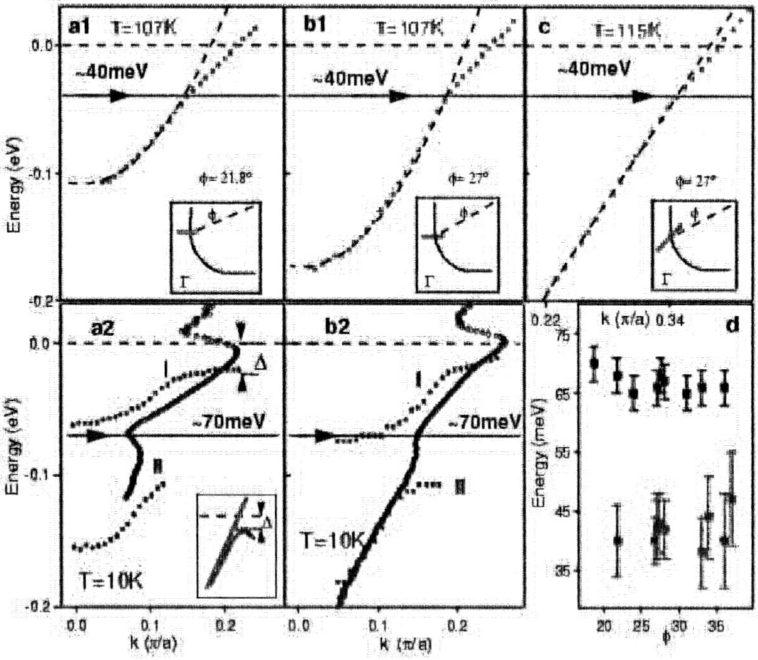

FIGURE 14. Quasiparticle dispersion $E(k)$ in the normal state (a1, b1, c), at 107 K and 115 K, along various directions ϕ around the anti-nodal point. The kink at $E = 40meV$ is shown by the horizontal arrow. (a2 and b2) is $E(k)$ in the superconducting state at 10 K with the shifted kink to $70meV$. (d) kink positions as a function of ϕ in the anti-nodal region. From [108].

2.6.2. Theory of the ARPES kink

We would like to point out that the breakthrough-experiments done by the Shen group [43], [108] shown in Fig. 13, Fig. 14, with the properties (**N6**), (**S3**) and (**S4**) - which we call the *ARPES non-shift puzzle*, are the *smoking-gun* experiments for the microscopic theory of HTSC oxides. Namely, any theory which reflects to explain the pairing in HTSC oxides must solve the *non-shift puzzle*.

In that respect the recent theory [44], which is based on the existence of the forward scattering peak in the EPI (which is due to strong correlations) - the *FSP model*, was able to explain this puzzle in a consequent way. The FSP model (see more in [2], [44]) contains the following basic ingredients: *(i)* the electron-phonon interaction is dominant in HTSC and its spectral function $\alpha^2F(\mathbf{k},\mathbf{k}',\Omega) \approx \alpha^2F(\varphi,\varphi',\Omega)$ (φ is the angle on the Fermi surface) has a pronounced forward scattering peak due to strong correlations. Its width is very narrow $\mid \mathbf{k}-\mathbf{k}' \mid_c \ll k_F$ even for overdoped systems [17], [18], [19]. In the leading order one can put $\alpha^2F(\varphi,\varphi',\Omega) \sim \delta(\varphi-\varphi')$; *(ii)* the dynamical part (beyond the Hartree-Fock) of the Coulomb interaction is characterized by the spectral function $S_C(\mathbf{k},\mathbf{k}',\Omega)$. The ARPES shift puzzle implies that S_C is *either peaked* at small transfer momenta $\mid \mathbf{k}-\mathbf{k}' \mid$, *or it is so small* that the shift is weakly affected and is beyond the experimental resolution of ARPES. We assume that the former case is realized, although this is not crucial at all because the ARPES spectra gives that the electron-phonon coupling is much larger than the Coulomb coupling, i.e. $\lambda_{ph} >> \lambda_C$; *(iii)* The scattering potential on non-magnetic impurities has pronounced forward scattering peak, which is also due to strong correlations [17], [18], [19]. The latter is characterized by two rates $\gamma_{1(2)}$. The case $\gamma_1 = \gamma_2$ mimics the extreme forward scattering, which does not affect pairing. On the other hand , $\gamma_2 = 0$ describes the isotropic exchange scattering - see discussion in

The Green's function is given by $G_k = 1/(i\omega_k - \xi_{\mathbf{k}} - \Sigma_k(\omega)) = -(i\tilde{\omega}_k + \xi_{\mathbf{k}})/(\tilde{\omega}_k^2 + \xi_{\mathbf{k}}^2 + \tilde{\Delta}_k^2)$ where in the $k = (\mathbf{k}, \omega)$. In the FSP model the equations for ω_k and $\tilde{\Delta}_k$ are [44]

$$\tilde{\omega}_{n,\varphi} = \omega_n + \pi T \sum_m \frac{\lambda_{1,\varphi}(n-m)\tilde{\omega}_{m,\varphi}}{\sqrt{\tilde{\omega}_{m,\varphi}^2 + \tilde{\Delta}_{m,\varphi}^2}} + \Sigma_{n,\varphi}^C, \tag{34}$$

$$\tilde{\Delta}_{n,\varphi} = \pi T \sum_m \frac{\lambda_{2,\varphi}(n-m)\tilde{\Delta}_{m,\varphi}}{\sqrt{\tilde{\omega}_{m,\varphi}^2 + \tilde{\Delta}_{m,\varphi}^2}} + \tilde{\Delta}_{n,\varphi}^C, \tag{35}$$

where

$$\lambda_{1(2),\varphi}(n-m) = \lambda_{ph,\varphi}(n-m) + \delta_{mn}\gamma_{1(2),\varphi}$$

with the electron-phonon coupling function

$$\lambda_{ph,\varphi}(n) = 2\int_0^\infty d\Omega \alpha_{ph,\varphi}^2 F_\varphi(\Omega) \frac{\Omega}{\Omega^2 + \omega_n^2}. \tag{36}$$

Since the EPI and $\Sigma_{n,\varphi}^C$ in Eq.(34-35) has a *local form* as a function of the angle φ, then the equation for $\tilde{\omega}_{n,\varphi}$ has also local form, which means that the different points on the Fermi surface are decoupled. In that case $\tilde{\omega}_{n,\varphi}$ depends on the local value of the gap $\tilde{\Delta}_{n,\varphi} \approx \Delta_0 \cos 2\varphi$. *Just this property is important in solving the ARPES shift puzzle*. So, in the *nodal point* ($\varphi = \pi/4$) one has $\tilde{\Delta}_{n,\varphi} = 0$ and the quasiparticle spectrum given by $E - \xi_{\mathbf{k}} - \Sigma_k(E, \tilde{\Delta}_{n,\varphi} = 0) = 0$ is *unaffected* by superconductivity, i.e. the kink is unshifted as shown in Fig. 15a. This is exactly what is seen in the experiment of the Shen group [43], [109] - see Fig. 13. The FSP model also predicts: (1) that the Σ_2 has a knee-like shape for $\omega < \omega_{ph}^{(70)}$ - see Fig. 15b, exactly what has been seen recently in ARPES spectra [43], [109]; (2) that for $\omega > \omega_{ph}^{(70)}$ the EPI contribution to Σ_2 is constant while its slope in this region is determined by the Coulomb scattering giving the small coupling constant $\lambda_C < 0.4$. This was also seen in ARPES spectra [43], [109], thus confirming the correctness of the FSP model.

In the case of the *antinodal point* ($\varphi \approx \pi/2$) there is a singularity at $40 meV$in the quasiparticle spectrum (E_{sing}) in the normal state - see Fig. 14. The analytic and numeric calculations of Eq.(34) show that this singularity is *shifted* by Δ_0 in superconducting state, i.e. $E_{sing} \to E_{sing} + \Delta_0$. This is exactly what is seen in the recent experiment on BISCO [108] - see Fig. 14, where the singularity of the normal state spectrum at 40 meV is shifted to $(65-70)$ meV in the superconducting state, since $\Delta_0 \approx (25-30)$ meV. The FSP model explains in the natural way also the *peak-dip-hump structure* in $A(\mathbf{k}, \omega)$ - for more details see [44].

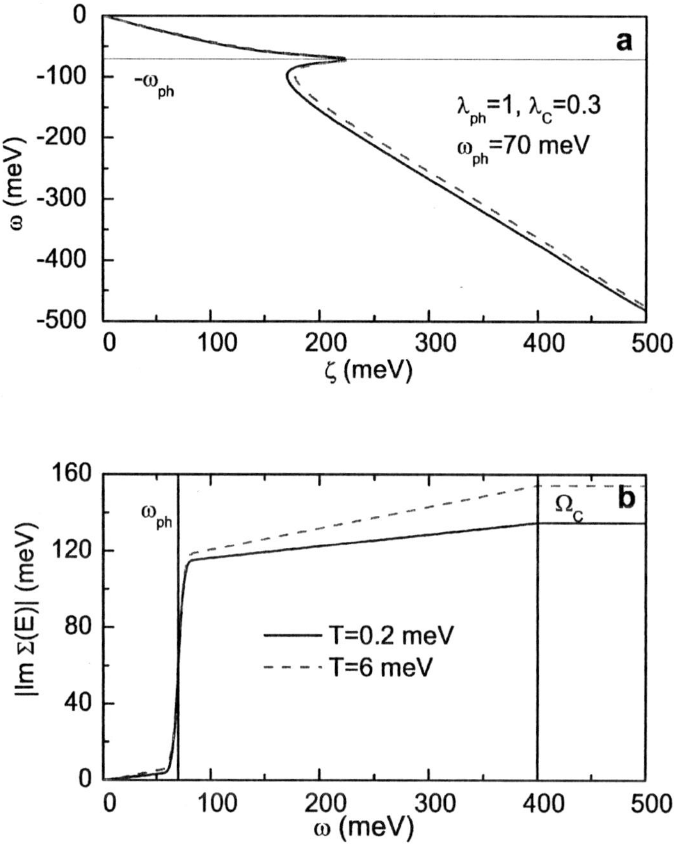

FIGURE 15. (a) The quasiparticle-spectrum $\omega(\xi_k)$ and (b) the imaginary self-energy $Im\Sigma(\xi = 0, \omega)$ in the nodal direction ($\varphi = \pi/4$) in the superconducting ($T = 0.2\ meV$) and normal ($T = 6\ meV$) state. $\Omega_C = 400\ meV$ is the cutoff in S_C. From [44].

2.6.3. ARPES and the EPI coupling constant λ

One can rise the question - is it possible to extract the coupling constant λ from ARPES measurements. As we have seen above, by assuming that the three-step model holds, where $I(\mathbf{k}, \omega) \sim A(\mathbf{k}, \omega)$, then one possibility is by measuring the kink in the quasiparticle renormalization, i.e. by measuring the real part of the self-energy $\Sigma_1(\mathbf{k}, \omega)$. These measurements [43], [108] give $\lambda_{ARPES}^{(1)} > 1$ in both nodal and antinodal direction. This is in accordance with the theoretical prediction in [44]. Another possibility is by measuring the width ($\Delta k_{FW}(\omega)$) of the momentum distribution curves (MDCs) which give the imaginary part $\Sigma_2(\omega, T)$ via

$$\Sigma_2(\omega, T) \approx \frac{1}{2} v_F \Delta k_{FW}(\omega). \tag{37}$$

In that respect very indicative are recent measurements [45] of $\Sigma_2(\omega, T)$ around the *nodal point* in a number of HTSC compounds, such as the superstructure free $Bi_{2-x}Pb_xSr_2CaCu_2O_{8+\delta}$ $(Bi(Pb) - 2212)$, $Bi_2Sr_2CaCu_2O_{8+\delta}$ $(Bi - 2212)$ and $Bi_2Sr_{2-x}La_xCu_2O_{8+\delta}$ $(Bi - 2201)$. In the analysis of ARPES spectra the authors in [45] have assumed that there are two ω-dependent contributions to Σ - the Fermi liquid contribution Σ_{FL} and the part Σ_B due to the interaction via bosonic excitations (let say phonons and spin-fluctuations).

The theoretical analysis in [46] shows that the procedure in [45] gives small value for the bosonic coupling constant $\lambda_B < 0.2$, which would give very small T_c and none of the bosonic pairing mechanisms would be effective. So, the division in [45] is inadequate. In fact the correct form of $\Sigma(\omega)$ is given in [44] where the EPI and Coulomb interaction are taken into consideration. In such a case, as we already explained, the imaginary part of the self-energy has a knee-like shape for $\omega < \omega_{ph}^{(70)}$ with the large slope proportional to the EPI coupling constant $\lambda_{ph} > 1$, as it is seen in ARPES spectra [43], [108], [109]. These experiments give also that in the high frequency region, i.e. for $\omega > \omega_{ph}^{(70)}$, the EPI contribution is constant, while from the slope of Σ_2 one extracts $\lambda_C < 0.4$.

In conclusion, the ARPES spectra, which are interpreted by the FSP theory, give that the EPI interaction dominates in the quasiparticle scattering in the frequency region responsible for pairing in cuprates. In that respect the ARPES kink and the knee-like shape of the spectral width are smoking-gun experiments for the theory of pairing in HTSC materials.

3. EPI IN HTSC OXIDES

In the following we present briefly some elements of the general theory of the strong EPI and its low-energy version. In the latter, high-energy processes are integrated out and the low-energy phenomena are governed by the high-energy vertex functions Γ_c, the excitation potential Σ_0 (part of the self-energy due to the Coulomb interaction) and the EPI coupling constants $g_{EP,ren} \sim \Gamma_c$ - see more in [2]. However, this procedure was never performed in its full extent, because of difficulties to calculate Σ_0 and Γ_c. Therefore, the EPI coupling constant was as a rule calculated by some other methods. Usually in *LTSC* materials the EPI is calculated by using the local-density functional (*LDA*) method, which is suitable for ground state properties of crystals (matter) and which is based on an effective electronic crystal potential V_g. Since in principle V_g may significantly differ from Σ_0 then the *LDA* calculated coupling constant $g_{EP}^{(LDA)}$ can be also very different from the real coupling constant g_{EP}. The calculation of $g_{EP}^{(LDA)}$ is complicated, even in the *LDA* method, and further approximations are necessary, like for instance the rigid-ion (*RI*) and rigid muffin-tin (*RMTA*) approximations. These approximation were justified in simple metals. However, these approximations are inadequate for HTSC oxides, because they fail to take into account correctly the *long-range forces* (Madelung energy - see below) and *strong electronic correlations*. Strictly speaking the EPI does not have meaning in the *LDA* method - see more in [2], because the latter treats the ground state properties of materials, while the EPI is due to excited states and inelastic processes in the system.

3.1. General strong coupling theory of the EPI

It is based on the fully microscopic electron-ion Hamiltonian for the interacting electrons and ions in a crystal - see for instance [110], [111], [112], and comprises electrons interacting between themselves as well as with ions and ionic vibrations. In order to describe superconductivity the Nambu-spinor $\hat{\psi}^{\dagger}(\mathbf{r})$ is introduced which operates in the electron-hole space $\hat{\psi}^{\dagger}(\mathbf{r}) = (\hat{\psi}_{\uparrow}^{\dagger}(\mathbf{r})\ \hat{\psi}_{\downarrow}(\mathbf{r}))$ (analogously for the column $\hat{\psi}(\mathbf{r})$) with $\hat{\psi}_{\uparrow}(\mathbf{r})$, $\hat{\psi}_{\uparrow}^{\dagger}(\mathbf{r})$ as annihilation and creation operators for spin up, respectively etc. The microscopic Hamiltonian which in principle should describe the normal and superconducting state of the system contains three parts $\hat{H} = \hat{H}_e + \hat{H}_i + \hat{H}_{e-i}$. The *electronic Hamiltonian* $\hat{H}_e$, which describes the kinetic energy and the Coulomb interactions of electrons, is given in the second-quantization by

$$\hat{H}_e = \int d^3r\hat{\psi}^{\dagger}(\mathbf{r})\hat{\tau}_3\varepsilon_0(\hat{p})\hat{\psi}(\mathbf{r}) +$$

$$+\frac{1}{2}\int d^3rd^3r'\hat{\psi}^{\dagger}(\mathbf{r})\hat{\tau}_3\hat{\psi}(\mathbf{r})V_c(\mathbf{r}-\mathbf{r}')\hat{\psi}^{\dagger}(\mathbf{r}')\hat{\tau}_3\hat{\psi}(\mathbf{r}'), \tag{38}$$

where $\varepsilon_0(\hat{p}) = \hat{p}^2/2m$ is the kinetic energy of electron and $V_c(\mathbf{r}-\mathbf{r}') = e^2/\mid \mathbf{r}-\mathbf{r}' \mid$ is the electron-electron Coulomb interaction. Note, that in the electron-hole space the pseudo-spin (Nambu) matrices $\hat{\tau}_i$, $i = 0,1,2,3$ are Pauli matrices.

The *lattice Hamiltonian* (describes lattice vibrations $\hat{u}_{\alpha n}$ of ions enumerated by n) is given by

$$\hat{H}_i = \frac{1}{2}\sum_n M(\frac{d\hat{\mathbf{u}}_n}{dt})^2 + \frac{1}{2}\sum_{n,m,\alpha} V_{ii}(\mathbf{R}_n^0 - \mathbf{R}_m^0) + \frac{1}{2}\sum_{n,m}(\hat{u}_{\alpha n} - \hat{u}_{\alpha m})\nabla_{\alpha}V_{ii}(\mathbf{R}_n^0 - \mathbf{R}_m^0) +$$

$$+\frac{1}{2}\sum_{n,m,\alpha,\beta}(\hat{u}_{\alpha n} - \hat{u}_{\alpha m})(\hat{u}_{\beta n} - \hat{u}_{\beta m})\nabla_{\alpha}\nabla_{\beta}V_{ii}(\mathbf{R}_n^0 - \mathbf{R}_m^0) + \hat{H}_i^{anh}. \tag{39}$$

The first term in Eq.(39) is the kinetic energy of vibrating ions (with charge Ze), $V_{ii}(\mathbf{R}_n^0 - \mathbf{R}_m^0) = Z^2e^2/\mid \mathbf{R}_n^0 - \mathbf{R}_m^0 \mid$ is the bare ion-ion interaction in equilibrium, while the third and fourth terms describe the change of V_{ii} by lattice vibrations with the ion-displacement is $\hat{u}_n = \mathbf{R}_n - \mathbf{R}_n^0$. The term $\hat{H}_i^{anh}$ describes higher anharmonic terms with respect to $\hat{u}_n^{\beta}$. The theory which we describe below holds for any kind of anharmonicity.

The *electron-ion Hamiltonian* describes the interaction of electrons with the equilibrium lattice and with its vibrations, respectively

$$\hat{H}_{e-i} = \sum_n \int d^3rV_{e-i}(\mathbf{r}-\mathbf{R}_n^0)\hat{\psi}^{\dagger}(\mathbf{r})\hat{\tau}_3\hat{\psi}(\mathbf{r}) + \int d^3r\hat{\Phi}(\mathbf{r})\hat{\psi}^{\dagger}(\mathbf{r})\hat{\tau}_3\hat{\psi}(\mathbf{r}), \tag{40}$$

$$\hat{\Phi}(\mathbf{r}) = \sum_n [V_{e-i}(\mathbf{r} - \mathbf{R}_n^0 - \hat{\mathbf{u}}_n) - V_{e-i}(\mathbf{r} - \mathbf{R}_n^0).$$

Here, $V_{e-i}(\mathbf{r} - \mathbf{R}_n^0)$ is the electron-ion potential - see [2]. The second term which depends on the lattice distortion operator $\hat{\Phi}(\mathbf{r})$ describes the interaction of electrons with harmonic ($\sim \hat{u}_{\alpha n}$) (or anharmonic $\sim \hat{u}_{\alpha n}^k$, $k = 2, 3...$) lattice vibrations.

Based on the above Hamiltonian one can in principle calculate the electron and phonon Green's functions

$$\hat{G}(1,2) = -\langle T\hat{\psi}(1)\hat{\psi}^\dagger(2)\rangle \tag{41}$$

and

$$\tilde{D}(1-2) = -\langle T\hat{\Phi}(1)\hat{\Phi}(2)\rangle, \tag{42}$$

respectively. The solution of these equations is written in the form of Dyson's equations

$$\hat{G}^{-1}(1,2) = \hat{G}_0^{-1}(1,2) - \hat{\Sigma}(1,2) \tag{43}$$

and

$$\tilde{D}^{-1}(1,2) = \tilde{D}_0^{-1}(1,2) - \tilde{\Pi}(1,2), \tag{44}$$

where $\hat{G}_0^{-1}(1,2)$ and $\tilde{D}_0^{-1}(1,2)$ are the bare inverse electron and phonon Green's function, respectively. The nontrivial effects of interactions are hidden in the self-energies $\hat{\Sigma}(1,2)$ and $\tilde{\Pi}(1,2)$. Here, $1 = (\mathbf{r}_1, \tau_1)$, where τ_1 is the imaginary time. The calculation of $\hat{\Sigma}$ is simplified by using the *Migdal adiabatic approximation* [139], which incorporates the experimental fact that in most metals the characteristic phonon (Debye) energy of lattice vibrations ω_D is much smaller than the characteristic electronic Fermi energy E_F ($\omega_D \ll E_F$). Using this fact Migdal formulated a theorem which claims that in the self-energy Σ one should keep explicitly terms linear in the phonon propagator $\tilde{D}$ only. As the result one obtains the *Migdal-Eliashberg theory* for

$$\hat{\Sigma} = \hat{\Sigma}_c + \hat{\Sigma}_{EP}, \tag{45}$$

where

$$\hat{\Sigma}_c(1,2) = -V_c^{sc}(1,\bar{1})\hat{\tau}_3\hat{G}(1,\bar{2})\hat{\Gamma}_c(\bar{2},2;\bar{1}). \tag{46}$$

$V_c^{sc}(1,2) = V_c(1,\bar{2})\varepsilon_e^{-1}(\bar{2},2)$ is the screened Coulomb interaction. The part which is due to the EPI has the following form

$$\hat{\Sigma}_{EP}(1,2) = -V_{EP}(\bar{1},\bar{2})\hat{\Gamma}_c(1,\bar{3};\bar{1})\hat{G}(\bar{3},\bar{4})\hat{\Gamma}_c(\bar{4},2;\bar{2}), \tag{47}$$

where

$$V_{EP}(1,2) = \varepsilon_e^{-1}(1,\bar{1})\tilde{D}(\bar{1},\bar{2})\varepsilon_e^{-1}(\bar{2},2)$$

is the screened EPI and ε_e is the electronic dielectric function. Note, $\hat{\Sigma}_{EP}(1,2)$ depends now quadratically on the vertex function $\hat{\Gamma}_c$, due to the adiabatic theorem. If $\hat{\Gamma}_c$ (which is a functional of $\hat{G}$) is known then the quasiparticle dynamics can be in principle determined. In that respect the central question is: **(1)** how to calculate $\hat{\Gamma}_c$ - which contains all information on Coulomb interaction and electronic correlations? This is a difficult task and practically never realized in its full extent for real systems. However, this program is realized recently in the t-J model with the EPI interaction in the framework of the X-method - see below and [2]; **(2)** how to calculate the effective EPI potential $V_{EP} \sim g_{EP}^2/\varepsilon_e^2$, or more precisely the coupling constant g_{EP} and the electronic dielectric function ε_e? In absence of a better theory these quantities are usually calculated in the framework of the LDA band-structure theory.

3.2. LDA calculations of λ in HTSC oxides

The LDA method considers electrons in the ground state (there is a generalization to finite T), whose energy can be calculated by knowing the spectrum $\{\varepsilon_k\}$ of the Kohn-Sham (Schrödinger like) equation

$$[\frac{\hat{\mathbf{p}}^2}{2m} + V_g(\mathbf{r})]\psi_k(\mathbf{r}) = \varepsilon_k \psi_k(\mathbf{r}), \tag{48}$$

which depends on the *effective one-particle potential*

$$V_g(\mathbf{r}) = V_{ei}(\mathbf{r}) + V_H(\mathbf{r}) + V_{XC}(\mathbf{r}). \tag{49}$$

Here. V_{ei} is the electron-lattice potential, V_H is the Hartree term and V_{XC} describes exchange-correlation effects - see [2]. Because the EPI depends on the excited states (above the ground state) of the system this means, that in principle the LDA method can not describe it - see [2]. However, by using an analogy with the microscopic Migdal-Eliashberg theory one can define the EPI coupling constant $g^{(Mig)} = g\Gamma_c/\varepsilon$ also in the LDA theory - see [2]. It reads

$$g_{\alpha,ll'}^{(LDA)}(\mathbf{k},\mathbf{k}') = \sum_n g_{\alpha,nll'}^{(LDA)}(\mathbf{k},\mathbf{k}') = \langle \psi_{\mathbf{k}l} \mid \sum_n \frac{\delta V_g(\mathbf{r})}{\delta R_{n\alpha}} \mid \psi_{\mathbf{k}'l'} \rangle, \tag{50}$$

where n means summation over the lattice sites, $\alpha = x,y,z$ and the wave function $\psi_{\mathbf{k}l}$ is the solution of the Kohn-Sham equation. Formally one has $\delta V_g/\delta \mathbf{R}_n = \Gamma_{LDA}\varepsilon_e^{-1}\nabla V_{ei}$. Even in such a simplified approach it is difficult to calculate $g_{\alpha,ll'}^{(LDA)} = g_{\alpha,n}^{RMTA} + g_{\alpha,n}^{nonloc}$ because it contains the short-range (local) coupling

$$g_{\alpha,n}^{RMTA} \sim g_{\alpha,n}^{RMTA}(\mathbf{k},\mathbf{k}') \sim \langle Y_{lm} \mid \hat{r}_\alpha \mid Y_{l'm'} \rangle \tag{51}$$

with $\Delta l = 1$, and the long-range coupling

$$g_{\alpha,n}^{nonloc}(\mathbf{k},\mathbf{k}') \sim \langle Y_{lm} \mid (\mathbf{R}_n^0 - \mathbf{R}_m^0)_\alpha \mid Y_{l'm'} \tag{52}$$

with $\Delta l = 0$. In most calculations the local term $g_{\alpha,n}^{RMTA}$ is calculated only, which is justified in simple metals only but not in the HTSC oxides. In HTSC oxides the latter gives very small EPI coupling $\lambda^{RMTA} \sim 0.1$, which is apparently much smaller than the experimental value $\lambda > 1$ giving rise to the pessimistically small T_c [113]. The small λ^{RMTA}

was also one of the reasons for abandoning the EPI as pairing mechanism in HTSC oxides. At the beginning of the HTSC era the electron-phonon spectral function $\alpha^2F(\omega)$ for the case $La_{2-x}Sr_xCuO_4$ was calculated in [114] by using the first-principles band structure calculations and the nonorthogonal tight-binding theory of lattice dynamics. It was obtained $\lambda = 2.6$ and for assumed $\mu^* = 0.13$ gave $T_c = 36\ K$. However, these calculations predict a lattice instability for the oxygen breathing mode near $La_{1.85}Sr_{0.15}CuO_4$ that is never observed. Moreover, the same method was applied to $YBa_2Cu_3O_7$ in [115] where it was found $\lambda = 0.5$ which leads at most to $T_c = (19-30)\ K$. In fact the calculations in [114], [115] do not take into account the Madelung coupling (i.e. neglect the matrix elements with $\Delta l = 0$).

However, because of the weak screening of the ionic (long-range) Madelung coupling in HTSC oxides - especially for vibrations along the c-axis, it is necessary to include the nonlocal term $g^{nonloc}_{\alpha,n}$. This goal was achieved in the LDA approach by the Pickett's group [51], where the EPI coupling for $La_{2-x}M_xCuO_4$ is calculated in the *frozen-in phonon* (*FIP*) method. They have obtained $\lambda = 1.37$ and $\omega_{\log} \approx 400\ K$ and for $\mu^* = 0.1$ one has $T_c = 49\ K$ ($T_c \approx \omega_{\log} \exp\{-1/[(\lambda/(1+\lambda)) - \mu^*]\}$). For more details see Ref. [2] and references therein. We point out, that some calculations which are based on the tight-binding parametrization of the band structure in $YBa_2Cu_3O_7$ gave rather large EPI coupling $\lambda \approx 2$ and $T_c = 90\ K$.

Recently, a new *linear-response full-potential linear-muffin-tin-orbital* (*LR-LMTO*) method for the calculation of λ^{LDA} was invented in [116]. It is very efficient in explaining the physics of elemental metals, like $Al, Cu, Mo, Nb, Pb, Pd, Ta$ and V with disagreements by only $10-30\%$ of theoretical and experimental results (obtained from tunnelling and resistivity measurements) for the EPI coupling constants λ and λ_{tr}. However, the *LR-LMTO* method applied to the doped HTSC oxide $(Ca_{1-x}Sr_x)_{1-y}CuO_2$ for $x \sim 0.7$ and $y \sim 0.1$ with $T_c = 110\ K$ gives surprisingly small *EPI* coupling $\lambda_s \approx 0.4$ for $s-wave$ pairing and $\lambda_d \leq 0.3$ for $d-wave$ pairing [117]. Although this finding, that λ_d is of the similar magnitude as λ_s ($\lambda_d \approx \lambda_s$), is interesting and encouraging it seems that this method misses some ingredients of the ionic structure of the layered structure [52], [53].

We point out, that the model calculations which take into account the long-range ionic Madelung potential appropriately [47], [48], [50] gave also rather large coupling constant $\lambda \sim 2$, what additionally hints to the importance of the long-range forces in the EPI.

Since in HTSC oxides the plasma frequency along the c-axis, ω^c_{pl}, is of the order (or even less) of some characteristic c-axis vibration mode, it is necessary to include the *nonadiabatic effects* in the EPI coupling constant, i.e. its frequency dependence $g_{\alpha,n} \sim g^0/\varepsilon_{cc}(\omega)$. This non-adiabaticity is partly accounted for in the Falter group [52], [53] by calculating the electronic dielectric function along the c-axis $\varepsilon_{cc}(\mathbf{k},\omega)$ in the RPA approximation. The result is that $g_{\alpha,n}$ is increased appreciable beyond its (well screened) metallic part, what gives a large increase of the EPI coupling not only in the phonon modes but also in the plasmon one. This question deserves much more attention than it was in the past.

3.3. Lattice dynamics and EPI coupling

The calculation of the phonon frequencies ω_{ph}, which are obtained from

$$D_0^{-1}(\mathbf{q},\omega_{ph}) - \hat{\Pi}(\mathbf{q},\omega_{ph}) = 0, \tag{53}$$

is in principle even more complicated problem than the calculations of the electronic properties. It lies on the difficulty to calculate the phonon polarization operator $\hat{\Pi}$ - see more in [2]. Schematically one has

$$\hat{\Pi} \sim (\nabla_\alpha V_{e-i})^2 \hat{\chi}_c, \tag{54}$$

where $\hat{\chi}_c$ is the *electronic charge susceptibility*. V_{e-i} is the bare electron-lattice interactions. $\hat{\chi}_c$ is schematically given by $\hat{\chi}_c = \hat{P}\hat{\varepsilon}_e^{-1}$ and the electronic *polarization operator* $\hat{P} = \hat{G}\hat{\Gamma}_c\hat{G}$. As we see the phonon frequencies depends crucially on the screening properties of electrons. The screening effects in HTSC oxides are determined by the specificity of the metallic-ionic structure and strong electronic correlations. At present there is a controllable theory for the electronic properties in the t-J model [17], [18], [19], [20], [2] only, where these two ingredients are successfully incorporated in the theory. However, until now there is no controllable theory for the lattice dynamics which incorporates these two ingredients, in spite the fact that the X-method (see below) offers well defined and procedure. There were a number of interesting attempts to calculate renormalization of some specific phonons [54], [55], [56], such as for instance of the half-breathing phonon mode along the $(1,0,0)$ direction - which is strongly softened. In spite of some alleged theoretical confirmation of the experimental softening in YBCO and LASCO, none of these calculations are reliable, because none of them take into account the screening due to strong correlations (the charge vertex $\hat{\Gamma}_c$ and the dielectric function $\hat{\varepsilon}_e$) in a controllable way. That is the reason that all attempts until were unable to extract the reliable magnitude of the coupling constant with a specific phonon. Even more, by playing only with a single phonon mode, and with a particular wave-vector in the Brillouin zone, *one can not get large EPI* and large λ - see [2]. The latter claim is confirmed by tunnelling experiments, which demonstrate that almost all phonons (for instance 39 modes in $YBCO$) contribute to λ. No particular mode can be singled out in the spectral function $\alpha^2F(\omega)$ as being the only one which dominates in pairing mechanism in HTSC oxides.

4. THEORY OF STRONG ELECTRONIC CORRELATIONS

The well established fact is that *strong electronic correlations* are pronounced in HTSC oxides, at least in underdoped systems. However, the LDA theory fails to capture effects of strong correlations by treating they as a local perturbation. This is, as we shall see later, an unrealistic approximation in HTSC oxides, where strong correlations introduce non-locality. The shortcoming of the *LDA* is that in the half-filling case (with $n = 1$ and one particle per lattice site) it predicts metallic state missing the existence of the *Mott insulating state*. In the latter, particles are localized at lattice sites independent of the (non)existence of the AF order and the localization is due to the large Coulomb repulsion U at a given lattice site, i.e. $U \gg W$ where W is the band width. Some properties in the metallic state can not be described by the simple canonical Landau-Fermi liquid concept. For instance, recent ARPES photoemission measurements [122] on the hole doped samples show a well defined Fermi surface in the one-particle energy spectrum, which contains $1-\delta$ electrons in the Fermi volume (δ is the hole concentration), but the band width is $(2-3)$ times smaller than the LDA band structure calculations predict. The latter is consistent with the Luttinger theorem as well as with the LDA band structure calculations. However, experimental data on the dynamical conductivity (spectral weight of the Drude peak), Hall measurements etc. indicate that in transport properties a low density of hole-like charge carriers (which is proportional to δ) participates predominantly. These carriers suffer strong scattering and their inverse lifetime is proportional to the temperature (at $T > T_c$) as we discussed earlier. It is worth of mentioning here that the local moments on the Cu sites, which are localized in the parent AF compound, are counted as part of the Fermi surface area when the system is doped by small concentration of holes in the metallic state. The latter fact gives rise to a *large Fermi surface* which scales with the number (per site) of electrons $1-\delta$. At the same time the conductivity sum-rule is proportional to the number of doped holes δ, instead of $1-\delta$ as in the canonical Landau-Fermi liquid. These two properties tell us that we deal with a *correlated state*, and the latter must be due to the specific electronic structure of HTSC oxides (cuprates). The common ingredient of all cuprates is the presence of the Cu atoms. In order to account for the absence of Cu^{3+} ionic configuration (the charge transfer $Cu^{2+} \rightarrow Cu^{3+}$ costs large energy $U \sim 10\ eV$, i.e. the occupation of the Cu site with two holes with opposite spins is unfavorable) P. W. Anderson [61] proposed the Hubbard model as the basic model for quasiparticle properties in these compounds. For some parameter values it can be derived from the (minimal) microscopic *three-band model*. Besides the hopping t_{pd} between the d-orbitals of Cu and p-orbitals of O ions (as well as t_{pp}) - the Emery model [123], it includes also the strong Coulomb interaction U_{Cu} on the Cu ions as well as interaction between p- and d-electrons. The main two parameters are $U_{Cu} \sim (6-10)\ eV$ and the charge transfer energy $\Delta_{pd} \equiv \varepsilon_d^0 - \varepsilon_p^0 \sim (2.5-4)$ eV, where $\varepsilon_d^0, \varepsilon_p^0$ are energies of the d- and p-level, respectively. In HTSC oxides the case $U_{Cu} >> \Delta_{pd}$ is realized, i.e. they belong to the class of *charge transfer materials*. This allows us to project the complicated three-band Hamiltonian onto the low-energy sector, and to obtain an effective single-band Hubbard Hamiltonian with an effective hopping parameter t and the effective repulsion $U \approx \Delta_{pd}$. It turns out that the case $U >> t$ is realized, since $\Delta_{pd} \gg t = t_{pd}^2/\Delta_{pd}$. The effective and minimal Hamiltonian which describes the low-energy physics of HTSC oxides comprises also the long-range Coulomb interaction $\hat{V}_C$ and the EPI $\hat{V}_{EPI}$ - see [7], [2]

$$\hat{H} = -\sum_{i,j,\sigma} t_{ij} c_{i\sigma}^{\dagger} c_{j\sigma} + U \sum_i n_{i\uparrow} n_{i\downarrow} + \hat{V}_C + \hat{V}_{EPI}. \tag{55}$$

The effective repulsion $U \approx 4\ eV$ has its origin in the charge-transfer gap of the three-band model, while the nearest neighbor and next-nearest neighbor hopping t and t', respectively are estimated to be $t = 0.3 - 0.5\ eV$ and t'/t equal -0.15 in La compounds and -0.45 $YBCO$. Since $(U/t) \gg 1$ the above Hamiltonian is again in the regime of strong electronic correlations, where the doubly occupancy of a given lattice site is strongly suppressed, i.e. $\langle n_{i\uparrow}n_{i\downarrow}\rangle \ll 1$. The latter restricts charge fluctuations of electrons (holes) on a given lattice site are allowed, since $n_i = 0, 1$ is allowed only, while processes with $n_i = 2$ are (practically) forbidden. Note, that in (standard) weakly correlated metals all charge fluctuation processes ($n_i = 0, 1, 2$) are allowed, since $U \ll W$ in these systems. From the Hamiltonian in Eq.(55), which is the 2D model for the low-energy physics in the CuO_2 plane, comes out that in the *undoped* system there is one particle per lattice - the so called half-filled case (in the band language) with $\langle n_i\rangle = 1$. It is an insulator because of large U and even antiferromagnetic insulator at $T = 0\ K$. The effective exchange interaction (with $J = 4t^2/U$) between spins is Heisenberg-like. By doping the system by holes (with the hole concentration $\delta(< 1)$), means that particles are taken out from the system in which case there is on the average $\langle n_i\rangle = 1 - \delta$ particles per lattice site. Above some (small) critical doping $\delta_c \sim 0.01$ the AF order is destroyed and the system is strongly correlated metal. For some *optimal* doping $\delta_{op}(\sim 0.1)$ the system is metallic with the large Fermi surface and can exhibit even high-T_c superconductivity in the presence of the EPI, as it will demonstrated below. The latter interaction and its interplay with strong correlations is the central subject in the following sections.

4.1. X-method for strongly correlated systems

Since $U >> t$ one can put with good accuracy $U \to \infty$, i.e. the system is in the *strongly correlated regime* where the doubly occupancy $n_i = 2$ is excluded. One of the ways to cope with such strong correlations is to introduce the (fermionic like) creation and annihilation operators ($\hat{X}_i^{\sigma 0}$ and $\hat{X}_i^{0\sigma} = (\hat{X}_i^{\sigma 0})^\dagger$)

$$\hat{X}_i^{\sigma 0} = c_{i\sigma}^\dagger(1 - n_{i,-\sigma}), \tag{56}$$

which respect the condition $n_{i,\sigma} + n_{i,-\sigma} \leq 1$ on each lattice site. The latter means that there is no more than one electron (hole) at a lattice site, i.e. the doubly occupancy is forbidden. The bosonic like operators

$$\hat{X}_i^{\sigma_1\sigma_2} = \hat{X}_i^{\sigma_1 0}\hat{X}_i^{0\sigma_2} \tag{57}$$

(with $\sigma_1 \neq \sigma_2$) create a spin fluctuation at the $i - th$ site. Here, the spin projection parameter $\sigma = \uparrow, \downarrow$ and $-\sigma = \downarrow, \uparrow$ and the operator $\hat{X}_i^{\sigma\sigma}$ has the meaning of the electron (hole) number on the i-th site. In the following we shall use the convention that when $\hat{X}_i^{\sigma\sigma} \mid 1\rangle = 1 \mid 1\rangle$ there is a fermionic particle ("electron") on the *i*-th site, while for $\hat{X}_i^{\sigma\sigma} \mid 0\rangle = 0 \mid 0\rangle$ the site is empty, i.e. there is a hole on it. It is useful to introduce the hole number operator

$$\hat{X}_i^{00} = \hat{X}_i^{0\sigma}\hat{X}_i^{\sigma 0} \tag{58}$$

at a given lattice site, i.e. if $\hat{X}_i^{00} \mid 0\rangle = 1 \mid 0\rangle$ the *i*-th site is empty - there is one hole on it, while for $\hat{X}_i^{00} \mid 1\rangle = 0 \mid 1\rangle$ it is occupied by an "electron" and there is no hole. They fulfill the non-canonical commutation relations

$$\left[\hat{X}_i^{\alpha\beta}, \hat{X}_j^{\gamma\lambda}\right]_\pm = \delta_{ij}\left[\delta_{\gamma\beta}\hat{X}_i^{\alpha\lambda} \pm \delta_{\alpha\lambda}\hat{X}_i^{\gamma\beta}\right]. \tag{59}$$

Here, $\alpha, \beta, \gamma, \lambda = 0, \sigma$ and δ_{ij} is the Kronecker symbol. The (anti)commutation relations in Eq.(59) are rather different from the canonical Fermi and Bose (anti)commutation relations.

Since $U \to \infty$ the doubly occupancy is excluded, i.e. $\hat{n}_{i\uparrow}\hat{n}_{i\downarrow} \mid \psi\rangle (= \hat{X}_i^{22} \mid\uparrow\downarrow\rangle) = 0$, and by construction the $\hat{X}$ operators satisfy the *local constraint* (the completeness relation)

$$\hat{C}_X(i) \equiv \hat{X}_i^{00} + \sum_{\sigma=1}^{N} \hat{X}_i^{\sigma\sigma} = 1\,. \tag{60}$$

This condition tells us that at a given lattice site there is either one hole ($\hat{X}_i^{00} \mid hole\rangle = 1 \mid hole\rangle$) ore one electron ($\hat{X}_i^{\sigma\sigma} \mid elec\rangle = 1 \mid elec\rangle$). Note, if Eq.(60) is obeyed then both commutation and anticommutation relations hold also in Eq.(59) at the same lattice site, which is due to the projection properties of the Hubbard operators $\hat{X}^{\alpha\beta}\hat{X}^{\gamma\mu} = \delta_{\beta\gamma}\hat{X}^{\alpha\mu}$.

For further purposes, i.e. for the study of low-energy excitations in a controllable way, the number of spin projections is *generalized* to be N , i.e. $\sigma = 1, 2, ...N$. This way of generalization was very useful in describing heavy fermion physics, where for some Ce compounds N means the number of projections of the total angular momentum, for instance when $j = 5/2$ then $N = 2j + 1 = 6$ ($N \gg 1$). For some Yb compounds one has $j = 7/2$, i.e. $N = 2j + 1 = 8$ ($N \gg 1$). By projecting out the doubly occupied (high energy) states from the Hamiltonian in Eq.(55) one obtains the generalized $t - J$ model. The details of this derivation are given in Appendix and here we give the final expression for the Hamiltonian which excludes the doubly occupancy

$$\hat{H} = \hat{H}_t + \hat{H}_J = -\sum_{i,j,\sigma} t_{ij}\hat{X}_i^{\sigma 0}\hat{X}_j^{0\sigma} + \sum_{i,j,} J_{ij}(\mathbf{S}_i \cdot \mathbf{S}_j - \frac{1}{4}\hat{n}_i\hat{n}_j) + \hat{H}_3. \tag{61}$$

The first term describes the hopping of the "electron" by taking into account that the doubly occupancy of sites are excluded. The second term describes the Heisenberg-like exchange energy of almost localized "electrons". $\hat{H}_3$ contains three-sites hopping which is usually omitted believing it is not important. For effects related to charge fluctuation processes it is plausible to omit it, while for spin-fluctuation processes it may be questionable approximation. The spin and number operators can be expressed via the Hubbard operators [125]

$$\mathbf{S} = \hat{X}_i^{\bar{\sigma}_1 0}(\vec{\sigma})_{\bar{\sigma}_1\bar{\sigma}_2}\hat{X}_i^{0\bar{\sigma}_2};\ \hat{n}_i = \hat{X}_i^{\bar{\sigma}\bar{\sigma}} \tag{62}$$

where summation over bar indices is understood.

The basic idea behind the *X-method* is that the Dyson's equation for the electron Green's function can be effectively obtained by introducing *external potentials* (sources) $u^{\sigma_1\sigma_2}(1)$. The source Hamiltonian $\hat{H}_s$ is used in the form

$$\int \hat{H}_s d\tau = \int \sum_{\sigma_1,\sigma_2} u^{\sigma_1\sigma_2}(1)\hat{X}^{\sigma_1\sigma_2}(1)\,d1 \equiv u^{\bar{\sigma}_1\bar{\sigma}_2}(\bar{1})\hat{X}^{\bar{\sigma}_1\bar{\sigma}_2}(\bar{1})\,, \tag{63}$$

where $1 \equiv (\mathbf{l}, \tau)$ and $\int(..)d1 \equiv \int(..)d\tau \sum_{\mathbf{l}}$ and τ is the Matsubara time. Here and in the following, integration over bar variables($\bar{1}, \bar{2}..$) and a summation over bar spin variables($\bar{\sigma}$..) is understood. The sources $u^{\sigma_1\sigma_2}(1)$ are useful in generating higher correlation functions entering the self-energy. The electronic Green's function is defined by [2] ($\hat{T}$ is the time-ordering operator)

$$G^{\sigma_1\sigma_2}(1,2) = \frac{-\langle\hat{T}\,(\hat{S}\hat{X}^{0\sigma_1}(1)\hat{X}^{\sigma_2 0}(2))\rangle}{\langle\hat{T}\hat{S}\rangle}, \tag{64}$$

where $\hat{S} = \hat{T}\exp\{-\int\hat{H}_s(1)d1\}$ and the corresponding Dyson's equation reads

$$\left[G_{0,u}^{-1,\sigma_1\bar{\sigma}_2}(1,\bar{2}) - \Sigma_G^{\sigma_1\bar{\sigma}_2}(1,\bar{2})\right]G^{\bar{\sigma}_2\sigma_2}(\bar{2},2) = Q^{\sigma_1\sigma_2}(1)\delta(1-2), \tag{65}$$

$$G_{0,u}^{-1,\sigma_1\sigma_2}(1,2) = (-\frac{\partial}{\partial t_1})\delta^{\sigma_1\sigma_2}\delta(1-2) - u^{\sigma_1\sigma_2}(1)\delta(1-2) \tag{66}$$

The so called Hubbard spectral weight is given by

$$Q^{\sigma_1\sigma_2}(1) = \delta^{\sigma_1\sigma_2}\langle\hat{X}^{00}(1)\rangle + \langle\hat{X}^{\sigma_1\sigma_2}(1)\rangle. \tag{67}$$

$\Sigma_G^{\sigma_1\sigma_2}(1,2)$ is a functional of the Green's function $G^{\sigma_1\sigma_2}(1,2)$. The latter describes the *composite* (correlated) particle (in the language of the *SB* theory it describes the combined "spinon + holon"). For further analysis it is useful to introduce the quasiparticle Green's $g^{\sigma_1\sigma_2}$ (something analogous to the "spinon" Green's function in the *SB* approach) and the vertex functions $\gamma_{\sigma_3\sigma_4}^{\sigma_1\sigma_2}(1,2;3)$, respectively

$$g^{\sigma_1\sigma_2}(1,2) = G^{\sigma_1\overline{\sigma_2}}(1,2)Q^{-1,\bar{\sigma}_2\sigma_2}(2) \tag{68}$$

$$\gamma_{\sigma_3\sigma_4}^{\sigma_1\sigma_2}(1,2;3) = -\frac{\delta g^{-1,\sigma_1\sigma_2}(1,2)}{\delta u^{\sigma_3\sigma_4}(3)}, \tag{69}$$

respectively. Note, that $\gamma_{\sigma_3\sigma_4}^{\sigma_1\sigma_2}(1,2;3)$ are the *three-point vertex function*, which also renormalizes the ionic EPI coupling constant - as we shall see below. $g^{\sigma_1\sigma_2}(1,2)$ is the solution of the equation

$$\left[G_{0,u}^{-1,\sigma_1\bar{\sigma}_2}(1,\bar{2}) - \Sigma_g^{\sigma_1\bar{\sigma}_2}(1,\bar{2})\right]g^{\bar{\sigma}_2\sigma_2}(\bar{2},2) = \delta^{\sigma_1\sigma_2}\delta(1-2), \tag{70}$$

where $\Sigma_g^{\sigma_1\sigma_2}(1,2)$ depends on the "quasiparticle" Green's function $g^{\sigma_1\sigma_2}(1,2)$. Note, that in Eq.(70) the Hubbard spectral weight $Q^{\sigma\sigma}$ disappears from the right hand side.

Since in the following we study *nonmagnetic (paramagnetic)* normal state one has $\Sigma_g^{\sigma\sigma}(1,2) \equiv \Sigma_g(1,2)$ for $\sigma = 1,..N$, as well as $Q^{\sigma\sigma}(\equiv Q)$. In [17], [18], [19], [20] it is shown that in that case Σ_g can be expressed via two vertex functions - the *charge vertex*

$$\gamma_c(1,2;3) \equiv \gamma_{\sigma\sigma}^{\sigma\sigma}(1,2;3) \tag{71}$$

and the *spin vertex*

$$\gamma_s(1,2;3) \equiv \gamma_{\bar{\sigma}\sigma}^{\bar{\sigma}\sigma}(1,2;3), \tag{72}$$

i.e. Σ_g is given by

$$\Sigma_g(1-2) = -\frac{t_0(1-2)}{N}Q(1) + \delta(1-2)\frac{J_0(1-\bar{1})}{N} < \hat{X}^{\sigma\sigma}(\bar{1}) > -$$

$$-\frac{t_0(1-\bar{1})}{N}g(\bar{1}-\bar{2})\gamma_c(\bar{2},2;1) +$$

$$+\frac{t_2(1,\bar{1},\bar{3})}{N}g(\bar{1}-\bar{2})\gamma_s(\bar{2},2;\bar{3}) + \Sigma_Q(1-2), \tag{73}$$

where $t_2(1,2,3) = \delta(1-2)t_0(1-3) - \delta(1-3)J_0(1-2)$. The notation $t_0(1-2)$ (and $J_0(1-2)$) means $t_0(1-2) = t_{0,i_1j_2}\delta(\tau_1-\tau_2)$. The first two terms in Eq.(73) represent an effective kinetic energy of quasiparticles in the lower Hubbard band. As we shall see below they give rise to the band narrowing and to the shift of the band center, respectively. The third and fourth terms describe the kinematic and dynamic interaction of quasiparticles with charge and spin fluctuations, respectively, while the very important term proportional to $\Sigma_Q(1,2)$ takes into account the counterflow of surrounding quasiparticles which takes place in order to respect the local constraint (absence of doubly occupancy). It reads

$$\Sigma_Q(1,2) = \frac{t_0(1-\bar{1})}{N}\frac{g(\bar{1}-\bar{2})}{Q}\left[\frac{\delta Q(\bar{2})}{\delta u^{\bar{\sigma}\bar{\sigma}}(1)} - \frac{\delta Q^{\bar{\sigma}\sigma}(\bar{2})}{\delta u^{\bar{\sigma}\sigma}(1)}\right]g^{-1}(\bar{2}-2).$$

Σ depends on the vertex functions $\gamma_c(1,2;3)$ and $\gamma_s(1,2;3)$ and it does not contain a small expansion parameter, like the interaction energy in weakly interacting systems, because the hopping parameter t describes at the same time the kinetic energy and *kinematic interaction* of quasiparticles. This means that there is no controllable perturbation technique, due to the lack of small parameter. There are various decoupling procedures and mean-field like techniques - the path integral method, or $1/N$ expansion in various slave-boson approaches [2].

What is the advantage of the *X-method* expressed by Eq.(68-73). It turns out that it allows to formulate a controllable $1/N$ expansion for Σ_g by including also the EPI [17], [18], [19], [20] - see below. For that purpose it is necessary to generalize the *local constraint* condition

$$\hat{C}_{X_N}(i) \equiv \hat{X}_i^{00} + \sum_{\sigma=1}^{N} \hat{X}_i^{\sigma\sigma} = \frac{N}{2}, \tag{74}$$

where $N/2$ replace the unity in Eq.(60). It is apparent from Eq.(60) that for $N=2$ it coincides with Eq.(60) and has the meaning that maximally half of all spin states at a given lattice site can be occupied.

The spectral function $A(\mathbf{k},\omega) = -\mathrm{Im}G(\mathbf{k},\omega)/\pi$ must obey the generalized Hubbard sum rule which respects the new local constraint in Eq.(74).

$$\int d\omega A(\mathbf{k},\omega) = \frac{1+(N-1)\delta}{2} \tag{75}$$

The $N>2$ generalization of the local constraint allows us to make a *controllable $1/N$ expansion* of the self-energy with respect to the small quantity $1/N$ (when $N \gg 1$).

Physically this procedure means that we select a *class of diagrams* in the self-energy and response functions which might be important in some parameter regime. By careful inspection of Eq.(73) one concludes that for large N there is $1/N$ expansion for various quantities - for instance

$$g = g_0 + \frac{g_1}{N} + ...; Q = Nq_0 + Q_1 + ..., \tag{76}$$

$$\Sigma_g = \Sigma_0 + \frac{\Sigma_1}{N} + ...; \gamma_c = \gamma_{c0} + \frac{\gamma_{c1}}{N} + ..., \tag{77}$$
$$\gamma_s(1,2;3) = N\delta(1-2)\delta(1-3) + \gamma_{s1} + ...$$

As a result of this expansion one obtains Σ_0 and g_0 in *leading* $O(1)$*-order* - see details in [2], [17], [18], [19], [20] (Σ_Q in Eq.(73) is of O(1/N) order)

$$g_0(\mathbf{k},\omega) \equiv \frac{G_0(\mathbf{k},\omega)}{Q_0} = \frac{1}{\omega - [\varepsilon_0(\mathbf{k}) - \mu]}, \tag{78}$$

where the *quasiparticle energy* $\varepsilon_0(\mathbf{k}) = \varepsilon_c - q_0 t_0(\mathbf{k}) - \sum_{\mathbf{p}} J_0(\mathbf{k}+\mathbf{p}) n_F(\mathbf{p})$, and the *level shift* $\varepsilon_c = \sum_{\mathbf{p}} t_0(\mathbf{p}) n_F(\mathbf{p})$. Here, $t_0(\mathbf{k})$ and $J_0(\mathbf{k})$ are Fourier transforms of $t_{0,ij}$ and $J_{0,ij}$, respectively. For $t' = 0$ one has $t_0(\mathbf{k}) = 2t_0(\cos k_x + \cos k_y)$ and $J_0(\mathbf{k}) = 2J_0(\cos k_x + \cos k_y)$. In the equilibrium state ($u^{\sigma_1\sigma_2} \to 0$) and in leading order one has

$$Q_0 = <\hat{X}_i^{00}> = Nq_0 = N\delta/2. \tag{79}$$

The chemical potential μ is obtained from the condition

$$1 - \delta = 2\sum_{\mathbf{p}} n_F(\mathbf{p}). \tag{80}$$

Let us summarize the main results which were obtained by the *X-method* in leading $O(1)$-order and compare these results with corresponding results of the *SB*-method [17], [18], [19], [20]: (**1**) In the $O(1)$ order the Green's function $g_0(\mathbf{k},\omega)$ describes the *coherent motion* of quasiparticles whose contribution to the total spectral weight of the Green's function $G_0(\mathbf{k},\omega)$ is $Q_0 = N\delta/2$. Note, $G_0(\mathbf{k},\omega) = Q_0 g_0(\mathbf{k},\omega)$ in leading order. The dispersion of the quasiparticle energy is dominated by the exchange parameter if $J_0 > \delta t_0$. In the case when $J_0 = 0$ there is a band narrowing by lowering the hole doping δ, where the band width is proportional to the hole concentration δ, i.e. $W = z \cdot \delta \cdot t_0$. (**2**) The X-method respects the local constraint at each lattice site and in each step of calculations. (**3**) In the important paper [126] - which is based on the theory elaborated in [17], [18], [19], [20], it is shown that in the superconducting state the anomalous self-energy (which is of $O(1/N)$-order in the $1/N$ expansion) of the *X*- and *SB*-methods *differ* substantially. As a consequence, the *SB*-method [127] predicts false superconductivity in the *t-J* model (for $J = 0$) with large T_c (due to the kinematical interaction), while the X-method gives extremely small $T_c (\approx 0)$ [126]. The reason for this discrepancy between the two methods is that calculations done in the *SB*-method miss a class of compensating diagrams, which are on the other hand taken automatically in the X-method. So, although the two approaches yield some similar results in leading $O(1)$-order their implementation in the next leading $O(1/N)$-order make that they are different. Note, that the $1/N$ expansion in the X-method is well-defined and transparent. **(4)** By explicit calculation and comparison of the two methods in [18] it is shown that the renormalization of the EPI coupling constant is different in the two approaches even in the large N-limit - see below; **(5)** Very interesting behavior as a function of doping concentration exhibits the optical conductivity $\sigma(\omega, \mathbf{q}=0) \equiv \sigma(\omega)$ which scales with the doping δ. Note, the volume below the Fermi surface in the case of strong correlations scales with $n = 1-\delta$, like in the usual Fermi liquid. The above analysis clearly demonstrate shows difference in response functions of strongly correlated systems and the canonical Landau-Fermi liquid.

4.2. Forward scattering peak in the charge vertex γ_c

The three-point *charge vertex* $\gamma_c(1,2;3)$ plays important role in the renormalization all charge processes, such as the EPI, Coulomb scattering and the scattering on non-magnetic impurities. It was shown in [17], [18], [19], [20], [2] that $\gamma_c(1,2;3)$ can be calculated exactly in the leading $O(1)$ order (and in all other orders) of the $t-J$ model. The integral equation in the $O(1)$ order reads

$$\begin{aligned}\gamma(1,2;3) = \delta(1-2)\delta(1-3) + t(1-2)g_0(1,\bar{1})g_0(\bar{2},1^+)\gamma(\bar{1},\bar{2};3) \\ +\delta(1-2)t(1-\bar{1})g_0(\bar{1},\bar{2}))g_0(\bar{3},1)\gamma(\bar{2},\bar{3};3 \\ -J(1-2)g_0(1,\bar{1}))g_0(\bar{2},2)\gamma(\bar{1},\bar{2};3.\end{aligned} \tag{81}$$

The analytical solution of Eq.(81) is given by [2], [20]

$$\gamma_c(\mathbf{k},q) = 1 - \sum_{\alpha=1}^{6}\sum_{\beta=1}^{6} F_\alpha(\mathbf{k})[\hat{1} + \hat{\chi}(q)]^{-1}_{\alpha\beta}\chi_{\beta 2}(q), \tag{82}$$

where

$$\chi_{\alpha\beta}(q) = \sum_p G_\alpha(p,q)F_\beta(\mathbf{p}), \tag{83}$$

$$G_\alpha(p,q) = [1, t(\mathbf{p}+\mathbf{q}), \cos p_x, \sin p_x, \cos p_y, \sin p_y]\Pi(p,q),$$

$$F_\alpha(\mathbf{k}) = [t(\mathbf{k}), 1, 2J_0\cos k_x, 2J_0\sin k_x, 2J_0\cos k_y, 2J_0\sin k_y],$$

where $\Pi(k,q) = -g(k)g(k+q)$ and $q = (\mathbf{q}, iq_n)$, $q_n = 2\pi nT$, $p = (\mathbf{p}, ip_m)$, $p_m = \pi T(2m+1)$. Note, the frequency sum over p_m in $\chi_{\alpha\beta}(q)$ in Eq.(82) involves only Π and can easily be carried out $\sum_{p_m}\Pi(p,q) = [n_F(\xi_{\mathbf{q}+\mathbf{p}}) - n_F(\xi_{\mathbf{p}})]/[\xi_{\mathbf{p}} - \xi_{\mathbf{q}+\mathbf{p}} - iq_n]$.

We stress that $\gamma_c(\mathbf{k},q)$ describes a specific screening of the charge potential due to of strong correlations. In the presence of perturbation (external source u) there is change of the band width, as well as of the local chemical potential, which comes from the suppression of doubly occupancy. The central result of the X-method is that for momenta $\mathbf{k}$ laying at (and near) the Fermi surface $\gamma_{c0}(\mathbf{k},\mathbf{q},\omega=0)$ has very pronounced *forward*

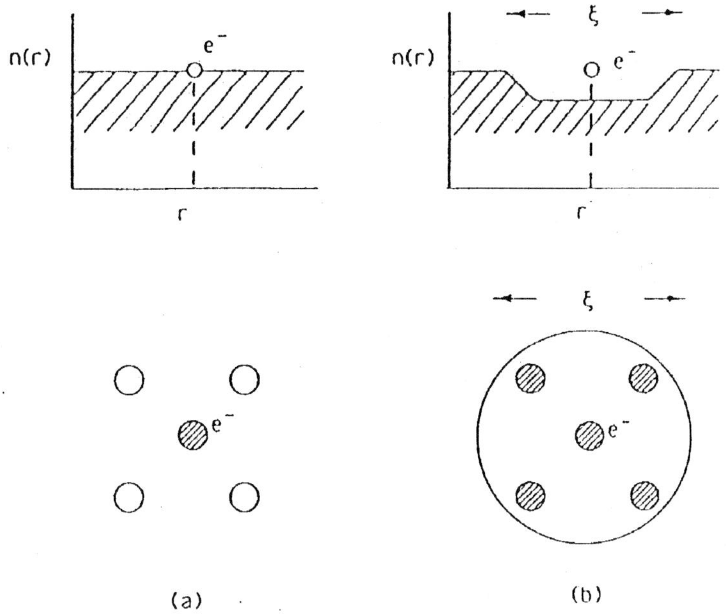

FIGURE 16. Schematic picture of electron correlation hole and the $E-P$ interaction for uncorrelated (weakly correlated) (a) and strongly correlated (b) electron. In the case (a) the electron does not perturb the electronic density $n(r)$ and it interacts with the vibrations of a single atom (shaded). In the case (b) an electron is accompanied by a large correlation hole of size $\xi \sim 1/\delta$ (δ is doping) and it will interact with atoms within this zone. From [120].

scattering peak at $\mathbf{q}=0$) at low doping concentration $\delta(\ll 1)$, while the backward scattering is substantially suppressed - see Fig. 17. The latter means that charge fluctuations are strongly suppressed (correlated) at small distances. Such a behavior of the vertex function means that a quasiparticle moving in the strongly correlated medium digs up a *giant correlation hole* with the radius $\xi_{ch} \approx a/\delta$, where a is the lattice constant - see Fig. 16.

However, in the highly doped systems with $\delta > 0.1$ - which corresponds to the overdoped HTSC oxides, the effects of strong correlations is progressively suppressed and the screening mechanism due to strong correlations is less effective. We stress that when $J < t$ then the last term in Eq.(81) for $\gamma_{c0}(\mathbf{k}_F, \mathbf{q})$ is unimportant. On the other hand both terms, the second (due to band narrowing) and the third (due band shifting) one, are in conjunction responsible for the development of the forward scattering peak at lower doping. If we omit in Eq.(81) the band shifting term (the third one) we get very weak forward scattering peak, while omitting the band narrowing term (the first one) $\gamma_{c0}(\mathbf{k}_F, \mathbf{q})$ is practically constant in a broad region of $\mathbf{q}$.

Finally, since the real physics is characterized by $N=2$ one can put the question - what is the *reliability* of the results for the quasiparticle properties obtained by the $1/N$ expansion (and $N \to \infty$) within the X-method? First, the exact diagonalization of the charge correlation function $N(\mathbf{k}, \omega)$ in the $t-J$ model [124] shows clearly that

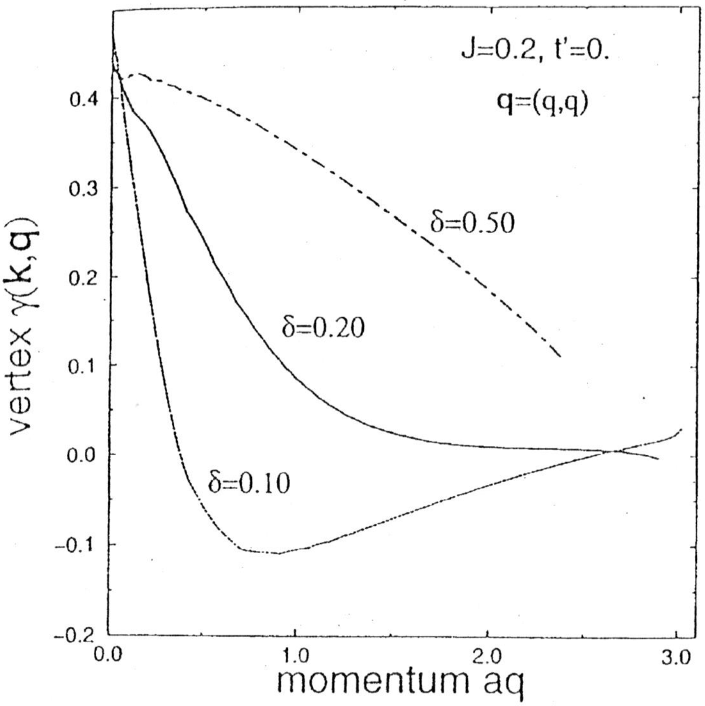

FIGURE 17. Zero-frequency vertex function $\gamma(\mathbf{k}_F,\mathbf{q})$ of the $t-J$ model as a function of the momentum aq with $\mathbf{q}=(q,q)$ for three different doping δ - from [19].

the low-energy charge scattering processes at large momenta $|\,\mathbf{k}\,|\approx 2k_F$ are strongly suppressed compared to the small transferred momenta ($|\,\mathbf{k}\,|\ll 2k_F$). These calculations confirm unambiguously the results obtained by the X-method in [17], [18], [19] on the suppression of the backward scattering in the vertex. Second, very recent Monte Carlo (numerical) calculations in the Hubbard model with finite U [21] show clear development of the forward scattering peak in $\gamma_c(\mathbf{k}_F,\mathbf{q})$ by increasing U, thus confirming the theoretical predictions in [17], [18], [19].

5. RENORMALIZATION OF THE EPI BY STRONG CORRELATIONS

In preceding Sections arguments we argued that because $\lambda_{tr}\ll\lambda$ the standard Migdal-Eliashberg EPI theory must be corrected in order to take into account screening properties of strongly correlated system. This analysis is done in the framework of the X-method in a series of papers [17], [18], [19] which we briefly discuss below. The renormalization of the EPI by strong correlations has been studied also by the *SB*-method [128], [121], [129], [130] by the $1/N$ expansion in the partition function, or by using the mean-field approach [131] where also the non-Migdal correction due to the EPI is considered. We stress that at present there are no systematic and controllable calculations within the *SB*-method for the EPI. From that point of view the *X*-method is of indispensable value.

5.1. The forward scattering peak in the EPI

The minimal model Hamiltonian for the HTSC oxides contains besides the $t-J$ terms also the EPI, i.e. $\hat{H}=\hat{H}_{tJ}+\hat{H}_{EP}$ where $\hat{H}_{EP}=\hat{H}_{EP}^{ion}+\hat{H}_{EP}^{cov}$

$$\hat{H}=-\sum_{i,j,\sigma}t_{ij}\hat{X}_i^{\sigma 0}\hat{X}_j^{0\sigma}+\sum_{i,j,}J_{ij}(\mathbf{S}_i\cdot\mathbf{S}_j-\frac{1}{4}n_in_j)+$$

$$+\sum_{i,\sigma}\varepsilon_{a,i}^0\hat{X}_i^{\sigma\sigma}+\hat{H}_{ph}+\hat{H}_{EP}+\hat{V}_{LC}, \tag{84}$$

where the ionic contribution to the EPI is

$$\hat{H}_{EP}^{ion}=\sum_{i,\sigma}\hat{\Phi}_i(\hat{X}_i^{\sigma\sigma}-\langle\hat{X}_i^{\sigma\sigma}\rangle)+\hat{H}_{EP}^{cov}. \tag{85}$$

Here, $\hat{\Phi}_i$ (given by Eq.(40)) describes the change of the atomic energy $\varepsilon_{a,i}^0$ due to the long-range Madelung energy, where L and κ enumerate unit lattice vectors and atoms in the unit cell, respectively. $Z_{L\kappa}$ is the effective charge of an ion at the site $L\kappa$. Note, in Eq.(85) we do not assume small displacement $\hat{\mathbf{u}}_i$ and the following analysis holds in principle also for an *anharmonic* EPI. The term proportional to $\langle\hat{X}_i^{\sigma\sigma}\rangle$ in Eq.(85) is introduced in order to have $\langle\hat{\Phi}_i\rangle=0$ in the equilibrium state. Note, that there is also covalent contribution to the *EPI* in Eq.(84) due to the change of the hopping (t) and exchange energy (J) by the ion displacements

$$\hat{H}_{EP}^{cov}=-\sum_{i,j,\sigma}\frac{\partial t_{ij}}{\partial(\mathbf{R}_i^0-\mathbf{R}_j^0)}(\hat{\mathbf{u}}_i-\hat{\mathbf{u}}_j)\hat{X}_i^{\sigma 0}\hat{X}_j^{0\sigma}+$$

$$+\sum_{i,j,}\frac{\partial J_{ij}}{\partial(\mathbf{R}_i^0-\mathbf{R}_j^0)}(\hat{\mathbf{u}}_i-\hat{\mathbf{u}}_j)\mathbf{S}_i\cdot\mathbf{S}_j. \tag{86}$$

The treatment of the first term is similar to the Madelung term in Eq.(85) although the equation for the *four-point vertex function* $\gamma_c(1,2;3,4)$ is different than Eq.(86). We stress, that the X-method has advantage also in the treatment of the covalent term, because it peaks up straightforwardly all important contributions in γ_c, due to strong correlations. On the other side the corresponding treatment by the *SB* method is complicated and not well defined, giving sometimes wrong results. For instance, in [121] several terms in the vertex equation are omitted leading to incorrect results for the covalent EPI coupling. We stress, that the covalent part contributes approximately $20-30$ K to the critical temperature in HTSC oxides as the band structure calculations in [114],

[115] have shown (partly discussed in Section 3.). Its renormalization by strong correlations will be studied elsewhere. The second term in $\hat{H}_{EP}^{cov}$ is due to the change of the exchange energy by phonon vibrations. Since it is second order with respect to t_{ij} it is much smaller than the first covalent term and accordingly contributes very little to the total EPI.

After technically lengthy calculations, which are performed in [17], [18] the expression for the ionic part of the EPI (frequency-dependent) part of the self-energy reads

$$\Sigma_{EP}^{(dyn)}(1,2) = -V_{EP}(\bar{1}-\bar{2})\gamma_c(1,\bar{3};\bar{1})g_0(\bar{3}-\bar{4})\gamma_c(\bar{4},2;\bar{2}), \tag{87}$$

where analogously to Eq.(47) one has $V_{EP}(1-2) = \varepsilon_e^{-1}(1-\bar{1})V_{EP}^0(\bar{1}-\bar{2})\varepsilon_e^{-1}(\bar{2}-2)$. The propagator of the bare EPI $V_{EP}^0(1-2) = -\langle T\hat{\Phi}(1)\hat{\Phi}(2)\rangle$ comprises in principle also the anharmonic contribution. From Eq.(87) it is seen that in strongly correlated systems the ionic part of the EPI is proportional to the square of the *three-point charge vertex* $\gamma_c(1,2;3)$ (due to correlations). The self-energy is given by

$$\Sigma_{EP}^{(dyn)}(\mathbf{k},\omega) = \int_0^\infty d\Omega \langle \alpha^2 F(\mathbf{k},\mathbf{k}',\Omega)\rangle_{\mathbf{k}'} R(\omega,\Omega), \tag{88}$$

where $R(\omega,\Omega)$ is given in [2], [17], [18]. The (momentum-dependent) Eliashberg spectral function is defined by

$$\alpha^2 F(\mathbf{k},\mathbf{k}',\omega) = N_{sc}(0)\sum_{\nu} \mid g_{eff}(\mathbf{k},\mathbf{k}-\mathbf{k}',\nu)\mid^2 \times$$

$$\times \delta(\omega-\omega_\nu(\mathbf{k}-\mathbf{k}'))\gamma_c^2(\mathbf{k},\mathbf{k}-\mathbf{k}'). \tag{89}$$

$n_B(\Omega)$ is the Bose distribution function and ψ is di-gamma function, while $g_{eff}(\mathbf{k},\mathbf{p},\nu)$ is the EPI coupling constant for the ν-the mode, where the renormalization by long-range Coulomb interaction is included, i.e. $g_{eff}(\mathbf{k},\mathbf{p},\nu) = g(\mathbf{k},\mathbf{p},\nu)/\varepsilon_e(\mathbf{p})$. $N_{sc}(0)$ is the density of states renormalized by strong correlations where $N_{sc}(0) = N_0(0)/q_0$ and $q_0 = \delta/2$ in the $t-t'$ model ($J=0$). In the *t-J* model $N_{sc}(0)$ has another form which does not diverge for $\delta \to 0$ but one has $N_{sc}(0)(\sim 1/J_0) > N_0(0)$, where the bare density of states $N_0(0)$ is calculated, for instance by the *LDA* scheme.

5.2. Pairing and transport EPI coupling constants

Depending on the symmetry of the superconducting order parameter $\Delta(\mathbf{k},\omega)$ ($s-$, $d-wave$ pairing) various averages (over the Fermi surface) of $\alpha^2 F(\mathbf{k},\mathbf{k}',\omega)$ enter the Eliashberg equations. Assuming that the superconducting order parameter transforms according to the representation Γ_i ($i=1,3,5$) of the point group $C_{4\nu}$ of the square lattice (in the CuO_2 planes) the appropriate symmetry-projected spectral function is given by

$$\alpha^2 F_i(\tilde{\mathbf{k}},\tilde{\mathbf{k}}',\omega) = \frac{N_{sc}(0)}{8}\sum_{\nu,j} \mid g_{eff}(\tilde{\mathbf{k}},\tilde{\mathbf{k}}-T_j\tilde{\mathbf{k}}',\nu)\mid^2 \times$$

$$\times\delta(\omega-\omega_\nu(\tilde{\mathbf{k}}-T_j\tilde{\mathbf{k}}'))\mid\gamma_c(\tilde{\mathbf{k}},\tilde{\mathbf{k}}-T_j\tilde{\mathbf{k}}')\mid^2 D_i(j). \tag{90}$$

$\tilde{\mathbf{k}}$ and $\tilde{\mathbf{k}}'$ are momenta on the Fermi line in the irreducible Brillouin zone which is $1/8$ of the total Brillouin zone. T_j , $j=1,..8$, denotes the eight point-group transformations forming the symmetry group of the square lattice. This group has five irreducible representations which we distinguish by the label $i=1,2,...5$. In the following the representations $i=1$ and $i=3$, which correspond to the $s-$ and $d-wave$ symmetry of the full rotation group, respectively, will be of importance. $D_i(j)$ is the representation matrix of the *j-th* transformation for the representation *i*. By assuming that the superconducting order parameter $\Delta(\mathbf{k},\omega)$ does not vary much in the irreducible Brillouin zone one can average over $\tilde{\mathbf{k}}$ and $\tilde{\mathbf{k}}'$ in the Brillouin zone. For each symmetry one obtains the corresponding spectral function $\alpha^2F_i(\omega)$

$$\alpha^2F_i(\omega)=\langle\langle\alpha^2F_i(\tilde{\mathbf{k}},\tilde{\mathbf{k}}',\omega)\rangle_{\tilde{\mathbf{k}}}\rangle_{\tilde{\mathbf{k}}'} \tag{91}$$

which (in the first approximation determines) the transition temperature for the order parameter with the symmetry Γ_i. In the case $i=3$ the electron-phonon spectral function $\alpha^2F_3(\omega)$ in the *d-channel* is responsible for $d-wave$ superconductivity represented by the irreducible representation Γ_3 (or sometimes labelled as B_{1g}).

Performing similar calculations (as above) for the phonon-limited resistivity one finds that the latter is related to the *transport spectral function* $\alpha^2F_{tr}(\omega)$ which is given by

$$\alpha^2F_{tr}(\omega)=\frac{\langle\langle\alpha^2F(\mathbf{k},\mathbf{k}',\omega)[\mathbf{v}(\mathbf{k})-\mathbf{v}(\mathbf{k}')]^2\rangle_{\mathbf{k}}\rangle_{\mathbf{k}'}}{2\langle\langle\mathbf{v}^2(\mathbf{k})\rangle_{\mathbf{k}}\rangle_{\mathbf{k}'}}. \tag{92}$$

$\mathbf{v}(\mathbf{k})$ is the Fermi velocity. The effect of strong correlations on the EPI was discussed in [17] and more extensively in [18] within the model where the phonon frequencies $\omega(\tilde{\mathbf{k}}-\tilde{\mathbf{k}}')$ and $g_{eff}(\mathbf{k},\mathbf{p},\lambda)$ are weakly momentum dependent - due to the long-range screening (RPA). In order to illustrate the effect of strong correlations on $\alpha^2F_i(\omega)$ we consider the latter functions at zero frequency ($\omega=0$) which are then reduced to the (so called) "enhancement" functions

$$\Lambda_i=\frac{1}{8}\frac{N_{sc}(0)}{N_0(0)}\sum_{j=1}^{8}\langle\langle\mid\gamma_c(\tilde{\mathbf{k}},\tilde{\mathbf{k}}-T_j\tilde{\mathbf{k}}')\mid^2\rangle_{\tilde{\mathbf{k}}}\rangle_{\tilde{\mathbf{k}}'}D_i(j) \tag{93}$$

Note, in the case $J=0$ one has $N_{sc}(0)/N_0(0)=q_0^{-1}$, where q_0 is related to the doping concentration, i.e. $q_0=\delta/2$. Similarly, the correlation effects in the resistivity $\rho(T)(\sim\Lambda_{tr})$ renormalize the transport coupling constant Λ_{tr}

$$\Lambda_{tr}=\frac{N_{sc}(0)}{N_0(0)}\frac{\langle\langle\mid\gamma_c(\tilde{\mathbf{k}},\tilde{\mathbf{k}}-T_j\tilde{\mathbf{k}}')\mid^2[\mathbf{v}(\mathbf{k})-\mathbf{v}(\mathbf{k}')]^2\rangle_{\mathbf{k}}\rangle_{\mathbf{k}'}}{2\langle\langle\mathbf{v}^2(\mathbf{k})\rangle_{\mathbf{k}}\rangle_{\mathbf{k}'}} \tag{94}$$

Note, that for quasiparticles with the isotropic band the absence of correlations implies that $\Lambda_1=\Lambda_{tr}=1$, $\Lambda_i=0$ for $i>1$.

The averages in Λ_1,Λ_3 and Λ_{tr} were performed numerically in [18] by using the realistic anisotropic band dispersion in the $t-t'-J$ model and the corresponding charge

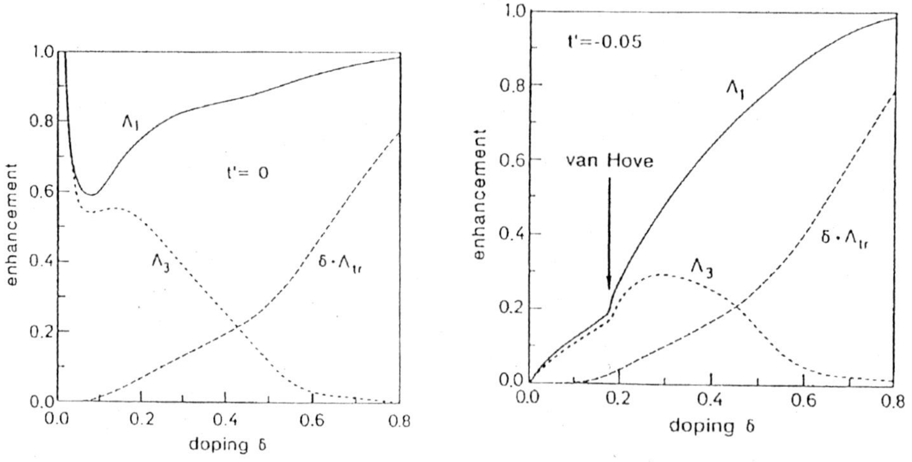

FIGURE 18. **(a)** - Enhancements Λ_1 and Λ_3 and $\delta \cdot \Lambda_{tr}$ as a function of doping δ for $t' = 0$ and $J = 0$ - from [18]. **(b)** Enhancements Λ_1 and Λ_3 and $\delta \cdot \Lambda_{tr}$ as a function of doping δ for $t' = -0.05$ and $J = 0$ - from [18].

vertex. The results for Λ_1, Λ_3 and Λ_{tr} are shown in Fig. 18 as functions of doping concentration in the t and $t - t'$ and $t - t' - J$ models, respectively. The three curves are multiplied with a common factor so that Λ_1 approaches 1 in the empty-band limit $\delta \rightarrow 1$, when strong correlations are absent. Note, that T_c in the weak coupling limit and in the *i-th* channel scales like

$$T_c^{(i)} \approx \langle \omega \rangle \exp(-1/(\Lambda_i - \mu_i^*), \tag{95}$$

where μ_i^* is the Coulomb pseudopotential in the i-th channel and $\langle \omega \rangle$ averaged phonon frequency.

Several interesting results, which are seen in Fig. 18, should be stressed.

First, in the empty-band limit $\delta \rightarrow 1$ the $d - wave$ coupling constant Λ_3 is much smaller than the $s - wave$ coupling constant Λ_1, i.e. $\Lambda_3 \ll \Lambda_1$. Furthermore, the totally symmetric function Λ_1 decreases with decreasing doping.

Second, in both models Λ_1 and Λ_3 meet each other (note $\Lambda_1 > \Lambda_3$ for all δ) at some small doping $\delta \approx 0.1 - 0.2$ where $\Lambda_1 \approx \Lambda_3$ but still $\Lambda_1 > \Lambda_3$. By taking into account a residual Coulomb repulsion of quasiparticles with $\mu_d^* \ll \mu_s^*$ one gets that the $s - wave$ superconductivity (which is governed by the coupling constant Λ_1) is suppressed, while the $d - wave$ superconductivity (governed by Λ_3) is only weakly affected. In that case the $d - wave$ superconductivity due to the EPI becomes more stable than the $s - wave$ superconductivity at sufficiently small doping δ, i.e. $T_c^{(d)} > T_c^{(s)}$. Experimentally, this occurs in underdoped, optimally doped an overdoped HTSC oxides [34]. This transition between s- and $d - wave$ superconductivity is triggered by electronic correlations because in the calculations it is assumed that the bare EPI coupling is momentum independent, i.e. the bare coupling constant contains the $s - wave$ symmetry only.

Third, the calculations are performed in the *adiabatic approximation*, where Λ_1 is less and Λ_{tr} much more suppressed by strong correlations. In the *nonadiabatic regime* $\omega > \mathbf{p} \cdot \mathbf{v}_F(\mathbf{p})$ i.e. for $\omega < \mathbf{p} \cdot \mathbf{v}_F(\mathbf{p})$ the vertex function grows by decreasing q finally reaching $\gamma_c(\mathbf{k}_F, \mathbf{p} = \mathbf{0}, \omega) = 1$. Due to the latter effect the enhancement function $\gamma_c^2(\mathbf{k}_F, \mathbf{p}, \omega)/q_0$ may be substantially larger compared to the adiabatic. This means that different phonons will be differently affected by strong correlations. For a given frequency the coupling to phonons with momenta $p < p_c = \omega/v_F$ will be *enhanced,* while the coupling to those with $p > p_c = \omega/v_F$ is substantially reduced due to the suppression of the backward scattering by strong correlations.

Fourth, the *transport coupling constant* Λ_{tr} (not properly normalized in Fig. 18 - see correction in [19]) is reduced in the presence of strong correlations, especially for lower doping where $\Lambda_{tr} < \Lambda/3$. This is very important result because it resolves the experimental puzzle that λ_{tr} (which enters resistivity $\rho(T) \sim \lambda_{tr} T$) is much smaller than the coupling constant λ (which enters the self-energy Σ and T_c), i.e. why $\lambda_{tr} << \lambda$. The answer lies in strong correlations which causes the forward scattering peak in charge scattering processes - the *FSP theory*.

As we already said, Monte Carlo (numerical) calculations if the Hubbard model at finite U - performed by Scalapino Group [21], show that the forward scattering peak in the EPI coupling constant (and the charge vertex) develops by increasing U. The latter effect is more pronounced at lower doping. The similar (to Monte Carlo) results were obtained quite recently in [22] in the framework of the Rückenstein-Kotliar (four slave-boson) model. These numerical results prove the *correctness of the EPI theory* based on the X-method.

We stress that contrary to the X-method, where the systematic $1/N$ calculations of the EPI self-energy is uniquely done, this is still a problem for the *SB* (Barnes) method where the $1/N$ expansion of the partition function $Z(T, \mu)$ is usually performed [121]. The existing expression (in the literature) for the vertex function in the SB method is different than that in the X-method [18], [18]. It seems that such a not well-controlled procedure omits a class of diagrams giving inadequate behavior of the coupling constant λ as a function of doping. Additionally, the vertex function in the *SB* approach is peaked not at $q = 0$ but at some finite $q_{\max}$, where $q_{\max} \to 0$ only for doping $\delta \to 0$ - see [23].

6. FSP THEORY AND NOVEL EFFECTS

There are a number of effects which are predicted by the FSP theory. We have already explained the effects of the forward scattering peak on the EPI. In Section 2. the non-shift puzzle in ARPES was also explained by the FSP theory (model). We discuss briefly some other predictions of the FSP theory containing parts not comprised in [2].

6.1. Nonmagnetic impurities and robustness of d-wave pairing

In the presence of strong correlations the impurity potential is also renormalized and the effective potential in the Born approximation is given by $u^2(\mathbf{q}) = \gamma_c^2(k_F, \mathbf{q}) u_0^2(\mathbf{q})$, where $u_0(\mathbf{q})$ is the single impurity scattering potential in the absence of strong correlations [142]. Since the charge vertex $\gamma_c(\mathbf{p}_F, \mathbf{q})$ is peaked at $\mathbf{q} = \mathbf{0}$ the potential $u(\mathbf{q})$ is also peaked at $\mathbf{q} = \mathbf{0}$. This means that the scattering amplitude contains not only the s-channel (as usually assumed in studying impurity effects in HTSC oxides), but also the *d-channel*, etc. Based on this property the FSP theory succeeded in explaining some experimental facts, such as: (**i**) the suppression of the residual resistivity ρ_i [17], [18]. It is observed in the optimally doped $YBCO$, where the resistivity $\rho(T)$ at $T = 0\ K$ has a rather small value $< 10\ \mu\Omega$cm.; (**ii**) the robustness of $d - wave$ pairing [142]. The previous theories [140], which assume $u(\mathbf{q}) = const$, i.e. the s-wave scattering channel only, predict that $T_c(\rho_{i,c}) = 0$ at much smaller residual resistivity $\rho_{i,c}^{(s)} \sim 50\ \mu\Omega$cm, while the experimental range is 200 $\mu\Omega$cm $< \rho_{i,c}^{\exp} < 1500\ \mu\Omega$cm [141]. The latter experimental fact means that d-wave pairing in HTSC is much more robust than the standard theory predicts, and it is one of the smoking gun experiments in testing the concept of the forward scattering peak in the charge scattering potential. It is worth of mentioning that in a number of papers the pair-breaking effect of non-magnetic impurities in HTSC was analyzed in terms of the impurity concentration n_i, i.e. the dependence $T_c(n_i)$. However, n_i is not the parameter which governs this pair-breaking effect. The more appropriate parameter for discussing the robustness of d-wave pairing is the impurity scattering amplitude $\Gamma(\theta, \theta')$, which can be related to the measured residual resistivity ρ_i which leads to the dependence of $T_c(\rho_i)$. The robustness of d-wave pairing in HTSC can be revealed only by studying the experimental curve $T_c(\rho_i)$, what has been first recognized experimentally in [141] and theoretically in [17], [142].

The theory of the robustness of d-wave pairing in HTSC was elaborated first in [142], where the FSP theory [17], [18] is applied to this problem. We shall not go into details - which are given in [142], [2], but we give here a general formula for the $T_c(\rho_i)$ dependence in anisotropic (including unconventional) superconductors, only. We assume that in anisotropic superconductivity the superconducting order parameter has the form $\Delta(\theta) = \Delta_0 Y(\theta)$ and generally one has $\langle Y(\theta) \rangle \neq 0$ ($\langle Y^*(\theta) \rangle Y(\theta) \rangle = 1$) where the momentum dependent impurity scattering amplitude is $\Gamma(\theta, \theta') = \Gamma_s(\theta, \theta') + \Gamma_d Y_d(\theta) Y_d(\theta') + ...$

$$\ln \frac{T_c}{T_{c0}} = \Psi(\frac{1}{2}) - \Psi(\frac{1}{2} + (1-\beta)x) - \langle Y(\theta) \rangle^2 [\Psi(\frac{1}{2}) - \Psi(x + \frac{1}{2})]. \quad (96)$$

Here, $x = \Gamma_s / 4\pi T_c$, $\beta = \Gamma_d / \Gamma_s$ and $\langle Y(\theta) \rangle$ means an averaging over the Fermi surface. Note, Eq.(96) holds independently of the scattering strength Γ_s,Γ_d, i.e. it holds in the Born as well as in the unitary limit. The residual resistivity ρ_i can be related to the transport scattering rate by $\rho_i = 4\pi \Gamma_{tr} / \omega_{pl}^2$, while the s-wave amplitude is related to Γ_{tr} by $\Gamma_s = p\Gamma_{tr}$. The parameter $p > 1$ can be obtained from the microscopic model (for instance in the t-J model $p \approx 2 - 3$) or can be treated as a fitting parameter - see more in [2]. In the case of an unconventional pairing one has $\langle Y(\theta) \rangle = 0$ and the last term drops. For the s-scattering only ($\Gamma_d = 0$) one has $\beta = 0$ and $T_c(\rho_i)$ should be suppressed very strongly contrary to the experimental results [141] - see Fig. 20.

The FSP theory of the impurity scattering in the t-J model [142] gives that the s-channel and d-channel almost equally contribute to the impurity scattering amplitude,

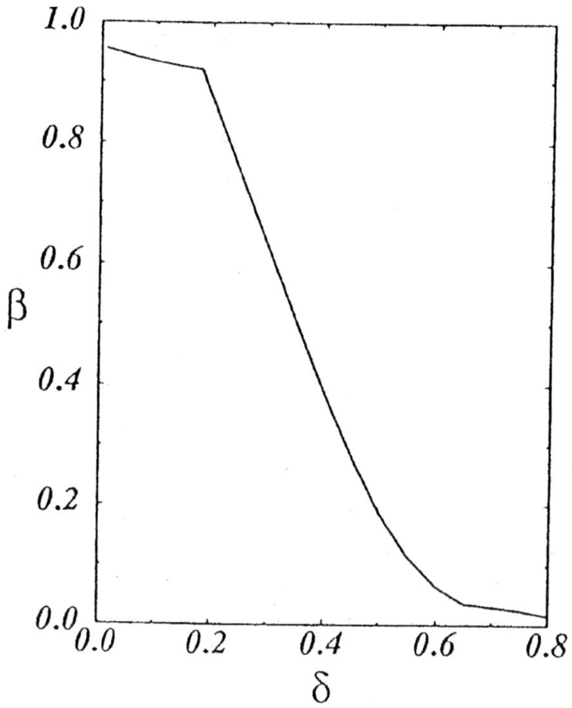

FIGURE 19. The anisotropy scattering parameter β as a function of doping δ in the t-J model. From [142].

since $\beta \approx 0.75 - 0.85$ for doping $\delta \approx 0.1 - 0.2$. The dependence of $\beta(\delta)$ is calculated for the t-J model - see Fig. 19.

Since the d-channel in scattering is not detrimental for d-wave pairing the FSP theory predicts that $T_c(\rho_i)$ vanishes at much larger $\rho_{i,c}$, i.e. $\rho_{i,c}^{(FSP)} >> \rho_{i,c}^{(s)}$, what is in the good agreement with experiments - as shown in Fig. 20.

6.2. Transport properties and superconductivity

The EPI was studied in the past in the *extreme limit* of the forward scattering peak in the Einstein model with the phonon frequency Ω [119], where in leading order the spectral function is singular, i.e. $\alpha^2 F(\mathbf{k},\mathbf{k}',\omega) \sim \delta(\mathbf{k}-\mathbf{k}')\delta(\omega - \Omega)$. Numerical calculations of the Eliashberg equations in the normal state [119] give very interesting behavior of the *density of states* $N(\omega)$, where a strong renormalization of $N(\omega)$ is present, but which is absent in the standard theory of the isotropic EPI. *First*, $N(\omega = 0) > N_{bare}(\omega = 0)$, where $N_{bare}(\omega = 0)$ is the density of states in the absence of the EPI

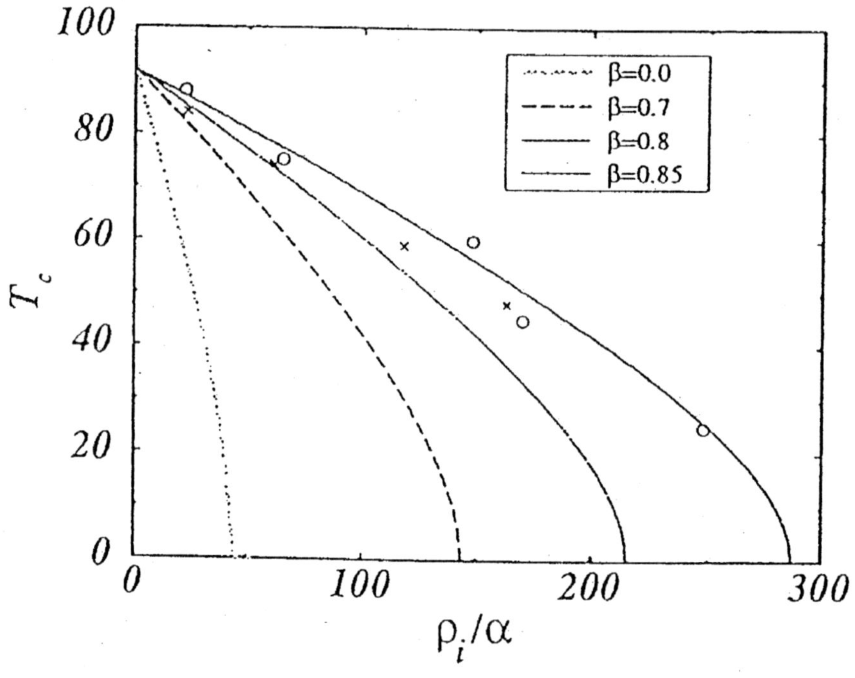

FIGURE 20. The critical temperature $T_c[K]$ of d-wave superconductor as a function of the experimental parameter $\rho_i/\alpha_c[K]$, ρ_i is the residual resistivity and α is defined in the text. The case $\beta = 0$ corresponds to the prediction of the standard d-wave theory with isotropic scattering [140]. The experimental data [141] are given by crosses - $YBa_2(Cu_{1-x}Zn_x)_3O_{7-\delta}$, and circles - $Y_{1-y}Pr_yBa_2Cu_3O_{7-\delta}$ - from [142].

- see Fig. 21.

There is a "pseudogap"-like feature in the region $(\Omega/5) < \omega \leq \Omega$ where $N(\omega) < N_{bare}(\omega)$. The "pseudogap" feature disappears at T comparable with the phonon energy Ω. Note, that the usual isotropic EPI does not renormalize the density of states in the normal state, i.e. $N(\omega) = N_{bare}(\omega)$. As a consequence of the pseudogap behavior of $N(\omega)$ the transport properties are very peculiar. For instance, the resistivity $\rho(T)$ is linear in T starting at very low temperatures, i.e. $\rho(T) \sim T$ for $(\Omega/30) \leq T$ and extends up to several Ω - as it is seen in Fig. 22. The dynamical conductivity $\sigma_1(\omega)$ shows the (extended) Drude-like behavior with the Drude width $\Gamma_{tr} \sim T$, for $\omega < T$ - see Fig. 22. The above numbered properties are in a qualitative agreement with experimental results in HTSC oxides, as discussed in Section 2.

In this extreme forward scattering peak limit one can calculate T_c. In leading order w.r.t. $(\Omega/T_c << 1)$ one has

$$T_{c0} \approx N(0)V_{EP} = \lambda N(0)/4, \tag{97}$$

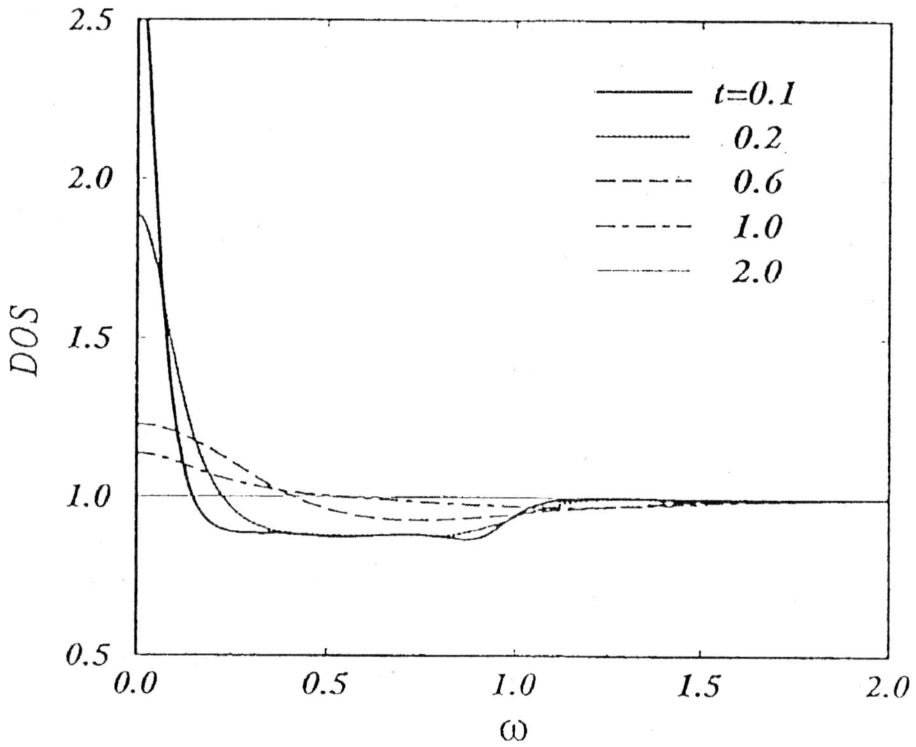

FIGURE 21. The density of states $N(\omega)$ in the FSP model for the EPI. with the dimensionless coupling $l(=V_{EP}/\pi\Omega)=0.1$ for various $t(=\pi T/\Omega)$. From [119].

where $\lambda = N(0)V_{EP}$. In that case the maximal superconducting gap is given by $\Delta_0 = 2T_c$ which is reached on the Fermi surface, while away from it the gap decreases, i.e.

$$\Delta_k = \Delta_0\sqrt{1-(\xi_k/\Delta_0)^2}. \tag{98}$$

The expression for T_c tells us that it can be large even for $\lambda < 0.1$, since in HTSC oxides the bare density of states is $N_{bare}(0) \sim 1 states/eV$. It is apparent that in this order there is no isotope effect, i.e. $\alpha = 0$. We stress that such an extreme limit is never realized in nature, but for the self-energy it is a good starting point, since the effects of the finite width (k_c) of $\alpha^2 F(\mathbf{k},\mathbf{k}',\omega)$, whenever $k_c \ll k_F$, change mainly the quantitative picture - see [119]. In case when $k_c v_F \ll \Omega$ the reduction of T_c is given by

$$T_c = T_{c0}(1-\frac{7\zeta(3)k_c v_F}{4\pi^2 T_{c0}}). \tag{99}$$

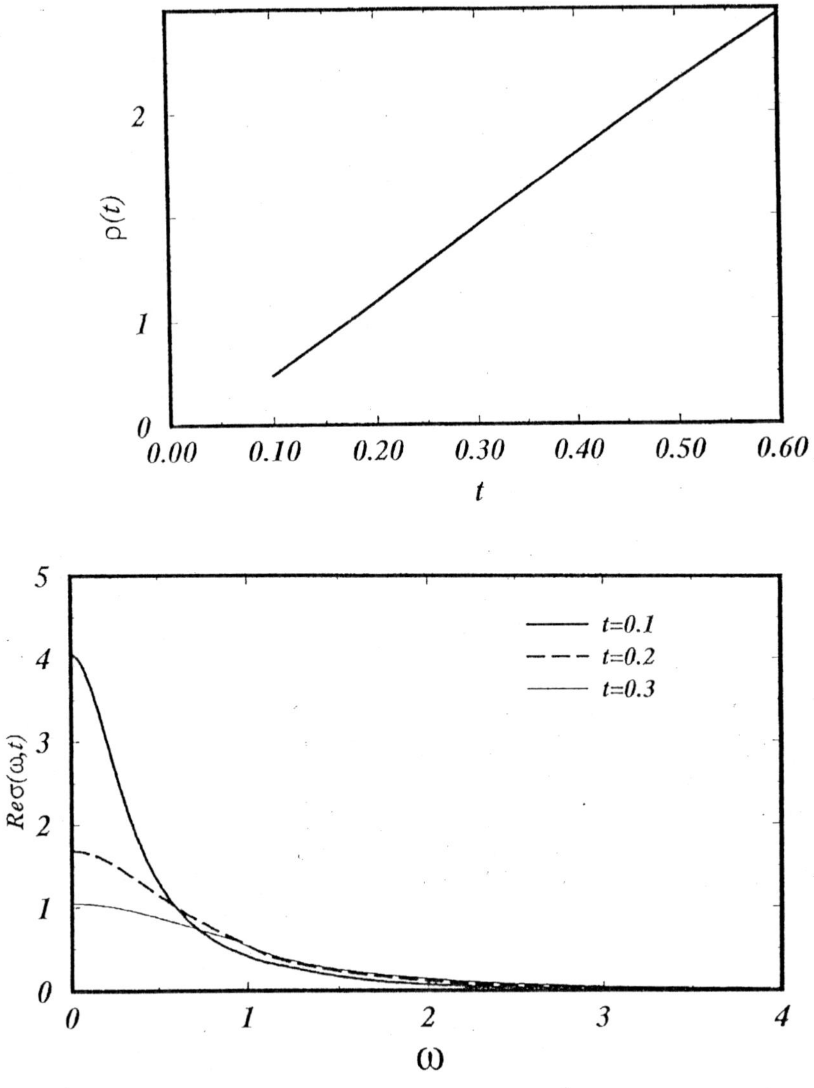

FIGURE 22. $\rho(T)$ (upper part) and $\sigma_1(\omega)$ in the FSP model for the EPI. with the dimensionless coupling $l(=V_{EP}/\pi\Omega)=0.1$ for various$t(=\pi T/\Omega)$. From [119].

Very interesting calculations in the more realistic FSP model with the finite width k_c, but $k_c \ll k_F$, were done in [134], where the FSP theory for the EPI and the SFI theory (based on spin-fluctuation mechanism of pairing) were compared. For instance, the FSP theory can explain the appreciable increase of the anisotropy ratio $R \equiv \Delta(\pi,0)/\Delta(\pi/2,\pi/2)$ when $T \to T_c$, while the SFI is unable. Furthermore, the FSP theory of the EPI can explain the pronounced orthorhombic ($a \neq b$) effect in $YBCO$ on the gap ratio Δ_a/Δ_b, penetration depth anisotropy λ_a^2/λ_b^2 and supercurrent ratio in the c-axis $Pb-YBCO$ junction. On the other hand, the SFI theory is ineffective, since it predicts at least one order of magnitude smaller effects - [134], [2].

6.3. Nonadiabatic corrections of T_c

HTSC oxides are characterized not only by strong correlations but also by relatively small Fermi energy E_F, which is not much larger than the characteristic (maximal) phonon frequency ω_{ph}^{max}, i.e. $E_F \simeq 0.1-0.3\ eV$, $\omega_{ph}^{max} \simeq 80\ meV$. The situation is even more pronounced in *fullerene compounds* A_3C_{60}, with $T_c = 20-35\ K$, where $E_F \simeq 0.2\ eV$ and $\omega_{ph}^{max} \simeq 0.16\ eV$. This fact implies a possible breakdown of the Migdal's theorem [139], [9], which asserts that the relevant vertex corrections due to the $E-P$ interaction are small if $(\omega_D/E_F) \ll 1$. In that respect a comparison of the intercalated graphite KC_8 and the fullerene A_3C_{60} compounds, given in [120], is very instructive, because both compounds have a number of similar properties. However, the main difference in these systems lies in the ratio ω_D/E_F, since $(\omega_D/E_F) \ll 1$ in KC_8, while it is rather large $(\omega_D/E_F) \sim 1$ in A_3C_{60}. Due to the appreciable magnitude of ω_D/E_F in the fullerene compounds and in HTSC oxides it is necessary to correct the Migdal-Eliashberg theory by vertex corrections due to the EPI. It is well-known that these vertex corrections lower T_c in systems with isotropic EPI. However, the vertex corrections in systems with the forward scattering peak and with the cut-off $q_c << k_F$ the increases of T_c appreciable. The calculations by the Pietronero group [120] gave two important results: **(1)** there is a drastic increase of T_c by lowering $Q_c = q_c/2k_F$, for instance $T_c(Q_c=0.1) \approx 4T_c(Q_c=1)$; **(2)** Even small values of $\lambda < 1$ can give large T_c. The latter results open a new possibility in reaching high T_c in systems with appreciable ratio ω_D/E_F and with the forward scattering peak. The difference between the Migdal-Eliashberg and non-Migdal theories can be explained qualitatively in the framework of an approximative McMillan formula for T_c (for not too large λ) which reads

$$T_c \approx \langle\omega\rangle e^{-1/[\tilde{\lambda}-\mu^*]}. \tag{100}$$

The *Migdal-Eliashberg theory* predicts

$$\tilde{\lambda} \approx \frac{\lambda}{1+\lambda}, \tag{101}$$

while the non-Migdal theory [120] gives

$$\tilde{\lambda} \approx \lambda(1+\lambda). \tag{102}$$

For instance $T_c \sim 100\ K$ in HTSC oxides can be explained by the Migdal-Eliashberg theory for $\lambda \sim 2$, while in the non-Migdal theory much smaller coupling constant is needed, i.e. $\lambda \sim 0.5$ as it is seen in Fig. 23. The pioneering approach done in [120] deserves more attention in the future.

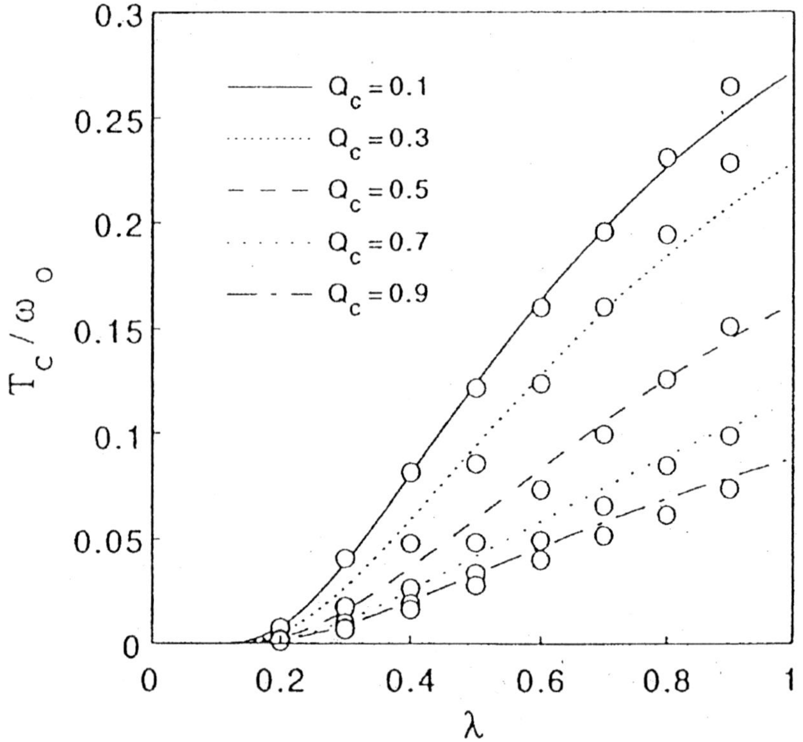

FIGURE 23. The approximative analytic (solid lines) and numerical (circles) solution of $T_c(\lambda)$ in the first nonadiabatic approximation for various cutoff Q_c and for $m = (\omega_D/E_F) = 0.2$. From [120].

6.4. Pseudogap behavior in the FSP model for the EPI

In this review we did not discuss a number of interesting topics such as the possible existence of stripes, the properties of the pseudogap state, etc.. There is a believe that the understanding of these properties might give some hints for pairing mechanism in HTSC oxides. Especially, the pseudogap (PG) problem is a very intriguing one and is not surprising at all, that a number of theoretical approaches were proposed for explaining the PG. We are not going to discuss it but only quote some of them. The *first* one is based on the assumption that the PG phase represents pre-formed pairs [143], and the true critical temperature T_c is smaller than the mean-field one T_c^{MF}. In the region $T_c<T<T_c^{MF}$ pre-formed pairs exist giving rise to the dip in the density of states $N(\omega)$. This approach is physically plausible having in mind that HTSC oxides are characterized by the short coherence length and quasi-two dimensionality. From the experimental side there are some supports. For instance, the specific heat measurements [144] point to the non-mean field character of the superconducting phase transition, particularly for the underdoped

systems. As we have already mentioned in the Introduction the ARPES measurements show that the PG has a d-wave like form

$$\Delta_{pg}(\mathbf{k}) \approx \Delta_{pg,0}(\cos k_x - \cos k_y) \tag{103}$$

(like the superconducting gap) and $\Delta_{pg,0}$ increases by lowering doping. The *second* approach assumes that the PG is due to a competing order, but usually without the long-range order, such as due to "spin-density wave" alias for strong antiferromagnetic fluctuations [145], [146]. There are other approaches which are based on the RVB and orbital current model, d-wave order, etc., but we shall not discussed it here.

However, the FSP theory, which predicts the long-range force due to the renormalization of the EPI by strong correlations, opens an additional possibility for the PG. As we discussed in Section 6.2, due to the forward scattering peak the critical temperature has a non-BCS dependence, i.e. $T_c^{MF} = V_{EP}/4$. However, this is the mean-field value, which is inevitably reduced by phase and internal Cooper pair fluctuations present in systems with long-range attractive forces, i.e. with the forward scattering peak.

The interesting problem of fluctuations in systems with long-range attractive forces was recently studied in [147]. It was shown there, that such a long-ranged superconductor exhibits a class of fluctuations in which the internal structure of the Cooper pair is soft. This leads to a "pseudogap" behavior in which the actual transition temperature T_c is greatly depressed from its mean-field value T_c^{MF}. We stress that these fluctuations are not the standard phase fluctuations in superconductors. Since the problem is very interesting and deserve much more attention in the future we discuss it here briefly. In the following the weak coupling limit is assumed, where the pairing Hamiltonian has the form

$$H = \sum_{\sigma} \int d\mathbf{x} \psi_{\sigma}^{\dagger}(\mathbf{x}) \xi(\hat{\mathbf{p}}) \psi_{\sigma}(\mathbf{x})$$

$$- \int d\mathbf{x} d\mathbf{x}' V(\mathbf{x}-\mathbf{x}') \psi_{\uparrow}^{\dagger}(\mathbf{x}) \psi_{\downarrow}^{\dagger}(\mathbf{x}') \psi_{\downarrow}(\mathbf{x}') \psi_{\uparrow}(\mathbf{x}). \tag{104}$$

In the MFA the order parameter $\Delta(\mathbf{x},\mathbf{x}')$ is given by

$$\Delta(\mathbf{x},\mathbf{x}') = V(\mathbf{x}-\mathbf{x}') \langle \psi_{\downarrow}(\mathbf{x}') \psi_{\uparrow}(\mathbf{x}) \rangle. \tag{105}$$

$\Delta(\mathbf{x},\mathbf{x}')$ depends in fact on the internal coordinate $\mathbf{r} = \mathbf{x} - \mathbf{x}'$ and the center of mass $\mathbf{R} = (\mathbf{x}+\mathbf{x}')/2$, i.e. $\Delta(\mathbf{x},\mathbf{x}') = \Delta(\mathbf{r},\mathbf{R})$. In usual superconductors with the short-range pairing potential $V_{sr}(\mathbf{x}-\mathbf{x}') \approx V_0 \delta(\mathbf{x}-\mathbf{x}')$ one has $\Delta(\mathbf{r},\mathbf{R}) = \Delta(\mathbf{R})$ and therefore there are practically the spatial ($\mathbf{R}$) fluctuations of the order parameter, only. In the case of long-range pairing potential there are additional fluctuations of the internal ($\mathbf{r}$) degrees of freedom. In the following we sketch the analysis given in [147].

When the range of the pairing potential is large, i.e. $r_c > \xi$ (the superconducting coherence length), fluctuations of the internal Cooper wave-function are important since they give rise to a tremendous reduction of the mean-field quantities. In order to make the physics of internal wave-function fluctuations we study much simpler Hamiltonian the so called reduced BCS Hamiltonian,

$$H = \sum_{\mathbf{k}\sigma} \xi_{\mathbf{k}} c_{\mathbf{k}\sigma}^{\dagger} c_{\mathbf{k}\sigma} - \sum_{\mathbf{k},\mathbf{k}'} V_{\mathbf{k}-\mathbf{k}'} c_{\mathbf{k}\uparrow}^{\dagger} c_{-\mathbf{k}\downarrow}^{\dagger} c_{-\mathbf{k}'\downarrow} c_{\mathbf{k}'\downarrow}. \tag{106}$$

Since we shall study excitations around the ground state we assume that there are no unpaired electrons which allows us to study the problem in the pseudo-spin Hamiltonian [148]

$$H = \sum_{\mathbf{k}\sigma} 2\xi_{\mathbf{k}} S^z_{\mathbf{k}\sigma} - \frac{1}{2}\sum_{\mathbf{k},\mathbf{k}'} V_{\mathbf{k}-\mathbf{k}'}(S^+_{\mathbf{k}} S^-_{\mathbf{k}'} + S^+_{\mathbf{k}'} S^-_{\mathbf{k}})$$

$$= \sum_{\mathbf{k}\sigma} 2\xi_{\mathbf{k}} S^z_{\mathbf{k}\sigma} - \sum_{\mathbf{k},\mathbf{k}'} V_{\mathbf{k}-\mathbf{k}'}(S^x_{\mathbf{k}} S^x_{\mathbf{k}'} + S^y_{\mathbf{k}'} S^y_{\mathbf{k}}), \tag{107}$$

where the pseudo-spin 1/2 operators $S^z_{\mathbf{k}\sigma}$, $S^+_{\mathbf{k}\sigma} = (S^-_{\mathbf{k}\sigma})^\dagger$ are given by

$$S^z_{\mathbf{k}\sigma} = \frac{1}{2}(c^\dagger_{\mathbf{k}\uparrow} c_{\mathbf{k}\uparrow} - c^\dagger_{-\mathbf{k}\downarrow} c_{-\mathbf{k}\downarrow} - 1),$$

$$S^+_{\mathbf{k}\sigma} = c^\dagger_{\mathbf{k}\uparrow} c^\dagger_{-\mathbf{k}\downarrow}. \tag{108}$$

We see that Eq.(107) belongs to the class of the Heisenberg ferromagnetic ($V_{\mathbf{k}-\mathbf{k}'} > 0$) Hamiltonian formulated on the lattice in the Brillouin zone. The mean-field approximation (MFA) for this Hamiltonian is given by

$$H_{MFA} = -\sum_k \mathbf{h}_{\mathbf{k}} \mathbf{S}_{\mathbf{k}} \tag{109}$$

with the mean-field $\mathbf{h}_{\mathbf{k}}$ given by

$$\mathbf{h}_{\mathbf{k}} = -2\xi_{\mathbf{k}}\mathbf{z} + \sum_{\mathbf{k}'} V_{\mathbf{k}-\mathbf{k}'} \langle S^x_{\mathbf{k}'}\mathbf{x} + S^y_{\mathbf{k}'}\mathbf{y}\rangle, \tag{110}$$

where $\mathbf{x},\mathbf{y}$ and $\mathbf{z}$ are unit vectors. Since x- and y-axis are equivalent one can searches $\mathbf{h}_{\mathbf{k}}$ in the form $\mathbf{h}_{\mathbf{k}} = -2\xi_{\mathbf{k}}\mathbf{z} + 2\Delta_{\mathbf{k}}\mathbf{x}$, where the order parameter $\Delta_{\mathbf{k}}$ is the solution of the equation

$$\Delta_{\mathbf{k}} = \sum_{\mathbf{k}'} V_{\mathbf{k}-\mathbf{k}'} \langle S^x_{\mathbf{k}'}\rangle = \sum_{\mathbf{k}'} V_{\mathbf{k}-\mathbf{k}'} \frac{V_{\mathbf{k}-\mathbf{k}'}\Delta_{\mathbf{k}'}}{2E_{\mathbf{k}}} \tanh\frac{\beta E_{\mathbf{k}}}{2}, \tag{111}$$

with $E_{\mathbf{k}} = \sqrt{\xi^2_{\mathbf{k}} + \Delta^2_{\mathbf{k}}}$.

In the case of *short-range BCS-like forces* $V_{BCS}(\mathbf{x}-\mathbf{x}') \approx V_0\delta(\mathbf{x}-\mathbf{x}')$ one has $V_{\mathbf{k}-\mathbf{k}'} = V_0$ for all momenta. This "long-range force" in the momentum space it is the "long-range force" means that the MFA is good approximation with the standard BCS solution of Eq.(111).

For the *long-range attractive forces* the function $V_{\mathbf{k}-\mathbf{k}'}$ is peaked at $\mid \mathbf{k}-\mathbf{k}' \mid = 0$, for instance in the extreme forward scattering peak case (see Section 6.2) one has $V_{\mathbf{k}-\mathbf{k}'} = V_0\delta(\mathbf{k}-\mathbf{k}')$. In the following we analyze s-wave pairing only where the solution of Eq.(111) gives $T_c^{MF} = V_0/4$ and $\Delta_0 = 2T_c^{MF}$. (Note, that in the BCS case one has $\Delta_0 = 1.76T_c^{MF}$.). The coherence length is defined by $\xi = v_F/\pi\Delta_0$. The important fact is that in the case of long-ranged superconductors the Heisenberg like Hamiltonian in the momentum space is short-ranged giving rise to low-laying spin-wave spectrum. The latter spectrum are in fact the low-energy bound states (excitons) which loosely correspond to the low-energy collective modes (in the true many-body theory based on Eq.(104)). This problem is studied in [147] for the long-range (but finite) potential $V(r) = V_0 \exp\{-r^2/2r_c^2\}$ (its Fourier transform is $V_k = (2\pi r_c^2)V_0 \exp\{-k^2 r_c^2/2\}$) where it was found a large number $N_{\rm cm} \sim \pi k_F r_c/6\xi$ (for $r_c \gg \xi$) of the excitonic like collective modes ω_{mn}^{exc} at zero momentum. These excitonic modes lie between the ground state and the two particle continuum for $\omega > 2\Delta_0$. Note, that since we assume that $\Delta_0 \ll E_F$ the system is far from the Bose-Einstein condensation limit.

The above analysis is useful for physical understanding, but the fully many-body fluctuation problem, which is based on the Hamiltonian in Eq.(104), is studied in [147] where the Ginzburg-Landau (G-L) equation is derived for the long-ranged superconductor. Due to the fluctuations of the internal wave-function the G-L free-energy functional $F\{\Delta(\mathbf{R},\mathbf{k})\}$ for the order parameter $\Delta(\mathbf{R},\mathbf{k}) = \int d\mathbf{r}\Delta(\mathbf{R}-\mathbf{r}/2,\mathbf{R}+\mathbf{r}/2)\exp\{-i\mathbf{kr}\}$ has much more complicated form

$$F\{\Delta(\mathbf{R},\mathbf{k})\} = \sum_k \int d\mathbf{R}\{A_{\mathbf{k}} \mid \Delta(\mathbf{R},\mathbf{k}) \mid^2 + B_{\mathbf{k}} \mid \Delta(\mathbf{R},\mathbf{k}) \mid^2$$

$$+\frac{1}{2M} \mid \partial_{\mathbf{k}}\Delta(\mathbf{R},\mathbf{k}) \mid^2 + \frac{1}{2m_{\mathbf{k}}} \mid \partial_{\mathbf{R}}\Delta(\mathbf{R},\mathbf{k}) \mid^2\}, \qquad (112)$$

where $M = r_c^2 V_0$ and

$$A_{\mathbf{k}} = \frac{1}{V_0} - \frac{\tanh(\beta\xi_{\mathbf{k}}/2)}{2\xi_{\mathbf{k}}}$$

$$\frac{1}{2m_{\mathbf{k}}} = \frac{\beta^2 v_F^2 \sinh(\beta\xi_{\mathbf{k}}/2)}{32\xi_{\mathbf{k}}\cosh^3(\beta\xi_{\mathbf{k}}/2)}$$

$$B_{\mathbf{k}} = \frac{\tanh(\beta\xi_{\mathbf{k}}/2)}{8\xi_{\mathbf{k}}^3} - \frac{\beta}{16\xi_{\mathbf{k}}^2\cosh^3(\beta\xi_{\mathbf{k}}/2)}. \qquad (113)$$

The term due to the partial derivative $\partial_{\mathbf{k}}$ is a direct consequence of the long-ranged pairing potential, and it describes of fluctuations of the internal Cooper wave-function. The effect of these fluctuations, described by the free-energy functional in Eq.(112), is studied in the Hartree-Fock approximation in the limit $r_c \gg \xi$, where it is found the large reduction of the mean-field critical temperature

$$T_c \sim \frac{T_c^{MF}}{(r_c/\xi)}. \qquad (114)$$

The latter result means that T_c in the long-ranged superconductors is *controlled by thermal fluctuations of collective modes* which is in contrast with the short-range (BCS-like) superconductivity. In the temperature interval $T_c < T < T_c^{MF}$ the system is in the pseudogap regime where the electrons are paired but there is no long-range phase coherence. The latter sets in only at $T < T_c$. We shall not further discuss this interesting approach but only stress that it can be generalized by including the repulsive interaction due to spin fluctuations, what shall be discussed elsewhere.

In conclusion, the forward scattering peak in the EPI gives rise to the long-ranged superconductivity in which the soft excitonic modes of the internal Cooper wave function reduce T_c strongly. In the region $T_c < T < T_c^{MF}$ the pseudogap (PG) phase is realized. In this approach the PG has the same symmetry as the superconducting gap.

7. ELECTRON-PHONON INTERACTION VS SPIN-FLUCTUATIONS

7.1. Interaction via spin fluctuations (SFI) and pairing

At present one of the possible candidates in explaining experimental results in HTSC oxides appears to be the theory based on the spin fluctuation pairing mechanism - the SFI theory. The latter is usually described by the single band Hubbard model, or on the phenomenological level by the postulated form of the self-energy (written below)[29], [136], [135], [83], [84], [137]. In the approach of Pines-school to the SFI the effective potential $V_{eff}(\mathbf{k},\omega)$ (see Eq.(5) in Sections 2.) depends on the imaginary part of the spin susceptibility $\mathrm{Im}\chi(\mathbf{k}-\mathbf{k}',\omega)$ (ω real). According to this school, the shape and the magnitude of $\mathrm{Im}\chi(\mathbf{q},\omega)$, which is peaked at $\mathbf{Q}=(\pi,\pi)$, plays an important role in obtaining T_c in this mechanism. There are two phenomenological approaches, which can be theoretically justified in a very weak coupling limit $g_{sf}\ll 1$ only, where $\mathrm{Im}\chi(\mathbf{q},\omega)$ is inferred from different experiments:

(**1**) From *NMR* experiments at very low ω - the *MMP model* [29], [136], [135], where $Im\chi_{MMP}$ is modelled by

$$Im\chi_{MMP}(\mathbf{q},\omega+i0^{+})=\frac{\omega}{\omega_{sf}}\frac{\chi_Q}{[1+\xi_M^2\mid\mathbf{q}-\mathbf{Q}\mid^2+(\omega/\omega_{sf})^2]^2}\Theta(\omega_c^{MMP}-\mid\omega\mid),\qquad(115)$$

with the frequency cutoff $\omega_c^{MMP}=400\ meV$. They fit the *NMR* experiments by assuming very large value for $\chi_Q\approx(30-40)\chi_0\sim 100\ eV^{-1}$. From Fig. 24 it is seen that the imaginary susceptibility is peaked at low frequency $\omega_{peak}\approx 5-10\ meV$.

(**2**) From the neutron scattering experiments [83], [84]) - the *RULN model*, where $Im\chi_{RULN}$ is modelled by

$$Im\chi_{RULN}(\mathbf{q},\omega+i0^{+})=C[\frac{1}{1+J_0[\cos q_x+\cos q_y]}]^2\times$$

$$\times\frac{3(T+5)\omega}{1.5\omega^2-60\mid\omega\mid+900+3(T+5)^2}\Theta(\omega_c^{RULN}-\mid\omega\mid),\qquad(116)$$

where $\omega_c^{RULN}=100\ meV$, $J_0=0.3$, $C=0.19\ eV^{-1}$ with T and ω measured in meV. From Fig. 25 it is seen that $Im\chi_{RULN}$ is peaked around 30 meV, which is much larger than in the MMP model.

By knowing $Im\chi$ one can calculate the effective pairing potential $V_{eff}(\mathbf{k},\omega)$ from Eq.(5) and the spectral function for the d-wave pairing $\alpha_d^2F(\omega)$

$$\alpha_d^2F(\omega)=-\frac{\langle\langle Y_d(\mathbf{k})Y_d(\mathbf{k}')V_{SF}(\mathbf{k}-\mathbf{k}',\omega+i0^{+})\rangle\rangle}{\langle Y_d^2(\mathbf{k})\rangle}.\qquad(117)$$

Here, $Y_d(\mathbf{k})=\cos k_x-\cos k_y$ is the $d-wave$ pairing function ($\Delta(\mathbf{k},\omega)\approx\Delta(\omega)Y_d(\mathbf{k})$). The bracket means an average over the Fermi surface. The spectral function $\alpha_d^2F(\omega)$ for two models is shown in Fig. 26, where it is seen that $\alpha_d^2F(\omega)^{RULN}$ is much narrower

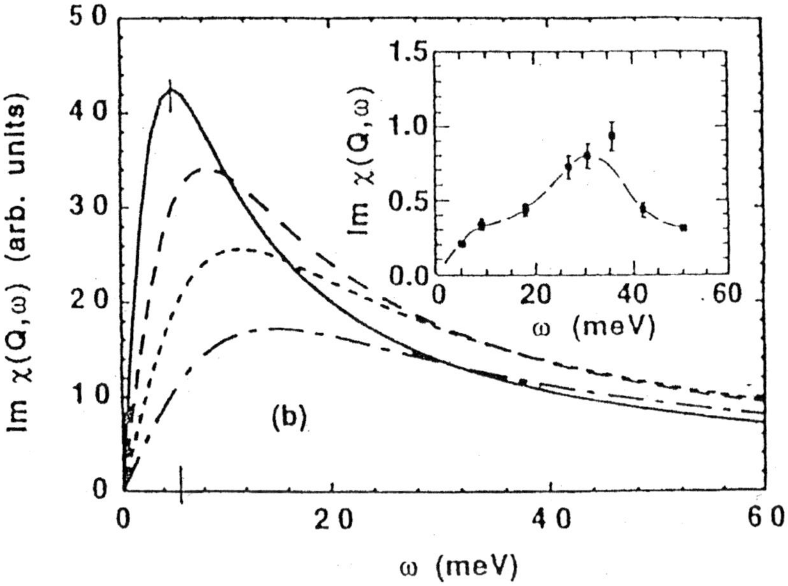

FIGURE 24. Spectral function $\text{Im}\chi(,\omega)$ for the *MMP model* of spin-mediated interactions $\text{Im}\chi(,\omega)$ in $YBa_2Cu_3O_{7-\delta}$. The spectral function is calculated at $=(\pi,\pi)$ and $T=0$ K (solid line), 100 K (long-dashed line), 200 K (short dashed line), and 300 K (dot-dashed line). Inset: experimental data of $YBa_2Cu_3O_{6.6}$ at $T=100$ K - the line is to guide the eye. From [83], [84].

function than $\alpha_d^2F(\omega)^{MMP}$. The latter is peaked almost at the same ω as $\alpha_d^2F(\omega)^{RULN}$, while $\alpha_d^2F(\omega)^{MMP}$ is much broader than $\alpha_d^2F(\omega)^{RULN}$.

Due to different shapes of the susceptibility and of $\alpha_d^2F(\omega)$ in these two approaches the calculated (from Eliashberg equations) critical temperatures are also very different. Since the MMP spectral function is much broader than the RULN one it turns out that $T_c^{(MMP)}$ can reach 100 K for rather large value of $g_{SF}\sim 0.64$ eV, while $T_c^{(RULN)}$ saturates already at 50 K even for $g_{SF}\gg 1$. From the physical pint of view the RULN model is more plausible than the MMP one, since the former is based on the neutron scattering measurements which comprise much larger frequencies than the NMR measurements. Note, that a valid model for HTSC oxides must be able to explain the high values of T_c (which needs $\lambda_{sf}(=2\int(\alpha_d^2F(\omega)/\omega)d\omega)\sim 2)$ and the resistivity $\rho(T)$ (and its slope ρ' with small $\lambda_{tr}\sim 0.6$). It turns out that $T_c^{(MMP)}$ can fit $T_c\approx 100$ K on the expense of large coupling $g_{sf}^{(MMP)}\sim 0.64$ eV and $\lambda_{sf}^{(MMP)}\sim 2.5$. However, the value $g_{sf}^{(MMP)}\sim 0.64$ eV gives much larger value for $\rho(T)$ and ρ' than the experiments do. On the other hand if one fits $\rho(T)$ and ρ' with the MMP model one gets very small $T_c<7$ K, thus making the MMP model ineffective in HTSC oxides.

In the physically more plausible *RULN model* $T_c^{(RULN)}$ saturates at 50 K even for $g_{sf}^{(RULN)}\gg 1$ eV. If one chooses an appropriate value for $g_{sf}^{(neut)}$ to fit $\rho(T)$ and ρ' one

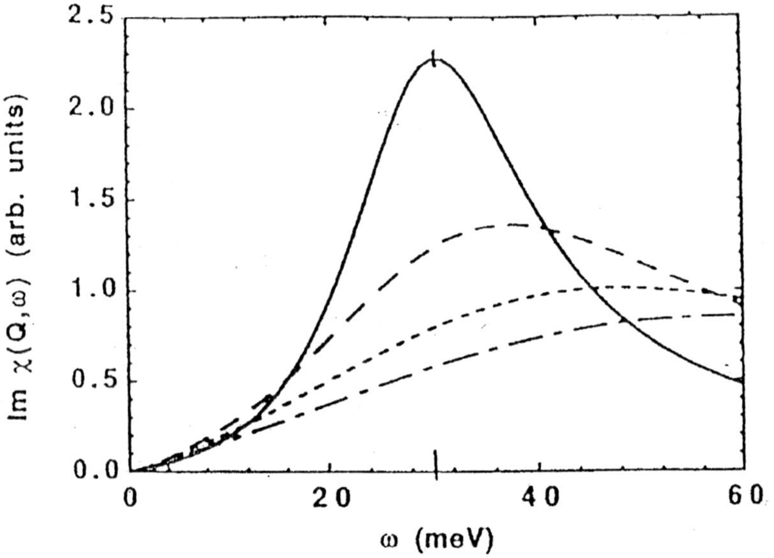

FIGURE 25. Spectral function $\mathrm{Im}\chi(,\omega)$ for the *RULN model* of spin-mediated interactions $\chi(,\omega)$ in $YBa_2Cu_3O_{7-\delta}$. The spectral function is calculated at $=(\pi,\pi)$ and $T=0\ K$ (solid line), 100 K (long-dashed line), 200 K (short dashed line), and 300 K (dot-dashed line). Inset: experimental data of $YBa_2Cu_3O_{6.6}$ at $T=100\ K$ - the line is to guide the eye. From [83], [84].

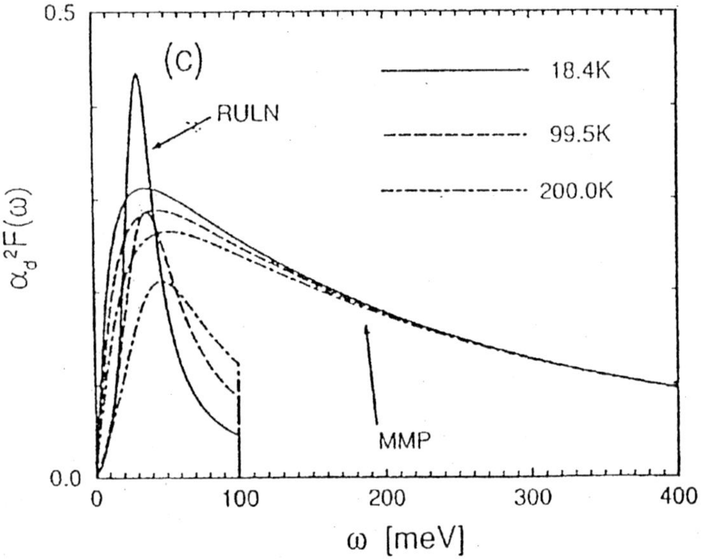

FIGURE 26. The spectral function $\alpha_d^2F(\omega)$ in the $d-wave$ pairing channel for the *MMP* and *RULN* model at different temperature - from [137].

gets $T_c^{(RULN)} \approx 7\ K$. This analysis gives a convincing evidence that the existing SFI theories are ineffective in HTSC oxides.

We stress again, that the large effective coupling constant, assumed in the SFI theories, $g_{sf} \sim 0.64\ eV$, is difficult to justify theoretically (if at all). By analyzing theoretically the possible strength of the coupling constant g_{sf}, in both the weak ($N(0)U \ll 1$) and strong ($N(0)U \gg 1$) coupling limit, one obtains that $g_{sf} < 0.2\ eV$ and $\lambda_{sf} < 0.2$ (note $\lambda_{sf} \sim g_{sf}^2$), which means that $\lambda_{sf} \ll \lambda_{sf}^{(MMP)}$ and $T_c^{(sf)} \ll T_c^{(MMP)}$ [2]. This analysis is supported by the recent theoretical results where $g_{sf} < 0.2\ eV$ is extracted from the calculation : **1.** of the width of the magnetic resonance peak at 41 meV [68]; **2.** of the small magnetic moment ($\mu < 0.1\ \mu_B$) in the antiferromagnetic order, which coexists with superconductivity in $La_{2-x}Sr_xCuO_4$ [69].

7.2. Are the EPI and SFI compatible in d-wave pairing?

The phenomenological SFI theories (MMP, RULN, FLEX approximation [149]] became popular because they can produce d-wave pairing, due to the repulsive character of spin-fluctuations (in the momentum space.) which are peaked in the backward ($q = Q$) scattering. However, as we have argued in previous Sections, a number of experiments point to a large EPI with $\lambda > 1$. On the other hand if one assumes that the EPI is momentum independent (isotropic), like in the standard Migdal-Eliashberg theory, then it is strongly pair-breaking for d-wave pairing. So, if one assumes (for the moment) that superconductivity in HTSC oxides is due to the SFI with $\lambda_{sf} \approx 2$, then in that case $T_c^{(sf)}$ would be drastically reduced (to almost zero) by the isotropic EPI even for moderate $\lambda_{EP} \sim 1$. The latter was shown in [138] where the Eliashberg equations are solved for the SFI treated in the FLEX approximation [149] and the EPI in the Einstein model with various momentum dependent $V_{EP}(\vec{k}, \omega)$. This result means, that if the SFI would be the basic pairing interaction in the presence of the isotropic and momentum independent EPI, then in order to reach $T_c \sim 100\ K$ the bare critical temperature should be $T_c^{(sf)} \sim (600 - 700)\ K$, which needs unrealistically large λ_{sf}. This is not only highly improbable, but would give enormous large resistivity and its slope, in contrast to experiments. The similar results were obtained in [11].

The calculations in [138] show that the SFI interaction is dominant in (d-wave) pairing if some strong constraints are realized, such as: **(1)** very large SFI coupling constant $\lambda_{sf} \approx 2$; **(2)** a strong forward scattering peak in the EPI with small EPI coupling $\lambda \ll 1$. Both these conditions are incompatible with experiments and theoretical analysis - see also Section 2.1. The way out from this controversy is that the EPI with the forward scattering peak is inevitably dominant interaction in the quasiparticle scattering and pairing in HTSC oxides. As we already discussed in Section 5. the forward scattering peak in the EPI gives rise to the large coupling constant in the d-wave channel, which is of the order of the one in the s-wave channel in the range of doping around the optimal one, i.e. $\lambda_{EP,d} \approx \lambda_{EP,s}$. This means that the residual Coulomb repulsion (by including also the SFI with the backward scattering peak (BSP)) with $\lambda_c < \lambda_{EP,d}$ *triggers* d-wave pairing.

8. IS THERE HIGH-TEMPERATURE SUPERCONDUCTIVITY IN THE HUBBARD AND t-J MODEL?

8.1. Hubbard model

There are a number of papers dealing with numerical calculations, such as Monte Carlo, exact Lanczos diagonalization, in the 2-dimensional (2D) single-band and three-band Hubbard model. One can say that the single-band Hubbard model does very well in describing the magnetic properties of HTSC oxides. Concerning the existence of superconductivity the situation is not definitely resolved. So far the calculations are done on finite clusters and rather high temperatures $T > 0.1t$ [150], [7] which show no tendency to superconductivity. It is worth of mentioning that most of these calculations deal with the pairing susceptibility - see Eq.(126) below, defined in terms of the bare electron operators $c_{mathbfk\sigma}$ in Eq.(128). Since superconducting pairing is realized on quasiparticles with the weight $z < 1$ there is a last hope that the accuracy of the present numerical calculations is not sufficient to pick up the suppressed pairing susceptibility. In that respect a very important approach to the problem of superconductivity (in any microscopic model), which is formulated without using any order parameter, was given by the Scalapino group [151]. The method of calculations is based on the two most important hallmarks of superconductivity: **(i)** *ideal diamagnetism* (the Meissner effect) and **(ii)** *ideal conductivity*. In that respect they study the superfluid density D_s (proportional to λ^{-2}, λ is the penetration depth) and the Drude weight D in the single-band n.n. (nearest neighbors) Hubbard model. The dynamical conductivity along the x-axis is given by

$$\sigma_{xx}(\omega) = -\frac{e^2}{i}\frac{\langle -T_x\rangle - \Lambda_{xx}(\mathbf{q}=0,\omega)}{\omega + i\delta}. \tag{118}$$

Here, $\langle -T_x\rangle = \langle -T\rangle/2$ where T is the kinetic energy in the n.n. tight-binding model - see Eq.(18), where the current-current response function $\Lambda_{xx}(\mathbf{q},\omega)$ is obtained from

$$\Lambda_{xx}(\mathbf{q},i\omega_m) = \frac{1}{N}\int_0^\beta d\tau e^{i\omega_m\tau}\langle j_x^p(\mathbf{q},\tau)j_x^p(-\mathbf{q},0)\rangle, \tag{119}$$

with $\omega_m = 2\pi mT$, by the standard analytic continuation $i\omega_m \to \omega + i\delta$ and

$$j_x^p(\mathbf{q},\tau) = it\sum_{l,\sigma} e^{-i\mathbf{q}\mathbf{l}}(c^\dagger_{l+x\sigma}c_{l\sigma} - c^\dagger_{l\sigma}c_{l+x\sigma}). \tag{120}$$

In the pure Hubbard model $\sigma_{xx}(\omega)$ contains the delta function contribution

$$\sigma_{xx}(\omega) = D\delta(\omega) + \sigma_{reg}(\omega), \tag{121}$$

where the Drude weight $D(\equiv (n/m)^*$, which measures the ratio of the density of the mobile charge carriers to their mass, is defined by

$$\frac{D}{\pi e^2} = \langle -T_x\rangle - \Lambda_{xx}(\mathbf{q}=0,\omega\to 0). \tag{122}$$

The Meissner effect is the current response to a static and transverse gauge potential $\mathbf{q}\cdot\mathbf{A}(\mathbf{q},\omega=0)=0$. In the small $\mathbf{q}$ limit one has

$$\langle j_\alpha(\mathbf{q}\to\mathbf{0})\rangle=-\frac{D_s}{\pi}(\delta_{\alpha\beta}-q_\alpha q_\beta/q^2)A_\beta \tag{123}$$

where $D_s(\equiv(n_s/m)^*$

$$D_s=\langle -T_x\rangle-\Lambda_{xx}(q_x=0,q_y\to 0,\omega=0). \tag{124}$$

Based on the above definitions of D and D_s we can study various phases of the system: (**1**) D=D_s=0 - an *isolator*; (**2**) D≠0 and D_s=0 - a *nonsuperconducting metal*; (**3**) $D_s \neq 0$, D≠0 - a *superconducting metal.*

The numerical Monte Carlo calculation in the repulsive Hubbard model (U=4t>0) [151] on an 8×8 lattice show that D_s=0 and D≠0 in a broad range of the filling 0.5<n<0.9 and for $T>0.1t$. This means that there is no tendency to high-temperature superconductivity in the single-band Hubbard model. This conclusion is supported by the projector-QMC calculations [152] for the quarter filling case n=0.5 and at T=0. That these results ($D_s=0$) are not a finite size effect confirm the calculations on the attractive Hubbard model (U=-4t<0), also on an 8×8 lattice, where the clear tendency to superconductivity is found, since $D_s \neq 0$, D≠0 already at T<0.2t.

The paper [151] is of great importance numerical studies of pairing in model systems, not only because it hints on the absence of superconductivity in the repulsive Hubbard model, but also because of the following two reasons: (**1**) It uses the general and unbiased criterion for superconductivity, which is independent on the type of the pairing amplitude; (**2**) It shows that the attractive interaction is more favorable for (high-temperature) superconductivity than the repulsion.

8.2. t-J model

The SFI phenomenological approaches root on their basic $t-J$ Hamiltonian Eq.(61). On can put a legitimate question - is there superconductivity in the t-J model? In the past there were various approaches confronting with this important problem. In spite of a number of controversial statements it seems that the results converge to the unique answer - *there is no superconductivity with appreciable T_c*. If superconductivity exists T_c is very low. As the strong support of this claim serve the recent calculations based on the high-temperature expansion in the t-J-V model [70],

$$\hat{H}=-\sum_{i,j,\sigma}t_{ij}\hat{X}_i^{\sigma 0}\hat{X}_j^{0\sigma}+\sum_{i,j=n.n.}[J(\mathbf{S}_i\cdot\mathbf{S}_j)+(V-\frac{J}{4})n_in_j]. \tag{125}$$

Here, the V-term mimic the screened Coulomb interaction which is always present in metals, where one expects that $V>J$. In [70] it was calculated the uniform susceptibility for the superconducting pairing

$$\chi_{SC}\equiv\frac{1}{N}\int_0^\beta d\tau\langle\langle\hat{T}_\tau e^{H\tau}O_{SC}e^{-H\tau}O_{SC}^\dagger\rangle\rangle \tag{126}$$

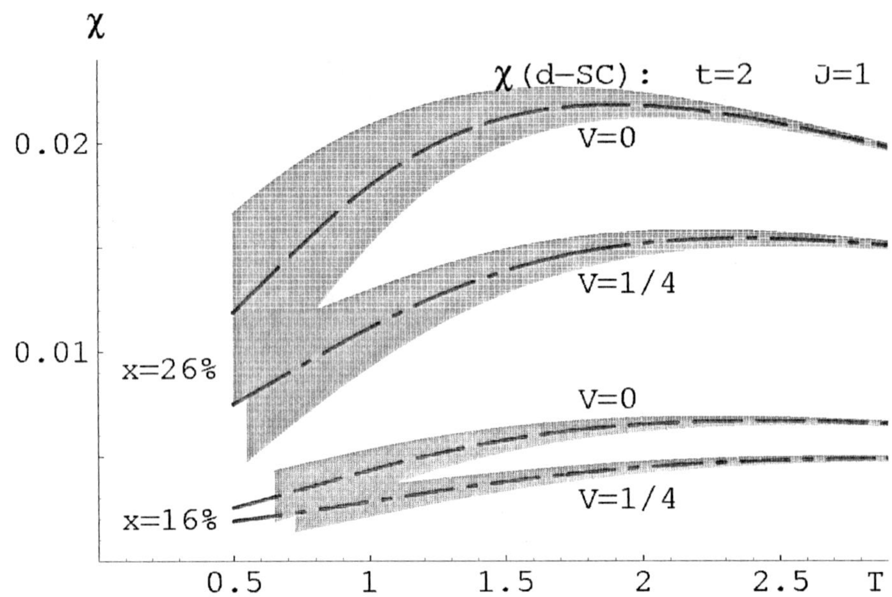

FIGURE 27. Superconducting (d-wave) susceptibility $\chi(d-Sc)(T)$ for $t = 2J$. Pairing correlations are already weak for $V = 0$, dashes, and decreases further by increasing V-n.n. repulsion (dot-dash). From [70]

$$O_{SC} = \frac{1}{2}\sum_{\mathbf{r}}(\Delta_{\mathbf{r},\mathbf{r}+\mathbf{x}} \pm \Delta_{\mathbf{r},\mathbf{r}+\mathbf{y}}) \tag{127}$$

$$\Delta_{ij} \equiv c_{i\uparrow}c_{j\downarrow} + c_{j\uparrow}c_{i\downarrow}. \tag{128}$$

where + sign holds for the s-Sc and the - sign for d-Sc.

In the physical region of parameters $t > J$ both χ_{s-SC} and χ_{d-SC} are small and further decrease by decreasing T. For rather small $V = J/4$, which is even much smaller than expected, the superconducting susceptibilities are drastically decreased as it is seen in Fig. 27. This means that in the more realistic models for HTSC oxides, such as the t-J-V, *there is no tendency to high-temperature superconductivity*. If superconductivity exists at all its T_c must be very low.

9. SUMMARY AND CONCLUSIONS

A number of experiments, such as optics (*IR* and Raman), transport, tunnelling, ARPES, neutron scattering, give convincing evidence that the electron-phonon interaction (EPI) in HTSC oxides is sufficiently strong and contributes to pairing. These experiments give also evidence for the presence of strong correlations which modify the EPI not only quantitatively but also qualitatively. The most spectacular result of the *EPI* theory in strongly correlated systems is the appearance of the *forward scattering peak* in the EPI, as well as in other charge scattering processes such as the residual Coulomb interaction and scattering on non-magnetic impurities - the FSP *theory* [17], [18], [19], [20] [2]. The forward scattering peak is especially pronounced at lower doping δ. This fundamental result allows us to resolve a number of experimental facts which can not be explained by the old theory based on th isotropic Migdal-Eliashberg equations for the EPI. The most important predictions of the FSP theory of the EPI and other charge scattering processes are: (**1**) the transport coupling constant λ_{tr} (entering the resistivity, $\rho \sim \lambda_{tr} T$) is much smaller than the pairing one λ, i.e. $\lambda_{tr} < \lambda/3$; (**2**) the strength of pairing in HTSC oxides is basically due to the EPI, while the residual Coulomb repulsion (including spin–fluctuations) *triggers* d-wave pairing; (**3**) d-wave pairing is very robust in the presence of non-magnetic impurities; (**4**) the nodal kink in the quasiparticle spectrum is unshifted in the superconducting state, while the anti-nodal singularity is shifted.

We stress the following two facts coming from the theoretical analysis: (**i**) the forward scattering peak in the EPI of strongly correlated systems is a general phenomenon by *affecting electronic coupling to all phonons*; (**ii**) the existence of the forward scattering peak in the EPI is confirmed numerically by the Monte Carlo calculations for the Hubbard-Holstein model with finite U [21], by exact diagonalization [124], as well as by some other methods [22].

Tunnelling experiments and ARPES measurements of the real part of the self-energy give evidence that the EPI coupling constant $\lambda > 1$. At present there are no reliable microscopic calculations of λ in HTSC oxides, which properly include (**a**) the ionic-metallic coupling (due to the long-range Madelung energy) and covalent coupling and (**b**) strong electronic correlations.

In the last several years a large number of published papers were devoted to the study of spin-fluctuation (SFI) interaction as a mechanism of pairing in HTSC oxides. In spite of many efforts and well financed theoretical projects (headed by the greatest authorities in the field), which have opened some new research directions in the theory of electron magnetism, there is no theoretical evidence for the effectiveness of the non-phononic mechanism of pairing. Until now superconductivity could not been proved in the repulsive single-band Hubbard model as well as in its derivative the $t-J$ model. Just opposite, quite recent numerical calculations [70] show in a convincing way, that there is no high-temperature superconductivity in the t-J model. If it exists its T_c is extremely low. Finally, the numerical calculations in the Hubbard model [151] show that the repulsive Hubbard interaction is unfavorable for high-temperature superconductivity, contrary to the attractive interaction which favors it.

The explanation of the high critical temperature in HTSC oxides should be searched in the electron-phonon interaction which is renormalized by strong electronic correlations.

To conclude, *one can not avoid unavoidable*.

ACKNOWLEDGMENTS

I am thankful to Professors Ferdinando Mancini and Adolfo Avella, the organizers of the "*VIII Training Course in the Physics of Correlated Electron Systems and High-Tc*

Superconductors" held in October 2004 in Vietri sul Mare (Salerno) Italy, for giving me the chance to inform the young and talented scientists on the fundamental problems in HTSC physics.

With the great honor, I express my deep gratitude and respect to Vitalii Lazarevich Ginzburg for supporting me for many years, for his deep understanding of superconductivity, sharing it generously with us - his students, collaborators and friends.

I am very thankful to Oleg V. Dolgov and Evgenii G. Maksimov for numerous elucidating discussions on superconductivity theory, on optical properties of HTSC oxides and on many-body theory, as well as for their permanent support. Discussions with Ivan Božović, A. V. Boris and N. N. Kovaleva on optics, with Z. -X. Shen and D. J. Scalapino on ARPES measurements and theory are acknowledged. I appreciate very much support of Ivan Božović, Ulrich Eckern and Igor M. Kulić.

10. APPENDIX: DERIVATION OF THE T-J MODEL

10.1. Hubbard model for finite U in terms of Hubbard operators

For simplicity we study the nearest neighbor (n.n.) Hubbard model [153]

$$H = -t \sum_{m \neq n\sigma} \hat{c}^{\dagger}_{m\sigma} \hat{c}_{n\sigma} + U \sum_{m} \hat{n}_{m\uparrow} \hat{n}_{m\downarrow}, \tag{129}$$

where the operator $\hat{c}^{\dagger}_{m\sigma}$ creates an electron at the m-th site with the spin projection σ.

The Hilbert space at the given lattice site contains four states $\{| \alpha > \Longrightarrow | 0 >, | 2 >, |\uparrow>, |\downarrow>\}$. Let us introduce the Hubbard projection operators $X^{\alpha\beta}$; $\alpha, \beta = 0, 2, \sigma$ (where $\sigma = \uparrow (+), \sigma = \downarrow (-)$)

$$X^{\alpha\beta} = | \alpha >< \beta | . \tag{130}$$

They fulfill the projection properties

$$X^{\alpha\beta} X^{\gamma\delta} = \delta_{\beta\gamma} X^{\alpha\delta}, \tag{131}$$

and rather "ugly" (anti)commutation algebra

$$X_i^{\alpha\beta} X_j^{\gamma\delta} \pm X_j^{\gamma\delta} X_i^{\alpha\beta} = \delta_{ij} (\delta_{\beta\gamma} X_i^{\alpha\delta} \pm \delta_{\delta\alpha} X^{\gamma\beta}. \tag{132}$$

The completeness relation of the Hilbert space reads

$$X_i^{00} + X_i^{22} + \sum_{\sigma} X_i^{\sigma\sigma} = 1. \tag{133}$$

The Hubbard operators describe the composite object. There is a connection between $\hat{c}_{i\sigma}$ and $X^{\alpha\beta}$ (if $\sigma = \uparrow \Longrightarrow \bar{\sigma} = \downarrow$)

$$\hat{c}_{i\sigma} = X_i^{0\sigma} + \sigma X_i^{\bar{\sigma}2}; \hat{c}^{\dagger}_{i\sigma} = X_i^{\sigma 0} + \sigma X_i^{2\bar{\sigma}} \tag{134}$$

$$n_i = 1 - X_i^{00} + X_i^{22} \tag{135}$$

$$S_i^+ = \hat{c}_{i\uparrow}^\dagger \hat{c}_{i\downarrow} = X_i^{+-} = (S_i^-)^\dagger = (X_i^{-+})^\dagger$$

$$S_i^z = \frac{1}{2}(\hat{c}_{i\uparrow}^\dagger \hat{c}_{i\uparrow} - \hat{c}_{i\downarrow}^\dagger \hat{c}_{i\downarrow}) = \frac{1}{2}(X_i^{++} - X_i^{--}), \tag{136}$$

and vice versa

$$X^{\sigma 0} = \hat{c}_\sigma^\dagger (1 - \hat{n}_{\bar{\sigma}}) \tag{137}$$

$$X^{\sigma\sigma} = \hat{n}_\sigma (1 - \hat{n}_{\bar{\sigma}}); \;\; X^{\sigma\bar{\sigma}} = \hat{c}_\sigma^\dagger \hat{c}_{\bar{\sigma}} \tag{138}$$

$$X^{00} = (1 - \hat{n}_\uparrow)(1 - \hat{n}_\downarrow) \tag{139}$$

$$X^{2\sigma} = \sigma \hat{c}_{\bar{\sigma}}^\dagger \hat{n}_\sigma; \;\; X^{20} = \sigma \hat{c}_{\bar{\sigma}}^\dagger \hat{c}_\sigma \tag{140}$$

$$X^{22} = n_\uparrow n_\downarrow \tag{141}$$

The Hubbard Hamiltonian $H = H_1 + H_{12} + H_0$ in terms of $X^{\alpha\beta}$ is given by

$$H_1 = -t \sum_{ij\sigma} (X_i^{\sigma 0} X_j^{0\sigma} + X_i^{2\sigma} X_j^{\sigma 2}) \tag{142}$$

$$H_{12} = -t \sum_{ij\sigma} \sigma (X_i^{\sigma 0} X_j^{\bar{\sigma} 2} + X_i^{2\bar{\sigma}} X_j^{0\sigma}) \tag{143}$$

$$H_0 = U \sum_i X_i^{22} \tag{144}$$

The first term in H_1 describes the motion of single electron in the lower (L) Hubbard band, while the second term describes the motion of the doubly occupied electrons from j-th to the i-th side in the upper (U) Hubbard band. The term H_0 is the repulsive energy of two electrons on the i-th site. Finally, H_{12} connects the two (lower and upper) bands.

10.2. Effective Hamiltonian for $U >> t$

There are various ways to obtain the effective Hamiltonian in the case $U >> t$. Because of its generality and simplicity we use here the canonical transformation method [153], where the operator S mixes lower and upper band. Under the action of S the Hamiltonian is transformed into $H_{eff} = e^S H e^{-S}$

$$H_{eff} = H + [S, H] + \frac{1}{2}[S, [S, H]] + .. \tag{145}$$

with S in the form

$$S = \kappa \sum_{ij\sigma} (X_i^{\sigma 0} X_j^{\bar{\sigma} 2} - X_i^{2\bar{\sigma}} X_j^{0\sigma}). \tag{146}$$

Now, we choose κ so that all first-order in t processes between the L- and U-band disappear from H_{eff}, i.e. one has

$$H_{12}+[S,H_2]=0. \tag{147}$$

The solution of Eq.(147) is $\kappa=-t/U$, and H_{eff} reads

$$H_{eff}=-t\sum_{ij\sigma}X_i^{\sigma 0}X_j^{0\sigma}+H_{3s}$$

$$+J\sum_{ij\sigma}(\mathbf{S}_i\mathbf{S}_j-\frac{1}{4}\hat{n}_i\hat{n}_j)+H_2, \tag{148}$$

where $J=2t^2/U$ is the exchange energy.

The term H_2 describes motion of "doublons" in the U-band

$$H_2=U\sum_i X_i^{22}-t\sum_{ij\sigma}X_i^{2\sigma}X_j^{\sigma 2}, \tag{149}$$

while H_{3s} describes the three-sites hopping.

$$H_{3s}=\frac{J}{2}\sum_{ijl\sigma}(X_i^{\bar{\sigma}0}X_l^{\sigma\bar{\sigma}}X_j^{0\sigma}-X_i^{\sigma 0}X_l^{\bar{\sigma}\bar{\sigma}}X_j^{0\sigma}). \tag{150}$$

Usually this term is neglected in the t-J model.

However, it may have tremendous effect on superconductivity by strongly suppressing it [154]. By projecting H_{eff} onto the lower Hubbard band one gets the famous t-J model Hamiltonian $H_{tJ}=PH_{eff}P$

$$H_{tJ}=-t\sum_{ij\sigma}X_i^{\sigma 0}X_j^{0\sigma}++J\sum_{ij\sigma}(\mathbf{S}_i\mathbf{S}_j-\frac{1}{4}\hat{n}_i\hat{n}_j)$$

$$=-t\sum_{ij\sigma}X_i^{\sigma 0}X_j^{0\sigma}+\frac{J}{2}\sum_{ij\sigma}(X_i^{\sigma\bar{\sigma}}X_j^{\bar{\sigma}\sigma}-X_i^{\sigma\sigma}X_j^{\bar{\sigma}\bar{\sigma}}). \tag{151}$$

Before we are going to discuss some representations for non-canonical operators $X_i^{\sigma 0}$ in terms of bosons and fermions let us stress that the so called "spin" operators $S^{\pm},S^z$ do not describe correctly the electron spin. Although they satisfy the correct spin-commutation relations

$$[S_i^+,S_j^-]=2\delta_{ij}S_i^z$$

$$[S_i^z,S_j^{\pm}]=\pm\delta_{ij}S_i^{\pm}, \tag{152}$$

they describe a particle with spins $S=0,1/2$ at the lattice site, since $\mathbf{S}_i^2$ fulfills

$$\mathbf{S}_i^2=\frac{3}{4}\hat{n}_i\neq\frac{3}{4}. \tag{153}$$

Since $X^{\alpha\beta}$ obey the non-canonical ("ugly") algebra the question is how to treat the Hamiltonian $H_{t,J}$? In Section 4.-5. we have shown that one can study directly with these operators by using the functional technique and 1/N expansion for the self-energy. However, there are very popular approaches which represent $X^{\alpha\beta}$ in terms of bosons and fermions with canonical commutation relations.

Slave boson (SB) method. Here, one introduces the fermion with spin (spinon) $F_{i\sigma}$ and the boson without spin (holon) B_i operators, where $X^{0\sigma} = F_\sigma B^\dagger$. The *constraint* on the SB Hilbert space (completeness), at the given lattice site, is given by $B^\dagger B + \sum_\sigma F_\sigma^\dagger F_\sigma = 1$ and the t-J Hamiltonian reads

$$H_{tJ} = -t\sum_{ij\sigma} F_{i\sigma}^\dagger F_{j\sigma} B_i B_j^\dagger + \frac{J}{2}\sum_{ij\sigma\sigma'} F_{i\sigma}^\dagger F_{j\sigma} F_{j\sigma'}^\dagger F_{i\sigma'}. \tag{154}$$

We stress that the *constraint* strongly limits the SB Hilbert space of bosons and fermions (at the given lattice site) which effectively means their strong interaction. In that respect any uncontrollable decoupling in the SB method (as in some RVB approaches) leads to a spin-charge (spinon-holon) separation, which is not realized in 2D and 3D systems. In order to correct this one introduces the so called gauge fields which keep the spin and charge together. We already discussed the difficulties of the SB method in studying the electron-phonon interaction.

Slave fermion (SF) method. In the SF method the boson has spin and fermion not, i.e. $X^{0\sigma} = B_\sigma^\dagger F$ with the constraint on the Hilbert space $F^\dagger F + \sum_\sigma B_\sigma^\dagger B_\sigma = 1$.

Spin fermion method. Here, the real fermion $\hat{c}^\dagger$ with the "spin" $\mathbf{S}$ is represented via the auxiliary fermion $F^\dagger$ and spin $\mathbf{s}$ by $\hat{c}_\uparrow^\dagger \hat{c}_\uparrow + \hat{c}_\downarrow^\dagger \hat{c}_\downarrow = 1 - F^\dagger F$ and $\mathbf{S} = \mathbf{s}(1 - F^\dagger F)$. The t-J Hamiltonian is rather complicated

$$H_{tJ} = 2t\sum_{ij} F_i^\dagger F_j(\mathbf{s}_i\mathbf{s}_j + \frac{1}{4})$$

$$+J\sum_{ij}(1 - F_i^\dagger F_i)(\mathbf{s}_i\mathbf{s}_j - \frac{1}{4})(1 - F_j^\dagger F_j). \tag{155}$$

Usually this method is used for analyzing motion of single hole in the half-filled system where the antiferromagnetic order is realized.

REFERENCES

1. J. G. Bednorz, K. A. Müller, Z. Phys. **B** 64, 189 (1986)
2. M. L. Kulić, Phys. Reports **338**, 1-264 (2003)
3. C. Thomsen and M. Cardona, in Physical Properties of High Temperature Superconductors I, ed. by D. M. Ginzberg (World Scientific, Singapore, 1989), pp. 409; R. Feile, Physica **C** 159 1 (1989); C. Thomsen, in Light Scattering in Solids VI, ed. by M. Cardona and G. Güntherodt (Berlin, Heidelberg, New York, Springer, 1991), pp. 285

4. V. G. Hadjiev, X. Zhou, T. Strohm, M. Cardona, Q. M. Lin, C. W. Chu, Phys. Rev. **B** 58, 1043 (1998)
5. V. G. Hadjiev,, T. Strohm, M. Cardona, Z. L. Du, Y. Y. Xue, C. W. Chu, preprint MPI, Stuttgart, 1998
6. D. J. Scalapino, Physics Reports **250**, 329 (1995)
7. A. P. Kampf, Physics Reports **249**, 220 (1994)
8. V. L. Ginzburg, E, G. Maksimov, Physica **C** 235-240, 193 (1994); Superconductivity (in Russian) **5**, 1505 (1992)
9. P.B. Allen, B. Mitrović, Solid State Physics, ed. H. Ehrenreich, F. Seitz, D. Turnbull, Academic, New York, **V** 37, p. 1, (1982)
10. M. L. Cohen, P. W. Anderson, Superconductivity in d and f band metals, AIP Conference Proceedings (ed. D. H. Douglass) New York, p.17 (1972)
11. E. G. Maksimov, Uspekhi Fiz. Nauk **170**, 1033 (2000)
12. D. A. Kirzhnits, in High Temperture Superconductivity, ed. V. L. Ginzburg and D. Kirzhnits, (Consultant Bureau New York, London 1982)
13. O. V. Dolgov, D. A. Kirzhnits, E. G. Maksimov, Rev. Mod. Phys. **53**, 81 (1981)
14. S. I. Vedeneev, A. G. M. Jansen, A. A. Tsvetkov, P. Wyder, Phys. Rev. **B** 51, 16380 (1995)
15. H. J. Kaufmann, O. V. Dolgov, E. K. Salje, Phys. Rev. **B** 58, 9479 (1998)
16. P. Samuely, N. L Bobrov, A. G. N. Jansen, P. Wyder, S. N. Barilo, S. V. Shiryaev, Phys. Rev. **B** 48, 13904 (1993); P. Samuely, P. Szabo, A. G. N. Jansen, P. Wyder, J. Marcus, C. Escribe-Filippini, M. Afronte, Physica **B** 194-196, 1747 (1994)
17. M. L. Kulić, R. Zeyher, Phys. Rev. **B** 49, 4395 (1994); Physica **C** 199-200, 358 (1994); Physica **C** 235-240, 358 (1994)
18. R. Zeyher, M. L. Kulić, Phys. Rev. **B** 53, 2850 (1996)
19. R. Zeyher, M. L. Kulić, Phys. Rev. **B** 54, 8985 (1996)
20. M. L. Kulić and R. Zeyher, Mod. Phys. Lett. **B** 11, 333 (1997)
21. Z. B. Huang, W. Hanke, E. Arrigoni, D. J. Scalapino , cond-mat/0306131 (2003)
22. E. Cappelluti, B. Cerruti, I. Pietronero, cond-mat/0312654 (2003)
23. M. L. Kulić, in preparation
24. H. Romberg, M. Alexander, N. Nücker, P. Adelmann, J. Fink, Phys. Rev. **B**42, 8768 (1990)
25. M. S. Hybersten, M. Schlüter, N. E. Christensen, Rev. **B** 39, 9028 (1989)
26. R. Claessen, R. Manzke, H. Carsten, B. Burandt, T. Buslaps, M. Skibowski, J. Fink, Phys. Rev. **B** 39, 7316 (1989); G. Mante, R. Claessen, T. Buslaps, S. Harm, R. Manzke, M. Skibowski, J. Fink, Z. Phys **B** 80, 181 (1990)
27. S. Uchida et al., Phys. Rev. **B** 43, 7942 (1991)
28. P. Lee, N. Nagaosa, Phys. Rev. **B** 46, 5621 (1992)
29. D. Pines, preprint CNSL Newsletter, LALP-97-010, No. 138, June 1997; Physica **B** 163, 78 (1990)
30. C. C. Tsuei, J.R. Kirtley, C. C. Chi, L. S. Yu-Jahnes, A. Gupta, T. Shaw, J. Z. Sun, M. B. Ketchen, Phys. Rev. Lett., **73**, 593 (1994)
31. C. C. Tsuei, J.R. Kirtley, M. Rupp, A. Gupta, J. Z. Sun, T. Shaw, M. B. Ketchen, C. Wang, Z. F. Ren, J. H. Wang, M. Bhushan Science **27**, 329 (1996)
32. C. C. Tsuei, J.R. Kirtley, J. Low Temp. Phys., **107**, 445 (1997)
33. C. C. Tsuei, J.R. Kirtley, Z. F. Ren, J. H. Wang, H. Raffy, Z. Z. Li, Nature **387**, 481 (1998)
34. C. C. Tsuei et al., cond-mat/0402655 (2004)
35. C.C. Tsuei, J.R. Kirtley, Rev. Mod. Phys. **72**, 969 (2000)
36. G. Hastreiter, U. Hofmann, J. Keller, K. F. Renk, Solid State Commun. **76**, 1015 (1990); G. Hastreiter, J. Keller, M. L. Kulić, in Proceedings of the first German-Soviet bilateral Conference on High-Temperature Superconductivity, 1990
37. S. I. Vedeneev, A. G. M. Jensen, P. Samuely, V. A. Stepanov, A. A. Tsvetkov and P. Wyder, Phys. Rev. **49**, 9823 (1994); S. I. Vedeneev, A. G. M. Jensen and P. Wyder, Physica **B** 218, 213 (1996)
38. D. Shimada, Y. Shiina, A. Mottate, Y. Ohyagi and N. Tsuda, Phys. Rev. **B 5**1, 16495 (1995); N. Miyakawa, A. Nakamura, Y. Fujino, T. Kaneko, D. Shimada, Y. Shiina and N. Tsuda, Physica **C** 282-287, 1519 (1997);
39. N. Miyakawa, Y. Shiina, T. Kaneko and N. Tsuda, J. Phys. Soc. Jpn. **62**, 2445 (1993); N. Miyakawa, Y. Shiina, T. Kido and N. Tsuda, J. Phys. Soc. Jpn. **58**, 383 (1989)
40. Y. Shiina, D. Shimada, A. Mottate, Y. Ohyagi and N. Tsuda, J. Phys. Soc. Jpn. **64**, 2577 (1995); Y.

Ohyagi, D. Shimada, N. Miyakawa, A. Mottate, M. Ishinabe, K. Yamauchi and N. Tsuda, J. Phys. Soc. Jpn. **64**, 3376 (1995)
41. R. S. Gonnelli, F. Asdente and D. Andoreone, Phys. Rev. **B** 49, 1480 (1994)
42. T. Schneider, cond-mat/0308595 (2003)
43. A. Lanzara et al., Nature **4**12, 510 (2001)
44. M. L. Kulić, O. V. Dolgov, cond-mat/0308597 (2003)
45. A. A. Kordyuk, et all, cond-mat/0402643 (2004)
46. M. L. Kulić, O. V. Dolgov, in preparation
47. T. Jarlborg, Solid State Comm., **6**7, 297 (1988); **7**1, 669 (1989)
48. S. Barišić, J. Zelenko, Solid State Comm., **74**, 367 (1990); S. Barišić, I. Batistić, Europhys. Lett. **8**, 765 (1989)
49. S. Barišić, Intern. J. of Mod. Phys. **B** 5, 2439 (1991)
50. R. Zeyher, Z. Phys. **B** 80, 187 (1990)
51. H. Krakauer, W. E. Picket, R. E. Cohen, Phys. Rev. **B** 47, 1002 (1993)
52. C. Falter, M. Klenner, G. A. Hoffmann, Phys. Rev. **B** 55, 3308 (1997); Phys. Stat. Sol. **(**b) 209, 235 (1998)
53. C. Falter, M. Klenner, G. A. Hoffmann, Phys. Rev. **B** 57, 14 444 (1998)
54. P. Horsch, G. Khaliullin, V. Oudovenko, Physica C **3**41, 117 (2000)
55. O. Rösch, O. Gunnarsson, cond-mat/0308035 (2003)
56. S. Ishihara, N. Nagaosa, cond-mat/0311200 (2003)
57. M. Gurvitch, A. T. Fiory, Phys. Rev. Lett. **5**9, 1337 (1987)
58. A. T. Fiory, S. Martin, R. M. Fleming, L. F. Schneemeyer, J. V. Waczak, A. F. Hebard, S. A. Sunshina, Physica **C** 162-164, 1195 (1989)
59. A. Mackenzie, E. Marseglia, I. Marsden, G. Lonzarich, C. Chen, B. Wanklyn, Physica **C** 162-164, 1029 (1989)
60. L. F. Mattheiss, Phys. Rev. Lett. **5**8, 1028 (1987)
61. P.W. Anderson, Science **2**35, 1196 (1987)
62. A. V. Puchkov, D. N. Basov, T. Timusk, J. Phys.: Condens. Matter **8**, 10049 (1996)
63. J. Hwang, T. Timusk, G. D. Gu, Nature **4**27, 714 (2004)
64. M. Norman, Nature **4**27, 692 (2004)
65. N. Bulut, D. J. Scalapino, Phys. Rev. Lett. **6**7, 2898 (1991); Phys. Rev. Lett. **6**8, 706 (1992); Phys. Rev. **B** 47, 3528 (1994)
66. Ph. Bourges, in *The Gap Symmetry and Fluctuations in High Temperature Superconductors*, J. Bok, G. Deutscher, D. Pavuna, S. A. Wolf, Eds. (Plenum, New York, 1998), pp. 349-371; preprint cond-mat/9901333 (1999)
67. M. Mehring, Appl. Magn. Resonance **3**, 383 (1992)
68. H. -Y. Kee, S. Kivelson, G. Aeppli, Phys. Rev. Lett. **8**8, 257002 (2002)
69. M. L. Kulić, I. M. Kulić, Physica **C** 391, 42 (2003)
70. L. Pryadko, S. Kivelson, O. Zachar, cond-mat/0306342 (2003)
71. J. Rossat-Mignod et al., Physica **C** 185-189, 86 (1991)
72. H. Mook et al., Phys. Rev. Lett. **7**0, 3490 (1993)
73. E. G. Maksimov, et al., Phys. Rev. Lett. **6**3, 1870 (1989)
74. H. J. Kaufmann, Ph. D. Thesis, Uni. Cambridge, February 1999
75. A. V. Boris et al., to be published in Science, 2004
76. I. Božović, Phys. Rev. **B**42, 1969 (1990)
77. M. Tinkham, Introduction to Superconductivity(McGrow-Hill, New York, ed. 2, 1996)
78. R. Kubo, J. Phys. Soc. Japan **1**2, 570 (1957)
79. H. J. Molegraaf, C. presura, D. van der Marel, P. H. Kes, M. Li, Science **2**95, 2239 (2002)
80. J. E. Hirsch, Physica **C** 199, 305 (1992); ibid **2**01, 347 (1992)
81. M. Norman, C. Pepin, Rep. Prog. Phys. **6**6, 1541-1610 (2003)
82. A. E. Karakozov, E. G. Maksimov, O. V. Dolgov, Sol. St. Comm. **1**24, 119 (2002)
83. R. J. Radtke, S. Illah, K. Levin, M. R. Norman, Phys. Rev. **B** 46, 11975 (1992)
84. R. J. Radtke, K. Levin, H. -B. Schüttler, M. R. Norman, Phys. Rev. **B** 48, 15957 (1993)
85. Z. Schlesinger, R. T. Collins, F. Holzberg, C. Field, V. Welp, Y. C. Chang, P. Z. Jiang, A. P. Paulikas, Phys. Rev. Lett. **6**5, 801 (1990)

86. O. V. Dolgov, E. G. Maksimov, S. V. Shulga, in Electron-Phonon Interaction in Oxide Superconductors, World Scien., p.30 (1991)
87. S. V. Shulga, O. V. Dolgov, E. G. Maksimov, Physica **C** 178, 266 (1989)
88. H. Takagi, B. Batlogg, H. L. Kao, J. Kwo, R. J. Cava, J. J. Krajewski, W. F. Peck, Jr., Phys. Rev. Lett. **6**9, 2975 (1992)
89. S. L. Cooper, K. E. Gray, in Physical Properties of High Temperature Superconductors IV, ed. by D. M. Ginzberg (World Scientific, Singapore, 1994)
90. S. N. Rashkeev and G. Wendin, Phys. Rev. **B** 47, 11603 (1993)
91. I. Božović, J. H. Kim, J. S. Harris, Jr., C. B. Eom, J. M. Phillips, J. T. Cheung, Phys. Rev. Lett., **73**, 1436 (1995)
92. M. Kranz, H. J. Rosen, R. M. Macfarlane and V. Y. Lee, Phys. Rev. **B** 38, 4992 (1988); C. Thomsen, M. Cardona, B. Gegenheimer, R. Liu and A. Simon, Phys. Rev. **B** 37, 9860 (1988)
93. C. Thomsen, in Light Scattering in Solids VI, ed. by M. Cardona and G. G. Güntherodt (Springer, Berlin, Heidelberg 1991)
94. W. Kress, U. Schröder, J. Prade, A. D. Kulkarni, F. W. de Wette, Phys. Rev. **B 38**, 2906 (1988)
95. R. Gajic, Thesis, Ubiversity of Belgrade (1992)
96. R. Zeyher and G. Zwicknagl, Z. Phys. **B** 78, 175 (1990)
97. C. Jiang, J. P. Carbotte, Phys. Rev. **B** 50, 9449 (1994)
98. A. Alligia, M. L. Kulić, V. Zlatić and K. H. Bennemann, Sol. State Comm. **6**5, 501 (1988); Proc. Adriatico Res. Conf. on High-T_c-Supercond., Proceedings ICTP Trieste, p.303, World Scien. Publish., Singapore (1987)
99. T. P. Deveraux, A. Virosztek and A. Zawadovski, Phys. Rev. **B** 51, 505 (1995)
100. E. L. Wolf, Principles of Electron Tunneling Spectroscopy, Oxford University Press, 1985
101. G. D. Mahan, Many-particle Physics, Plenum Press, New York 1990, pp. 796,
102. D. Shimada, N. Tsuda, U. Paltzer and F. W. de Wette, preprint (1998)
103. Q. Huang, J. F. Zasadinski, N. Tralshawala, K. E. Gray, D. Hinks, J. L. Peng and R. L. Greene, Nature (London) **3**47, 389 (1990)
104. J. F. Franck, in Physical Properties of High Temperature Supercoductors V , ed. D. M. Ginsberg, (World Scientific, Singapore, 1994)
105. J. P.Franck, Physica **C** 282-287, 198 (1997)
106. A. Damascelli, Z. Hussain, Z. -X. Shen, Rev. Mod. Phys. **7**5, 473 (2003)
107. J. C. Campuzano, M. R. Norman, M. Randeria, cond-mat/0209476 (2002)
108. T. Cuk et al., cond-mat/0403521 (2004); Z.-X. Shen, talk at the Int. Symp. "*Competing phases in novel condensed-matter systems, Würzburg*, Germany, July 9-11 (2003)
109. X. J. Zhou et al. , cond-mat/0405130 (2004)
110. D. J. Scalapino, in Superconductivity ed. R. D. Parks, **V** 1, Chapter 11, Dekker, New York, 1969
111. O. V. Dolgov, E. I.Maksimov, Tr. Fiz. Inst. Akad. Nauk SSSR, **1**48 3 (1983),
112. D. Rainer, in Progress in Low-Temperature Physics (Ed. D. F. Brewer) (Amsterdam, Elsevier, 1986) p. 371
113. I. I. Mazin et al., Phys. Rev. **B** 42, 366 (1990)
114. W. Weber, Phys. Rev. Lett. **5**8, 2154 (1987)
115. W. Weber, L. F. Mattheiss, Phys. Rev. **B**37, 599 (1988)
116. S. Y. Savrasov, Phys. Rev. Lett. **6**9, 2819 (1992); S. Y. Savrasov, D. Y. Savrasov, Phys. Rev. **B** 46, 12181 (1992)
117. S. Y. Savrasov, K. Andersen, Phys. Rev. Lett. **7**7, 4430 (1996)
118. E. I.Maksimov, D. Y. Savrasov, S. Y. Savrasov, Physics - Uspekhi **4**0, 337 (1997)
119. O. V. Danylenko, O. V. Dolgov, M. L. Kulić, V. Oudovenko, cond-mat/9710234 (1998); cond-mat/9710234 (1997); Europ. Phys. Jour. **B**9 - Cond. Matter, 201 (1999)
120. C. Grimaldi, L. Pietronero, S. Sträßler, Phys. Rev. **B** 52, 10516 (1995); ibid **B** 52, 10530 (1995)
121. J. H. Kim and Z. Tešanović, Phys. Rev. Lett. **7**1, 4218 (1993)
122. C. G. Olson, R. Liu, A. Yang, D. W. Lynch, A. J. Arko, R. S. List, B. Veal, Y. Chang, P. Jiang and A. Paulikas, Science **2**45 (1989)
123. V. J. Emery, Phys. Rev. Lett. **5**8, 2794 (1987)
124. T. Tohyama, P. Horsch and S. Maekawa, Phys. Rev. Lett. **7**4, 980 (1995)
125. J. Hubbard, Proc. Roy. Soc., **A**276 238 (1963); **A**277 237 (1963); **A**281 401 (1964); **8**4 455 (1964)

126. A. Greco, R. Zeyher, Europh. Lett., **35**, 115 (1996); R. Zeyher, A. Greco, Z. Phys. **B** 104, 737 (1997)
127. M. Grilli, G. Kotliar, Phys. Rev. Lett. **64**, 1170(1990)
128. Ju H. Kim, K. Levin, R. Wentzovitch and A. Aurbach, Phys. Rev. **B** 44, 5148 (1991)
129. M. Grilli and C. Castellani, Phys. Rev. **B** 50, 16880 (1994)
130. M. Grilli and C. Castellani, Phys. Rev. Lett. **74**, 1488 (1995)
131. M. Mierzejewski, J. Zielinski, P. Entel, Phys. Rev. **B** 57, 590 (1998)
132. A. A. Abrikosov, Physica C **244**, 243 (1995); J. Ruvalds et al., Phys. Rev. B **51**, 3797 (1995); G. Santi, T. Jarlborg, M. Peter, M. Weger, J. of Supercond. **8**, 215 (1995)
133. M. Weger, B. Barbelini, M. Peter, Z. Phy. **B**94, 387 (1994); M. Weger, M. Peter, L. P. Pitaevskii, Z. Phy. **B**101, 573 (1996); M. Weger, J. of Supercond. **10**, 435 (1997); M. Weger, M. Peter, First Euroconf. on Anomal. Complex Superc., Crete Sept. 26-Oct.3, 1998
134. G. Varelogiannis, Phys. Rev. B **57**, 13743 (1998)
135. A. J. Millis, H. Monien, D. Pines, Phys. Rev. **B** 42, 167 (1990)
136. P. Monthoux and D. Pines, Phys. Rev. Lett. **69**, 961 (1992); Phys. Rev. B **47**, 6069 (1993); R. J. Radtke, K. Levin, H. -B. Schüttler and M. R. Norman, Phys. Rev. B **48**, 15957 (1993)
137. H. -B. Schüttler, M. R. Norman, Phys. Rev. B **54**, 13 295 (1996)
138. A. I. Lichtenstein, M. L. Kulić, Physica C **245**, 186 (1995)
139. A. B. Migdal, Sov. Phys. JETP **34**, 996 (1958); G. M. Eliashberg, Sov. Phys. JETP **11**, 696 (1960)
140. R. Ferenbacher, Phys. Rev. Lett. **77**, 1849 (1996); R. Ferenbacher, M. Norman, Phys. Rev. **B** 50, 3495 (1994)
141. S. K. Tolpygo et al., Phys. Rev. **B**53, 12454 (1996); ibid 12462 (1996)
142. M. L. Kulić, V. Oudovenko, Solid State Comm. **104**, 731 (1997); M. L. Kulić, O. V. Dolgov, Phys. Rev.**B**60, 13062(1999)
143. J. R. Engelbracht, M. Randeria, C. A. R. Sa de Melo, Phys. Rev. **55**, 15 153 (1997)
144. A. Junod, A. Erb, C. Renner, Physica **C 317**, 333 (1999)
145. M. V. Sadovskii, Physics-Uspekhi **44**(5), 515 (2001)
146. A, M. Gabovich, A. I. Voitenko, M. Ausloos, Phys. Rep. **367**, 583 (2002)
147. K. Yang, S. L. Sondhi, Phys. Rev. **62**, 11 778 (2000)
148. Phys. Rev. **112**, 1900 (1957)
149. P. Monthoux, D. J. Scalapino, Phys. Rev. Lett. **72,** 1874 (1994); Chien-Hua Pao, N. E. Bickers, Phys. Rev. Lett. **72,** 1870 (1994)
150. E. Dagotto, Rev. Mod. Phys. **66**, 763 (1994)
151. D. J. Scalapino, S. R. White, S. C. Zhang, Phys. Rev. Lett. **68**, 2830 (1992)
152. F. F. Assad, W. Hanke, D. J. Scalapino, Phys. Rev. Lett. **71**, 1915 (1993)
153. Yu. A. Izyumov, Physics-Uspekhi **40**(5). pp. 445-476 (1997)
154. A. Alligia, private communication

Monte Carlo Simulations of Quantum Systems with Global Updates

Alejandro Muramatsu

Institut für Theoretische Physik III, Universität Stuttgart, Pfaffenwaldring 57, D-70550 Stuttgart, Germany

Abstract. We start this series of lectures on quantum Monte Carlo methods with the simplest model for a strongly correlated system, namely an antiferromagnetic Heisenberg S-$1/2$ chain. We will review methods for its simulation starting with the world-line algorithm and then introducing the loop-algorithm with global updates. We discuss next a model for doped antiferromagnets, with emphasis on the strong correlation limit that is central for high temperature superconductors and related materials, namely the so-called t-J model. An exact canonical transformation for this model that leads to a formulation with separated charge and spin degrees of freedom will be discussed. Results for a single hole in antiferromagnetic chains and planes will be summarized, giving an initial picture of charge dynamics in a quantum antiferromagnet. An additional algorithmic element, namely the determinantal method, will be introduced next, in order to deal with a finite number of charge degrees of freedom. Merging the loop- and the determinantal methods leads to a new one, the hybrid-loop algorithm, that combines the efficiency of the loop-algorithm for spins with the one of the determinantal method for fermions. Finally we discuss a first application of the hybrid-loop algorithm, where the spectral function of the t-J model in one dimension with finite doping is determined. There we can see how charge spin separation shows up in the one-particle spectral function, and a detailed description can be achieved by comparison with an analitically soluble model, namely the t-J model with $1/r^2$ interaction. The results show that in addition to spinons and holons expected in one-dimensional metals, antiholons with charge $Q = 2e$, spin $S = 0$, and twice the mass of the holons are necessary to describe the inverse photoemission spectra at finite doping.

INTRODUCTION

Triggered by the discovery of high temperature superconductors [1] (HTS), a plethora of new systems were discovered [2], where the correlated dynamics of the spin and charge degrees of freedom of the electrons resists up to now a theoretical understanding.

The essential physical properties of correlated materials are in essence represented by the Hubbard or the t-J model [3]. The Hubbard model describes electrons on a lattice that can tunnel from site to site with an amplitude t, and with an on-site interaction U. Although rather simple, this model shows a metal-insulator transition at commensurate fi llings and magnetic phases, as observed in many correlated materials. In the limit of strong correlations ($U \gg t$), electrons will avoid occupied sites and a new energy scale $J \sim t^2/U$ appears, that corresponds to an antiferromagnetic exchange interaction between electrons on neighboring sites. This is the regime described by the t-J model, that constitutes the minimal model for HTS. Away from density $n = 1$ (i.e. on averge one electron per site), and in spatial dimensions larger than one, not much is known about these models.

CP715, *Lectures on the Physics of Highly Correlated Electron Systems VIII: Eighth Training Course*, edited by A. Avella and F. Mancini

Even in one dimension, where exact solutions are available for the Hubbard model for all values of U [4], and for the t-J model at the so-called supersymmetric point $J = 2t$ [5, 6], correlation and spectral functions are very diffi cult to compute. The situation is worse in two dimensions, where no controlled theoretical methods are yet available. However, progress was made recently for the Hubbard model in infi nite dimensions or within the so-called dynamical mean-fi eld theory (DMFT) [7].

Due to the avoidance of doubly occupied sites, the t-J model has a reduced Hilbert space with respect to the Hubbard model, such that larger system sizes can be exactly diagonalized [8]. However, since the dimension of the Hilbert space increases exponentially with the system size, such studies are limited to very small systems (approx. 30 sites), whereas the anomalies observed experimentally correspond to the thermodynamic limit. It is therefore very important to develop accurate methods able to deal with rather large systems, such that fi nite size extrapolations become possible, in order to access the physics on the macroscopic scale.

Quantum Monte Carlo (QMC) methods offer an alternative, since in this case, the idea of *importance sampling* is used. This idea underlying Monte Carlo simulations in general [9], allows to avoid the problem of considering the full confi guration space, but ensures that the most important states contributing to physical observables are taken into account. In QMC algorithms a mapping is made of the original quantum mechanical problem into an equivalent classical one, that is simulated with the Monte Carlo method.

We start the next section by considering a $S-1/2$ Heisenberg antiferromagnetic chain, where an explicit mapping to the equivalent classical system will be given, leading to world-lines. Next, we will introduce the loop-algorithm [10, 11] that allows for global updatings of the confi gurations of world-lines.

THE WORLD-LINE AND LOOP-ALGORITHMS

The world-line (WL) algorithm introduced by Hirsch *et al.* [12] provides a very easy way of simulating lattice fermions in one dimension and non-frustrated spin systems in one or more dimensions. Its main advantages are given by the use of Ising-like moves, such that its implementation is easy and fast, as well as by the large freedom in the choice of systems that can be simulated, in contrast to determinantal methods (see below and Refs.[13, 14]), where it is essential that the interaction terms in the Hamiltonian can be transformed into a square of bilinear forms. In the following we give a short description of the method where we explicitly deal with the $S-1/2$ Heisenberg antiferromagnetic chain. We then proceed to discuss on the same model the loop-algorithm devised by Evertz *et al* [10, 11].

Brief description of world-lines

We start by considering an anisotropic antiferromagnetic Heisenberg chain (AFHC),

$$H_H = J\sum_i \left(S_i^x S_{i+1}^x + S_i^y S_{i+1}^y + \Delta S_i^z S_{i+1}^z\right) , \qquad (1)$$

where $\Delta = J_z/J$ characterizes the anisotropy, in such a way that $\Delta = 1$ corresponds to the isotropic case. S_i^α with $\alpha = x,y$ or z are $S-1/2$ operators fulfi lling the SU(2) algebra

$$\left[S_i^\alpha, S_j^\beta\right] = i\,\delta_{ij}\,\varepsilon^{\alpha\beta\gamma} S_i^\gamma\,, \tag{2}$$

where $\varepsilon^{\alpha\beta\gamma}$ is the Levi-Civita symbol in three dimensions.

Equation (1) can be mapped by means of a Jordan-Wigner transformation [15] into a model of spinless fermions with nearest neighbor interactions

$$H = -t\sum_i \left(c_i^\dagger c_{i+1} + \text{h.c.}\right) + V\sum_i \left(n_i - \frac{1}{2}\right)\left(n_{i+1} - \frac{1}{2}\right)\,, \tag{3}$$

where $c_i^\dagger$ and c_i are creation and annihilation operators respectively, for a fermion on site i and $n_i = c_i^\dagger c_i$. The transformation maps the state $|\uparrow>$ ($|\downarrow>$) of the AFHC into $|\,1>$ ($|\,0>$) of spinless fermions and the coupling constants are related by $t = J/2$ and $V = J\Delta$, such that the isotropic point of the AFHC corresponds to $V = 2t$. We should remark that the sign of the kinetic term in eq. (3) was changed by performing a canonical transformation $c_i \to (-1)^i c_i$, that can be performed in a bipartite lattice.

In the following we use the Hamiltonian (1) in order to discuss the world-line algorithm. The partition function at an inverse temperature $\beta = 1/k_B T$ is given by

$$\begin{aligned} Z &= \mathrm{Tr}\,\mathrm{e}^{-\beta H} = \mathrm{Tr}\prod_{\ell=1}^{L} \mathrm{e}^{-\Delta\tau H} \\ &= \sum_{\{i_\ell\}} < i_1|\mathrm{e}^{-\Delta\tau H}|i_L >< i_L|\mathrm{e}^{-\Delta\tau H}|i_{L-1} > \cdots < i_2|\mathrm{e}^{-\Delta\tau H}|i_1 >, \end{aligned} \tag{4}$$

where $\Delta\tau = \beta/L$, and $\{|i_\ell>\}$ are complete sets of states introduced at each time-slice. In order to evaluate the matrix elements in (4), a Trotter-Suzuki decomposition [16, 17] is introduced. In the simplest case where the Hamiltonian is splitted into two parts, $H = H_1 + H_2$, we have

$$\mathrm{e}^{-\Delta\tau H} = \mathrm{e}^{-\Delta\tau H_1}\mathrm{e}^{-\Delta\tau H_2} + \mathcal{O}\left[(\Delta\tau)^2\right]\,, \tag{5}$$

In our example one can conveniently choose

$$H_{1\,(2)} = \sum_{i\,\mathrm{odd}\,(\mathrm{even})} H_{i,i+1}\,, \tag{6}$$

such that H_1 and H_2 consist each of a sum of mutually commuting pieces. In this way, the matrix elements are reduced to a product of 2-site matrix elements that can be easily calculated:

$$\begin{aligned} &< i_\ell\,|\,\mathrm{e}^{-\Delta\tau H}\,|\,i_{\ell+1}> \\ &\quad =< i_\ell\,|\,\mathrm{e}^{-\Delta\tau H_1}\,|\,i_{\tilde\ell}>< i_{\tilde\ell}\,|\,\mathrm{e}^{-\Delta\tau H_2}\,|\,i_{\ell+1}> + \mathcal{O}\left[(\Delta\tau)^2\right] \end{aligned}$$

$$= \prod_{i\,\mathrm{odd}} < i_\ell \,|\, \mathrm{e}^{-\Delta\tau H_{i,i+1}} \,|\, i_{\tilde{\ell}} >$$

$$\times \prod_{i\,\mathrm{even}} < i_{\tilde{\ell}} \,|\, \mathrm{e}^{-\Delta\tau H_{i,i+1}} \,|\, i_{\ell+1} > + \mathcal{O}\left[(\Delta\tau)^2\right] . \qquad (7)$$

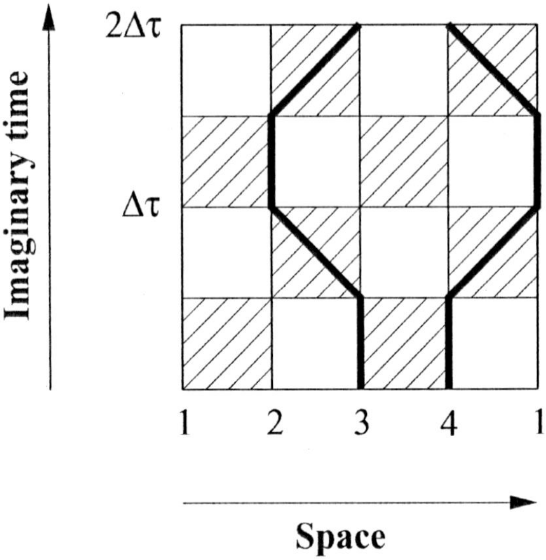

FIGURE 1. Checkerboard decomposition of space-time for a one-dimensional problem with only nearest neighbor contributions and periodic boundary conditions. The thick lines are fermion world-lines.

A graphical representation leads to a checkerboard for space-time, where exchange processes take place only within a shaded square (Fig. 1). The calculation of matrix elements for the two-sites problem leads in our case to

$$\begin{aligned}
<\uparrow\uparrow|\, \mathrm{e}^{-\Delta\tau H_{i,i+1}} \,|\uparrow\uparrow> &= <\downarrow\downarrow| \cdots |\downarrow\downarrow> = \mathrm{e}^{-\Delta\tau J\Delta/4} , \\
<\uparrow\downarrow|\, \mathrm{e}^{-\Delta\tau H_{i,i+1}} \,|\uparrow\downarrow> &= <\downarrow\uparrow| \cdots |\downarrow\uparrow> = \mathrm{e}^{+\Delta\tau J\Delta/4} \cosh \Delta\tau J/2 , \\
<\uparrow\downarrow|\, \mathrm{e}^{-\Delta\tau H_{i,i+1}} \,|\downarrow\uparrow> &= <\downarrow\uparrow| \cdots |\uparrow\downarrow> = -\mathrm{e}^{+\Delta\tau J\Delta/4} \sinh \Delta\tau J/2 . \qquad (8)
\end{aligned}$$

A problem appears in the last matrix element, where a minus sign is present. In principle, since the product of matrix elements give the weight of a confi guration in eq. (4), this could lead to negative weights. In a bipartite lattice, it is possible to perform the following canonical transformation

$$\begin{aligned}
S_i^x &\rightarrow (-1)^i S_i^x , \\
S_i^y &\rightarrow (-1)^i S_i^y , \\
S_i^z &\rightarrow S_i^z ,
\end{aligned}$$

that reverses the sign in such a way that the weights are positive. However, this shows that simulations of spin systems on a geometrically frustrated lattice, where the transformation (9) is not well defi ned, will suffer under the so-called minus sign problem. This problem refers to the situation where changes in the sign of the weights during a simulation lead to large statitical fluctuations, such that a bad signal to noise ratio results. After the canonical transformation above and shifting the energy per bond by $J/4$, such that the energy of the Heisenberg model coincides with that of the t-J model at half-filling (see the corresponding section below), we have finally the following simple forms for the matrix elements:

$$
\begin{aligned}
<\uparrow\uparrow|\, \mathrm{e}^{-\Delta\tau H_{i,i+1}}\, |\uparrow\uparrow> &= <\downarrow\downarrow| \cdots |\downarrow\downarrow> = 1\,, \\
<\uparrow\downarrow|\, \mathrm{e}^{-\Delta\tau H_{i,i+1}}\, |\uparrow\downarrow> &= <\downarrow\uparrow| \cdots |\downarrow\uparrow> = \mathrm{e}^{+\Delta\tau J\Delta/2} \cosh \Delta\tau J/2\,, \\
<\uparrow\downarrow|\, \mathrm{e}^{-\Delta\tau H_{i,i+1}}\, |\downarrow\uparrow> &= <\downarrow\uparrow| \cdots |\uparrow\downarrow> = \mathrm{e}^{+\Delta\tau J\Delta/2} \sinh \Delta\tau J/2\,.
\end{aligned}
\qquad (9)
$$

Possible moves for the problem considered here are shown in Fig. 2. Additional moves have to be considered in other models, like the so-called t-J model in order to ensure ergodicity [18]. The acceptance of a move is decided either in the frame of a Metropolis

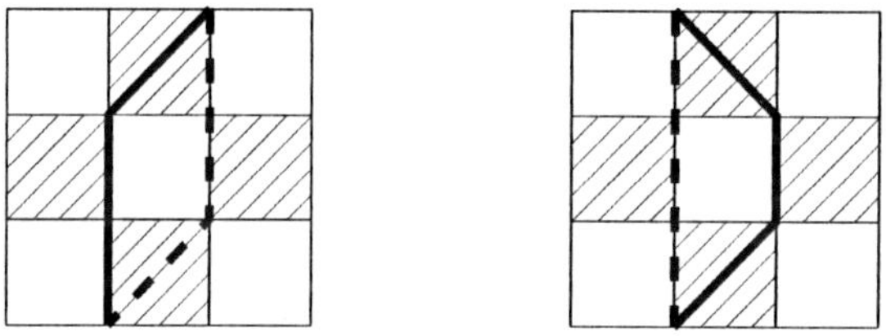

FIGURE 2. Possible moves for spinless fermions with nearest neighbor interactions. Thick (dashed) lines are world-lines before (after) the move.

or a heat bath algorithm [9] by considering the ratio of weights of the new configuration with respect to the old one. This ratio is given in our case by

$$R = \frac{W_{\text{new}}}{W_{\text{old}}} = [\tanh \Delta\tau J/2]^{su} \left[\cosh(\Delta\tau J/2)\, \mathrm{e}^{\Delta\tau J\Delta/2}\right]^{sv} \qquad (10)$$

with

$$s \equiv n(i,j) + n(i,j+1) - n(i+1,j) - n(i+1,j+1)\,, \qquad (11)$$

and

$$u = 1 - n(i+1,j-1) - n(i+1,j+2),\; v = n(i-1,j) - n(i+2,j)\,. \qquad (12)$$

We can consider now measurements in the WL-algorithm, where we have to distinguish three different classes of observables. First we have operators that are diagonal in occupation number. In such a case,

$$< \mathcal{O} > = \lim_{M\to\infty} \frac{1}{N2LM} \sum_{k=1}^{M} \sum_{j=1}^{2L} \sum_{i=1}^{N} \mathcal{O}(n_i^k(j))\,, \qquad (13)$$

where N is the number of sites, $2L$ the number of time-slices, and M the number of samples taken in the simulation.

Secondly, we consider operators that conserve spin or number of particles on two sites, like the spin-exchange on nearest neighbor sites, or kinetic energy or current-operator for fermions or bosons. In such a case, it is possible to split the operators as done before with the Hamiltonian,

$$\mathcal{O} = \sum_i \mathcal{O}_{i,i+1} = \overbrace{\sum_{i\,\mathrm{odd}} \mathcal{O}_{i,i+1}}^{\mathcal{O}_1} + \overbrace{\sum_{i\,\mathrm{even}} \mathcal{O}_{i,i+1}}^{\mathcal{O}_2} . \tag{14}$$

Then, the expectation value is given by

$$\begin{aligned} <\mathcal{O}> &= \frac{1}{Z}\mathrm{Tr}\,\mathcal{O}\left[\mathrm{e}^{-\Delta\tau H_1}\mathrm{e}^{-\Delta\tau H_2}\right]^L \\ &= \frac{1}{Z}\sum_{\{i_1\}} < i_1 \mid \mathcal{O}\left[\mathrm{e}^{-\Delta\tau H_1}\mathrm{e}^{-\Delta\tau H_2}\right]^L \mid i_1 > \\ &= \sum_{\{i_1\cdots i_{2L}\}} P(i_1,\ldots,i_{2L})\left\{\frac{< i_1 \mid \mathcal{O}_1\mathrm{e}^{-\Delta\tau H_1} \mid i_2 >}{< i_1 \mid \mathrm{e}^{-\Delta\tau H_1} \mid i_2 >}\right. \\ &\qquad\left. + \frac{< i_{2L} \mid \mathrm{e}^{-\Delta\tau H_2}\mathcal{O}_2 \mid i_1 >}{< i_{2L} \mid \mathrm{e}^{-\Delta\tau H_2} \mid i_1 >}\right\}, \end{aligned} \tag{15}$$

where

$$P(i_1,\ldots,i_{2L}) \equiv \frac{1}{Z} < i_1 \mid \mathrm{e}^{-\Delta\tau H_1} \mid i_2 > \cdots < i_{2L} \mid \mathrm{e}^{-\Delta\tau H_2} \mid i_1 > \tag{16}$$

is the weight for a given confi guration of world-lines.

Finally, there are operators that do not conserve spin or particle number locally, like transverse correlation functions $< S_i^+ S_j^- >$, where $S^\pm = S^x \pm iS^y$, or the one-particle Green's function in the case of a fermionic or a bosonic model

$$G_{ij} = - < Tc_i(\tau)c_j^\dagger(0) > , \tag{17}$$

where $c_i(\tau) = \exp(\tau H)c_i\exp(-\tau H)$ is the operator in the Heisenberg representation, and T is the time ordering operator. The diffi culty with such propagators can be easily illustrated by considering $\tau = 0$. Proceeding as in the previous case (we illustrate this in the case of the fermionic model), we have

$$\begin{aligned} G_{ij} &= \frac{1}{Z}\mathrm{Tr}\left\{c_j^\dagger c_i\left[\mathrm{e}^{-\Delta\tau H_1}\mathrm{e}^{-\Delta\tau H_2}\right]^L\right\} \\ &= \frac{1}{Z}\sum_{\{i_1\cdots i_{2L}\}} P(i_1,\ldots,i_{2L})\frac{< i_1 \mid c_j^\dagger c_i \mathrm{e}^{-\Delta\tau H_1} \mid i_2 >}{< i_1 \mid \mathrm{e}^{-\Delta\tau H_1} \mid i_2 >} . \end{aligned} \tag{18}$$

Such an expression is not well defi ned, since for $\mid i - j \mid > 1$, we have confi gurations where the denominator vanishes whereas the numerator does not. These confi gurations correspond to a world-line that terminates at i and starts at j. This can be cured by inserting additional states as follows

$$\begin{aligned} G_{ij} &= \frac{\sum < i_1 \mid c_j^\dagger c_i \mid i'_1 >< i'_1 \mid \mathrm{e}^{-\Delta\tau H_1} \mid i_2 > \cdots}{\sum < i_1 \mid i'_1 >< i'_1 \mid \mathrm{e}^{-\Delta\tau H_1} \mid i_2 > \cdots} \\ &\equiv \frac{<< i_1 \mid c_j^\dagger c_i \mid i'_1 >>_{\tilde{P}}}{<< i_1 \mid i'_1 >>_{\tilde{P}}}, \end{aligned} \tag{19}$$

where the new probability distribution is given by

$$\tilde{P} \equiv \frac{< i'_1 \mid \mathrm{e}^{-\Delta\tau H_1} \mid i_2 > \cdots < i_{2L} \mid \mathrm{e}^{-\Delta\tau H_2} \mid i_1 >}{\sum < i'_1 \mid \mathrm{e}^{-\Delta\tau H_1} \mid i_2 > \cdots < i_{2L} \mid \mathrm{e}^{-\Delta\tau H_2} \mid i_1 >}. \tag{20}$$

However, a deterioration in the statistics is expected since now one has to deal with a ratio of expectation values and disconnected world-lines have to be sampled. For time displaced correlation functions two additional intermediate states have to be incorporated, leading in general to an even worse statistics.

Apart from the difficulty in the measurement of off-diagonal correlation functions, the world-line algorithm has several other drawbacks. Since the update of configurations takes place locally, it is expected that such an algorithm will suffer under critical slowing down [9], when a correlation length in the system becomes large or even diverges, as it is expected in many correlated systems in low dimensions. This point will be addressed again below.

Another, even more serious problem is the fact that, when working with periodic boundary conditions, simulations are restricted to a given winding number of the world-lines. The winding number denotes how many times a world-line winds up around the system. Due to periodic boundary conditions in the imaginary time direction, and due to the indistinguishability of identical particles, the possible configurations of world-lines can be grouped in topological sectors that differ in their winding numbers. Local moves are not able to change from one topological sector to another, so that such a simulation is not ergodic. Such problems can be avoided by the loop-algorithm, since as discussed below, global updates are able to reduce autocorrelation times and are not restricted to a given winding number.

The loop algorithm

As already mentioned above, the local update of world-lines as described in the previous section suffers from long autocorrelation times [19] in a similar way as classical simulations with local updates do. A reduction of long autocorrelations can in general be reached by global moves in cluster algorithms as introduced by the pioneering work of Swendsen and Wang [20] for the classical Ising model. Such an improvement was developed for the six-vertex model by Evertz *et al.* [10, 11], and subsequently extended to other quantum spin and fermion models [19,21-25]. Extensive calculations with high accuracy were performed for the 1/5 depleted two dimensional Heisenberg system [26], that is realized in newly discovered materials.

Apart from reducing the autocorrelation time, the loop algorithm removes a number of limitations mentioned at the end of the previous section. Since it can in principle change the number of world-lines, the simulation is made in a grand canonical ensemble, and hence, neither the number of particles, nor the total spin is conserved. This is certainly crucial in order to be able to describe a transition towards a ferromagnetic state. Furthermore, exchange of particles is possible, since the topological constraint for the linking number of world-lines can be removed when the sites along a loop are flipped. In the following we focus on an intuitive approach to the loop algorithm, in contrast to a more rigorous treatment in the frame of vertex models, that can be found in the original papers [10, 22, 23]. For a recent review on loop algorithms we refer to [11].

It is assumed, that the weight of a given confi guration of the system $W(\vec{i})$, with $\vec{i} \equiv (i_1, \ldots, i_{2L})$ can be decomposed in the sum of some graphs $\mathcal{G}$ with weights $V(\mathcal{G})$, where we anticipate that those graphs are loops. This is natural since the objects we are dealing with, namely the world-lines, are one-dimensional (although in general embeded in dimensions higher or equal than 2), and the graphs should give a prescription to change those objects without changing their nature, i.e. after flipping the graphs (by this we mean that the states of the sites belonging to the graph are flipped from one state to another), the new confi guration should also be made out of world-lines. Then, we have

$$W(\vec{i}) = \sum_{\mathcal{G}} V(\mathcal{G}) \Delta(\vec{i}, \mathcal{G}) , \tag{21}$$

where $\Delta(\vec{i}, \mathcal{G})$ tells whether the graph $\mathcal{G}$ is compatible with the confi guration $\vec{i}$, i.e.

$$\Delta(\vec{i}, \mathcal{G}) \begin{cases} 1 & \text{if graph } \mathcal{G} \text{ compatible with } \vec{i} \\ 0 & \text{otherwise} . \end{cases} \tag{22}$$

The construction should be such that the relationship above is also fulfi lled at each plaquette, and therefore, the loops can be built up on the basis of local decisions. Consequently, eq. (21) should be valid also per plaquette:

$$w(u) = \sum_{g} v(g) \Delta(u, g) , \tag{23}$$

where the new notation refers to the weight w for a confi guration u on a plaquette. Accordingly, v is the weight of a graph g on a plaquette. From the equation above, a probability for a graph given a confi guration can be calculated on a plaquette,

$$p(g \mid u) = \frac{v(g) \Delta(u, g)}{w(u)} , \tag{24}$$

whenever eq. (23) can be inverted such that $v(g)$ can be determined univocally. The next step is to determine the possible graphs per plaquette. The plaquettes under consideration should be only the shaded ones in the checkerboard, since it is there, where the dynamics of the particles is described. Furthermore, since the aim is to construct loops, i.e. one-

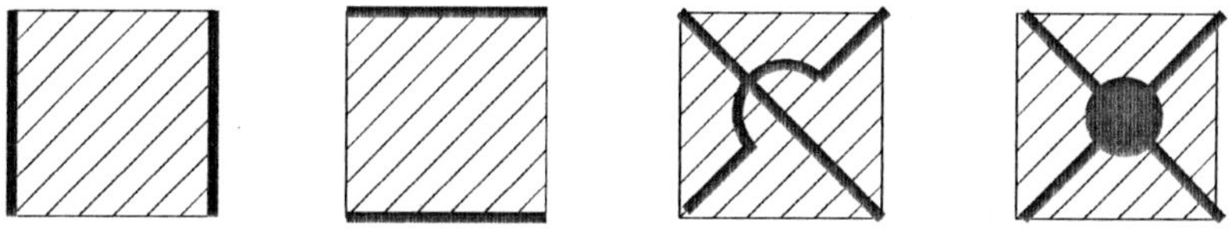

$v(||)$ $v(=)$ $v(X)$ $v(\otimes)$

FIGURE 3. Possible local graphs with the symbols for their weights $v(g)$

dimensional objects, each point of the shaded plaquette can be connected with only one point in the same plaquette, such that taking into account the neighboring plaquettes, each point belongs to a single line, i.e. the loops do not branch. There is however, a further possibility, namely that all points in the plaquette are flipped, a move that is called "freezing" [11]. Such a graph links loops into a cluster. The possible linkings are depicted in Fig. 3. For the "freezing" graphs, the freedom to introduce different weights depending on the confi guration is denoted by the subindex i.

Once the possible graphs were determined, the compatibility of such graphs with the possible confi gurations of a plaquette should be analyzed. Compatibility is dictated by the fact that flipping a loop should allways change a given confi guration into an allowed one. Such a compatibility table is depicted in Fig. 4. After having determined the possible graphs and their compatibility with confi gurations on a plaquette, we see from eq. (9) that the six non-vanishing matrix elements have pairwise the same weight, such that only three independent weights per plaquette exist. In fact, those confi gurations with the same weight are those connected by flipping the "freezing" graph. From eqs. (23) and (9), the following equations connecting the weight for each graph on a plaquette with the weights for the confi gurations are obtained.

$$\begin{aligned} v(\times)+v(||)+v_1(\otimes) &= 1 &\equiv w_1 \\ v(\times)+v(=)+v_2(\otimes) &= \sinh(\Delta\tau J/2)\, e^{\Delta\tau J\Delta/2} &\equiv w_2\ . \\ v(=)+v(||)+v_3(\otimes) &= \cosh(\Delta\tau J/2)\, e^{\Delta\tau J\Delta/2} &\equiv w_3 \end{aligned} \tag{25}$$

Since we have only three equations and have to determine six weights v for the graphs, further restrictions are needed. At this point we should recall that the formation of clusters that extend through the whole system renders the simulation useless. Therefore, the weights for "freezing" graphs should be minimized. The best possibility is to set them to zero. In such a way we have no clusters and the system of equations above is invertible, with the following solution,

$$\begin{aligned} v(\times) &= \frac{1}{2}\left\{1-\exp\left[-\Delta\tau\left(\frac{J}{2}-\frac{J\Delta}{2}\right)\right]\right\}, \\ v(||) &= \frac{1}{2}\left\{1+\exp\left[-\Delta\tau\left(\frac{J}{2}-\frac{J\Delta}{2}\right)\right]\right\}, \\ v(=) &= \frac{1}{2}\left\{\exp\left[\Delta\tau\left(\frac{J}{2}+\frac{J\Delta}{2}\right)\right]-1\right\}. \end{aligned} \tag{26}$$

The solutions above show that in the case of the isotropic Heisenberg model ($\Delta = 1$), $\nu(\times)$ vanishes, so that only two graphs are necessary.

g \ u	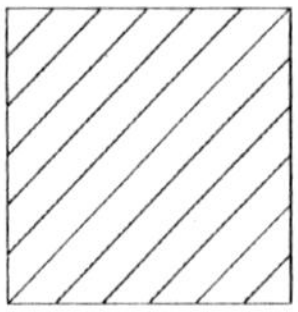	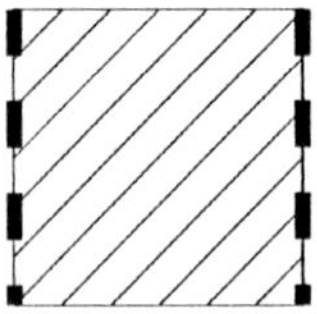	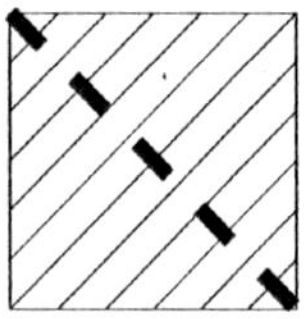	
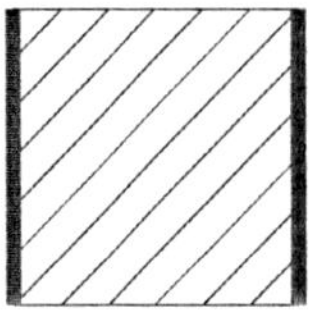	1	1	0	1
	0	0	1	1
	1	1	1	0
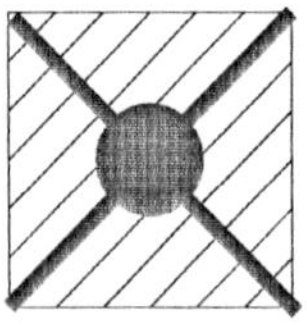	1	1	1	1

FIGURE 4. Compatible graphs g for configurations u. Dashed lines denote world-lines.

Once the weights of the graphs are obtained, it is straightforward to use eq. (24) in order to obtain the probabilities for a graph g given a certain confi guration u. In the case of an isotropic Heisenberg antiferromagnet, the probabilities $p(g \mid u)$ are as follows:

$$
\begin{aligned}
p\left(\parallel \middle| \begin{matrix}\uparrow\uparrow\\ \uparrow\uparrow\end{matrix}\right) &= p\left(\parallel \middle| \begin{matrix}\downarrow\downarrow\\ \downarrow\downarrow\end{matrix}\right) = 1\,, \\
p\left(\parallel \middle| \begin{matrix}\uparrow\downarrow\\ \uparrow\downarrow\end{matrix}\right) &= p\left(\parallel \middle| \begin{matrix}\downarrow\uparrow\\ \downarrow\uparrow\end{matrix}\right) = \frac{e^{-\Delta\tau J/2}}{\cosh\left(\frac{\Delta\tau J}{2}\right)}\,,
\end{aligned}
$$

$$
\begin{aligned}
p\left(\parallel\middle|\begin{matrix}\uparrow\downarrow\\ \downarrow\uparrow\end{matrix}\right) &= p\left(\parallel\middle|\begin{matrix}\downarrow\uparrow\\ \uparrow\downarrow\end{matrix}\right) = 0\,,\\
p\left(=\middle|\begin{matrix}\uparrow\uparrow\\ \uparrow\uparrow\end{matrix}\right) &= p\left(=\middle|\begin{matrix}\downarrow\downarrow\\ \downarrow\downarrow\end{matrix}\right) = 0\,,\\
p\left(=\middle|\begin{matrix}\uparrow\downarrow\\ \uparrow\downarrow\end{matrix}\right) &= p\left(=\middle|\begin{matrix}\downarrow\uparrow\\ \downarrow\uparrow\end{matrix}\right) = \tanh\left(\frac{\Delta\tau J}{2}\right)\,,\\
p\left(=\middle|\begin{matrix}\uparrow\downarrow\\ \downarrow\uparrow\end{matrix}\right) &= p\left(=\middle|\begin{matrix}\downarrow\uparrow\\ \uparrow\downarrow\end{matrix}\right) = 1\,. \qquad (27)
\end{aligned}
$$

In this special case, only for one type of confi guration, namely $\left(\begin{smallmatrix}\uparrow\downarrow\\ \uparrow\downarrow\end{smallmatrix}\right)$, or $\left(\begin{smallmatrix}\downarrow\uparrow\\ \downarrow\uparrow\end{smallmatrix}\right)$ there is a stochastic assignment of graphs. For all other confi gurations of world-lines in a plaquette, the assignment is deterministic. From the discussion above, it is clear that the probabilities obtained are positive defi nite and smaller than or equal one. It should also become clear that this is not allways the case, but depends on the model. In our case it was possible to do so, but different schemes may be needed depending on the values of the parameters.

After discussing the construction of loops, we consider the change in confi guration on the basis of loops. Let us fi rst remark that by flipping the sites on a given line of a graph in a plaquette, we reach another confi guration that is compatible with the same graph. That is, we make a transition between confi guration along a row in Fig. 4 (and to confi gurations not depicted but obtained by reflexion in the third and fourth column). This being valid per plaquette, is also valid on a graph $\mathscr{G}$, such that the flipping of all the sites on a loop bring us from a confi guration$\vec{i}$ to another $\vec{i}'$ with the same graph $\mathscr{G}$. Following the notation in eqs. (21) and (23), the weight of a given confi guration with a given graph is

$$
W(\vec{i},\mathscr{G}) = V(\mathscr{G})\Delta(\vec{i},\mathscr{G})\,, \qquad (28)
$$

such that

$$
W(\vec{i},\mathscr{G}) = W(\vec{i}',\mathscr{G}) \qquad (29)
$$

as long as the confi gurations$\vec{i}$ and $\vec{i}'$ are compatible with $\mathscr{G}$. This is not the more general prescription in constructing loops (see e.g. Refs. [24, 25] for the case of a pure loop-algorithm for the t-J model). However, for the example we are considering and a variety of related problems like spin 1/2 systems, it leads to a very effective update, as shown immediately below.

Next we ask for detailed balance for such a flip

$$
W(\vec{i},\mathscr{G})\, p(\vec{i}\to\vec{i}',\mathscr{G}) = W(\vec{i}',\mathscr{G})\, p(\vec{i}'\to\vec{i},\mathscr{G})\,. \qquad (30)
$$

This requirement ensures in fact, that detailed balance is respected by the original weight $W(\vec{i})$ [11]. The probability for a change of confi guration for a given graph can be solved with the following form

$$
p(\vec{i}\to\vec{i}',\mathscr{G}) = \frac{W(\vec{i}',\mathscr{G})}{W(\vec{i},\mathscr{G})+W(\vec{i}',\mathscr{G})}\,. \qquad (31)
$$

Together with eq. (29) this leads to $p(\vec{i} \to \vec{i}', \mathcal{G}) = 1/2$ for flipping a loop. For more complicated models like the t-J model, a more general prescription results [24, 25], such that $p(\vec{i} \to \vec{i}', \mathcal{G})$ departs from 1/2.

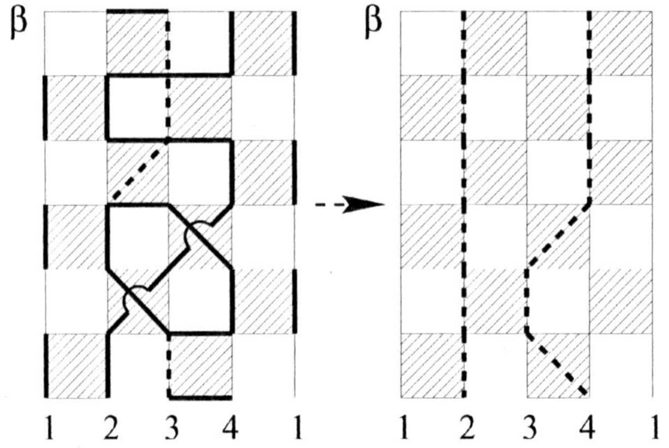

FIGURE 5. Example for loops and the result of flipping them. The dashed lines correspond to world-lines whereas the full lines denote the loops. The largest loop has been flipped in this example

Figure 5 shows an example, where by flipping all the sites along the largest loop, one obtains two world-lines out of one. It can be seen also that the local assignement of graphs leads to closed loops taking into account periodic boundary condition in space and imaginary time.

There are a number of further algorithmic recent advances, like the formulation of loops in continuous imaginary time [27], eliminating thus the systematic errors introduced by the Trotter-Suzuki splitting, and the so-called worm algorithm, also in continuous imaginary time, that allows the evaluation of one-particle Green's functions [28] and also allows effective simulations of spin systems in a magnetic field [29]. Other recent developments include the use of a single loop for the simulation of the infinite lattice and/or zero temperature [30], the solution of the minus sign problem in specific cases [31], and the stochastic series expansion [32], that allows for global updates in the frame of a series expansion. We just mention them here and refer to the original work and other review articles [11], in order to be able to discuss in later sections other algorithms for lattice fermions.

LOOP-ALGORITHM AND SINGLE HOLE DYNAMICS

In this section we discuss a first algorithm, where global updates are used to study charge dynamics in an antiferromagnet, in the frame of the t-J model. This algorithm is restricted to deal with the dynamics of a single hole in an antiferromagnet. As will be discussed in more detail in the next subsection, such a case, that may appear academic on a first glance, has a direct connection with photoemission experiments in cuprates.

The t-J model describes the limit of extremely strong correlation and because of its simplicity, is one of the most studied correlated systems nowadays, mostly in connection with high-T_c superconductors. In order to be able to discuss variants of the t-J model later, we write the Hamiltonian as follows,

$$H_{tJ} = -\sum_{\substack{i,j\\ \sigma}} t_{ij}\, \tilde{c}^{\dagger}_{i\sigma} \tilde{c}_{j\sigma} + \sum_{i,j} J_{ij} \left(\vec{S}_i \cdot \vec{S}_j - \frac{1}{4} \tilde{n}_i \tilde{n}_j \right) , \tag{32}$$

with $\tilde{c}^{\dagger}_{i\sigma} = (1 - n_{i,-\sigma}) c^{\dagger}_{i\sigma}$, $\tilde{n}_i = \sum_\alpha \tilde{c}^{\dagger}_{i\alpha} \tilde{c}_{i\alpha}$, and $\vec{S}_i = \frac{1}{2} \sum_{\alpha,\beta} c^{\dagger}_{i\alpha} \vec{\sigma}_{\alpha\beta} c_{i\beta}$, where $\vec{\sigma}$ are the usual Pauli-matrices. The operators $\tilde{c}^{\dagger}_{i\sigma}$ and $\tilde{c}_{i\sigma}$ prevent a site from doubly-occupancy, reflecting thus the strong local Coulomb repulsion. We will center our attention, like most of the existing studies, on the nearest neighbor t-J model, where $t_{ij} = t$ and $J_{ij} = J$ for i and j on nearest neighbor sites, $t_{ij} = J_{ij} = 0$ otherwise.

Most of the numerical studies were until now performed by means of exact diagonalizations [8], such that only very small systems ($\sim$ 20 - 30 sites) could be considered. An exception is the one-dimensional case, where the world-line algorithm with local moves, as described at the beginning, was extensively applied [18]. However, the world-line algorithm with local moves is not able to simulate in an effective way exchange processes among particles. In one dimension exchange of particles can only arise through the boundaries when periodic boundary conditions are present, and hence such processes become unimportant for large systems. On the contrary, exchange processes can become very frequent on small regions of the system in higher dimensions, such that the lack of ergodicity of world-lines with local moves does not allow for the simulation of fermionic systems at all. A determinantal algorithm was also developed [33] but no applications are known so far. Green's function Monte Carlo simulations were performed [34-36] for the single and two holes case, however, until now only ground state energies and related quantities but no correlation-functions could be obtained. Further progress in numerical simulations of the t-J model will be discussed in the next section, where we address the case of finite doping.

In order to prepare for the two new algorithms developed for the t-J model, we discuss next a different representation for the model. Furthermore, it is interesting to mention this transformation on its own, since there a form of charge-spin separation appears exactly, without necessity of enlarging the Hilbert space, as done with e.g. slave-bosons [37]. In the transformation introduced by Khaliullin [38, 39], the four possible states for an electron on a site are mapped into the two states of a spinless fermion and the two states of a $S-1/2$ pseudospin:

$$\begin{array}{rcrcl}
c^{\dagger}_{i\uparrow}|0>_i & \rightarrow & |v>_i & \leftrightarrow & |0,\Uparrow>_i\,, \\
|0>_i & \rightarrow & f^{\dagger}_i|v>_i & \leftrightarrow & |1,\Uparrow>_i\,, \\
c^{\dagger}_{i\downarrow}|0>_i & \rightarrow & \sigma^{-}_i|v>_i & \leftrightarrow & |0,\Downarrow>_i\,, \\
c^{\dagger}_{i\downarrow}c^{\dagger}_{i\uparrow}|0>_i & \rightarrow & f^{\dagger}_i\sigma^{-}_i|v>_i & \leftrightarrow & |1,\Downarrow>_i\,.
\end{array} \tag{33}$$

Such a mapping corresponds to the following canonical transformation

$$c^{\dagger}_{i\uparrow} = \gamma^{+}_i f_i - \gamma^{-}_i f^{\dagger}_i\,, \quad c^{\dagger}_{i\downarrow} = \sigma^{-}_i (f_i + f^{\dagger}_i)\,, \tag{34}$$

where $\gamma^{\pm}_i = (1 \pm \sigma^z_i)/2$ and $\sigma^{\pm}_i = (\sigma^x_i \pm i\sigma^y_i)/2$. The spinless fermion operators fulfill the canonical anticommutation relations $\{f^{\dagger}_i, f_j\} = \delta_{i,j}$, and σ^a_i, $a = x, y$, or z, are the

Pauli matrices. As discussed by Angelucci [39], the transformation above can be used in general for the Hubbard model or any other lattice model. In the case of the t-J model, we have to eliminate the doubly occupied state, that corresponds to the last state in (33). In this language, it means that a state with a fermion and a spin-down are not allowed. This leads to the constraint

$$\gamma_i^- f_i^\dagger f_i = 0\,, \tag{35}$$

for every site.

Using the transformation (34), the Hamiltonian (32) in its nearest neighbor form becomes

$$\tilde{H}_{t-J} = +t \sum_{<i,j>} P_{ij} f_i^\dagger f_j + \frac{J}{2} \sum_{<i,j>} \Delta_{ij}(P_{ij} - 1), \tag{36}$$

where $P_{ij} = (1 + \vec{\sigma}_i \cdot \vec{\sigma}_j)/2$ is the permutation operator for the pseudospins, $\Delta_{ij} = (1 - n_i - n_j)$ and $n_i = f_i^\dagger f_i$. In order to obtain the form above, the constraint (35) is used. To apply it on all sites, we need to require that

$$\sum_i \gamma_i^- f_i^\dagger f_i = 0\,. \tag{37}$$

This operator commutes with the Hamiltonian, such that it is a constant of motion. This in turn implies that if a state fulfi lling (37) is initially constructed, then the evolution by (36) will lead only to states in the physical subspace. A very important feature of the Hamiltonian (36) is the fact that although it describes strongly correlated electrons, it is bilinear in the operators $f_i^\dagger$, f_i. This shows that we are dealing with *non-interacting* holes under the influence of a fluctuating quantum mechanical spin fi eld. As we will see next, this is a key feature that allows an effi cient simulation of the model.

Single hole dynamics

As a fi rst step in order to study charge dynamics in a quantum antiferromagnet we consider the dynamics of a single hole in quantum antiferromagnets. Such studies were pioneered by Brinkman and Rice [40], who investigated the propagation of a hole neglecting quantum fluctuations of the spin background. After the discovery of high temperature superconductors and the suggestions by Anderson [41] on the possibility of a non-Fermi liquid state in those materials, the question whether the quasiparticle weight of a hole vanishes due to the interaction with an antiferromagnetic background became central in the fi eld of strongly correlated fermions.

Although the single hole case may appear academic, it can be realized in photoemission experiments in quantum antiferromagnets. They were in fact carried out in compounds such as $SrCuO_2$ [42], $Na_{0.96}V_2O_5$ [43] as examples for chains, $Sr_{14}Cu_{24}O_{41}$ [44] for ladders and $Sr_2CuO_2Cl_2$ [45] for planes.

Since the spectral function in photoemission experiments is determined by the one-particle Green's function, we consider in the following an algorithm suited for its calculation [46]. Due to the form of the canonical transformation (34), it is convenient to focus on the one-particle Green's function for spin up.

$$G(i-j,\tau) = -\langle T\tilde{c}_{i,\uparrow}(\tau)\tilde{c}^{\dagger}_{j,\uparrow}\rangle = -\langle T f_i^{\dagger}(\tau) f_j\rangle \tag{38}$$

where T corresponds to the time ordering operator. The propagation of down spin electrons cannot be easily considered, since the operators $\sigma_{i,\pm}$ cut world-lines. This is certainly not a problem for fi nite-size systems, where SU(2) spin symmetry is conserved.

Since we restrict ourselves to the case where a hole is created in a pure antiferromagnet, we can calculate the Green's function (38) as follows (we display here the case $d=1$ for simplicity):

$$G(i-j,-\tau) = \frac{\sum\limits_{\{\sigma_0\}} \langle v| \otimes \langle\sigma_0| e^{-(\beta-\tau)\tilde{H}} f_j e^{-\tau\tilde{H}} f_i^{\dagger} |\sigma_0\rangle \otimes |v\rangle}{\sum\limits_{\{\sigma_0\}} \langle v| \otimes \langle\sigma_0| e^{-\beta\tilde{H}} |\sigma_0\rangle \otimes |v\rangle}, \tag{39}$$

where $\{\sigma_0\}$ is a complete set of spin states, and $|v\rangle$ is the vacuum state for holes. $\tilde{H}$ stands for the Hamiltonian (36). However, both in the denominator and in the numerator after the destruction of the hole, it reduces to the Heisenberg Hamitonian, since we are considering the case without holes. Therefore, after a Trotter decomposition, where the Hamiltonian is splitted into contributions containing only two-site terms, we have

$$G(i-j,-\tau) = \sum_{\vec{\sigma}} P(\vec{\sigma}) G(i,j,\tau,\vec{\sigma}) + \mathcal{O}(\Delta\tau^2), \tag{40}$$

where

$$P(\vec{\sigma}) = \frac{\langle\sigma_0| e^{-\Delta\tau\tilde{H}} |\sigma_{m-1}\rangle \cdots \langle\sigma_1| e^{-\Delta\tau\tilde{H}} |\sigma_0\rangle}{\sum\limits_{\{\vec{\sigma}\}} \langle\sigma_0| e^{-\Delta\tau\tilde{H}} |\sigma_{m-1}\rangle \cdots \langle\sigma_1| e^{-\Delta\tau\tilde{H}} |\sigma_0\rangle} \tag{41}$$

is the probability distribution of a Heisenberg antiferromagnet for the confi guration $\vec{\sigma}$, where $\vec{\sigma}$ is a vector containing all intermediate states $(\sigma_{m-1},\ldots,\sigma_0)$, with $m\Delta\tau = \beta$, and $\beta = 1/k_B T$ the inverse temperature. Furthermore, we have defi ned the propagator $G(i,j,\tau,\vec{\sigma})$ of a hole *for each configuration* $\vec{\sigma}$ as

$$G(i,j,\tau,\vec{\sigma}) \;=\; \frac{\langle v| f_j \langle\sigma_n| e^{-\Delta\tau\tilde{H}} |\sigma_{n-1}\rangle \cdots \langle\sigma_1| e^{-\Delta\tau\tilde{H}} |\sigma_0\rangle f_i^{\dagger} |v\rangle}{\langle\sigma_n| e^{-\Delta\tau\tilde{H}} |\sigma_{n-1}\rangle \cdots \langle\sigma_1| e^{-\Delta\tau\tilde{H}} |\sigma_0\rangle}, \tag{42}$$

where $n\Delta\tau = \tau$, $\Delta\tau t \ll 1$. Whereas the matrix elements in the numerator describe the evolution of a hole created by $f_i^{\dagger}$ on the vacuum $|v\rangle$, the matrix elements in the denominator correspond to the ones of a Heisenberg model for the same spin confi guration. With such a reweighting we obtain the probability distribution $P(\vec{\sigma})$ in (40).

The sum over spins is performed by using the world-line loop-algorithm for a Heisenberg antiferromagnet with discretized imaginary time, as discussed in the previous sections. In general we have $\Delta\tau = 0.05$, such that the extrapolation to $\Delta\tau = 0$ leads to values of the observables within the statistical error bars.

As the evolution operator for the holes is a bilinear form in the fermion operators, $G(i,j,\tau,\vec{\sigma})$ can be calculated exactly, in contrast to a direct implementation in the loop algorithm [28], where fermion paths are sampled stochastically. $G(i,j,\tau,\vec{\sigma})$ contains a sum over all possible fermion paths between $(i,0)$ and (j,τ). The numerical effort to calculate $G(i,j,\tau,\vec{\sigma})\,\forall i,\tau$ scales as $N\tau$, where N is the number of lattice points in space. Therefore, the present method is more effi cient for large systems than e.g. projector algorithms for the Hubbard model [14], that scale with the system size cubed.

We now address the explicit calculation of $G(i,j,\tau,\sigma)$. After the Trotter splitting, we have to deal with the Hamiltonian in each bond. From (36), we see that for each nearest neighbor bond $<i,j>$ we have

$$\tilde{H}_{<i,j>} = \hat{\mathscr{A}}\left(1+\vec{\sigma}_i\cdot\vec{\sigma}_j\right) + \left[\frac{J}{4} - \hat{\mathscr{B}}\right]\left(\vec{\sigma}_i\cdot\vec{\sigma}_j - 1\right) , \tag{43}$$

where

$$\hat{\mathscr{A}} \equiv \frac{t}{2}\left(f_i^\dagger f_j + f_j^\dagger f_i\right) , \tag{44}$$

and

$$\hat{\mathscr{B}} \equiv \frac{J}{4}\left(f_i^\dagger f_i + f_j^\dagger f_j\right) . \tag{45}$$

Furthermore, $\left[\hat{\mathscr{A}},\hat{\mathscr{B}}\right] = 0$. This property is useful to calculate the matrix elements of $\exp\left(-\Delta\tau\tilde{H}_{<i,j>}\right)$ with respect to spin-states. They have the following form:

$$\begin{aligned}
i)\ <\uparrow\uparrow|\,\mathrm{e}^{-\Delta\tau\tilde{H}_{<i,j>}}\,|\uparrow\uparrow> &= \mathrm{e}^{-\Delta\tau 2\hat{\mathscr{A}}} ,\\
ii)\ <\downarrow\downarrow|\,\mathrm{e}^{-\Delta\tau\tilde{H}_{<i,j>}}\,|\downarrow\downarrow> &= i) ,\\
iii)\ <\uparrow\downarrow|\,\mathrm{e}^{-\Delta\tau\tilde{H}_{<i,j>}}\,|\uparrow\downarrow> &= \frac{1}{2}\left\{\mathrm{e}^{-\Delta\tau 2\hat{\mathscr{A}}} + \exp\left[\Delta\tau\left(2\hat{\mathscr{A}} + J - 4\hat{\mathscr{B}}\right)\right]\right\} ,\\
iv)\ <\uparrow\downarrow|\,\mathrm{e}^{-\Delta\tau\tilde{H}_{<i,j>}}\,|\downarrow\uparrow> &= \frac{1}{2}\left\{-\mathrm{e}^{-\Delta\tau 2\hat{\mathscr{A}}} + \exp\left[\Delta\tau\left(2\hat{\mathscr{A}} + J - 4\hat{\mathscr{B}}\right)\right]\right\} ,\\
v)\ <\downarrow\uparrow|\,\mathrm{e}^{-\Delta\tau\tilde{H}_{<i,j>}}\,|\uparrow\downarrow> &= iv) ,\\
vi)\ <\downarrow\uparrow|\,\mathrm{e}^{-\Delta\tau\tilde{H}_{<i,j>}}\,|\downarrow\uparrow> &= iii) .
\end{aligned} \tag{46}$$

Next, we consider the action of the fermionic operators above on a single hole state. We start with

$$\begin{aligned}
\exp\left(-4\Delta\tau\hat{\mathscr{B}}\right) &= \exp\left[-\Delta\tau J\left(f_i^\dagger f_i + f_j^\dagger f_j\right)\right]\\
&= \left[1+\left(\mathrm{e}^{-\Delta\tau J}-1\right)f_i^\dagger f_i\right]\left[1+\left(\mathrm{e}^{-\Delta\tau J}-1\right)f_j^\dagger f_j\right] ,
\end{aligned} \tag{47}$$

such that for a general single-hole state

$$|\Psi> = \sum_{\ell=1}^{N} P_\ell f_\ell^\dagger\,|\nu> , \tag{48}$$

with N the number of sites, we have

$$| \Psi' >= \exp\left(-4\Delta\tau\hat{\mathscr{B}}\right) | \Psi > , \tag{49}$$

with $| \Psi' >= \sum_\ell P'_\ell f^\dagger_\ell | \nu >$, and

$$P'_\ell = \begin{cases} \mathrm{e}^{-\Delta\tau J} P_\ell & \text{for } \ell = i, \text{ and } \ell = j \\ P_\ell & \text{else.} \end{cases} \tag{50}$$

Therefore, the action of this operator on single-hole states for the considered two sites is given by

$$\mathrm{e}^{-4\Delta\tau\hat{\mathscr{B}}} \quad \to \quad \mathrm{e}^{-\Delta\tau J}\begin{pmatrix} 1 & 0 \\ 0 & 1 \end{pmatrix} . \tag{51}$$

For the operator $\mathscr{A}$, one can see in a similar way that

$$\mathrm{e}^{\pm\Delta\tau 2\hat{\mathscr{A}}} \quad \to \quad \begin{bmatrix} \cosh(\Delta\tau t) & \pm\sinh(\Delta\tau t) \\ \pm\sinh(\Delta\tau t) & \cosh(\Delta\tau t) \end{bmatrix} , \tag{52}$$

such that for the operators in (46) we have

$$\begin{aligned} i) \quad &\to \quad \begin{bmatrix} \cosh(\Delta\tau t) & -\sinh(\Delta\tau t) \\ -\sinh(\Delta\tau t) & \cosh(\Delta\tau t) \end{bmatrix} , \\ iii) \quad &\to \quad \begin{bmatrix} \cosh(\Delta\tau t) & 0 \\ 0 & \cosh(\Delta\tau t) \end{bmatrix} , \\ iv) \quad &\to \quad \begin{bmatrix} 0 & \sinh(\Delta\tau t) \\ \sinh(\Delta\tau t) & 0 \end{bmatrix} . \end{aligned} \tag{53}$$

With the above in mind, we go back to the propagator of a hole for a given pseudospin confi guration (42). There we defi ne evolution operators

$$U(\sigma_\ell, \sigma_{\ell-1}) \equiv \frac{< \sigma_\ell | \mathrm{e}^{-\Delta\tau\tilde{H}} | \sigma_{\ell-1} >}{< \sigma_\ell | \mathrm{e}^{-\Delta\tau H_H} | \sigma_{\ell-1} >} , \tag{54}$$

where in the denominator we write explicitly the Heisenberg-Hamiltonian in order to make clear that these matrix elements do not include a hole. For the evolution from one time-slice to the next one, the matrix elements of the operators $U(\sigma_\ell, \sigma_{\ell-1})$ between one-particle states leads to $N \times N$ matrices that we denote as follows:

$$U(\sigma_\ell, \sigma_{\ell-1}) \to B_\ell . \tag{55}$$

They are built up by 2×2 matrices as given by (53), and operate on amplitudes P as introduced in (48). The possible values for these entries are summarized in Table 1.

Then, $G(i, j, \tau, \vec{\sigma})$ becomes

$$G(i, j, \tau, \vec{\sigma}) \quad = \quad \sum_{k_n, \ldots, k_0} \delta_{j,k_n} (B_n)_{k_n,k_{n-1}} \cdots (B_1)_{k_1,k_0} \delta_{k_0,i} , \tag{56}$$

TABLE 1. Contributions for the propagation of the hole on one plaquette. The first column shows the weight for a propagation where the hole stays on the same site x, whereas in the second column the weight corresponds to the propagation to the adjacent site. The third column represents the spin background on the plaquette.

$x \to x$	$x + \vec{\delta} \to x$	spin configuration
0	0	$\begin{smallmatrix}\downarrow & \downarrow \\ \downarrow & \downarrow\end{smallmatrix}$
$\cosh(\Delta\tau t)$	$-\sinh(\Delta\tau t)$	$\begin{smallmatrix}\uparrow & \uparrow \\ \uparrow & \uparrow\end{smallmatrix}$
$\frac{\cosh(\Delta\tau t)}{\exp(\Delta\tau J/2)\cosh(\Delta\tau J/2)}$	0	$\begin{smallmatrix}\uparrow & \downarrow \\ \uparrow & \downarrow\end{smallmatrix}, \begin{smallmatrix}\downarrow & \uparrow \\ \downarrow & \uparrow\end{smallmatrix}$
0	$\frac{\sinh(\Delta\tau t)}{\exp(\Delta\tau J/2)\sinh(\Delta\tau J/2)}$	$\begin{smallmatrix}\downarrow & \uparrow \\ \uparrow & \downarrow\end{smallmatrix}, \begin{smallmatrix}\uparrow & \downarrow \\ \downarrow & \uparrow\end{smallmatrix}$

where the indices $k_n, \ldots, k_0$ run over the site indices $1, \ldots, N$. As we are only interested in the Hilbert space with no double occupancy, we have to enforce the constraint at one single position of the propagation by projecting out the fermionic states which do not respect the constraint. We do so at $\tau = 0$ corresponding to the fi rst propagation.

It is possible to extract spectral functions from the imaginary time propagators since the spectral function

$$A(\vec{k}, \omega) = \sum_{f,\sigma} \left| \langle f, N-1 | c_{\vec{k},\sigma} | 0, N \rangle \right|^2 \delta\left(\omega - E_0^N + E_f^{N-1}\right), \tag{57}$$

is connected with the Green's function in imaginary time at $T = 0$, by the spectral theorem

$$G(\vec{k}, \tau) = \int_{-\infty}^{\infty} d\omega \frac{\exp(-\tau\omega)}{\pi} A(\vec{k}, \omega). \tag{58}$$

Here $|0, N\rangle$ is the ground-state at half fi lling with energy E_0^N and $|f, N-1\rangle$ are states in the $N-1$ particle Hilbert space with energy E_f^{N-1}. We perform the inversion of Eq. (58), that due to the statistical errors of $G(\vec{k}, \tau)$ is an extremely ill-posed problem, by means of MaxEnt, where the $A(\vec{k}, \omega)$ obtained is the one that maximizes the probability $P(A|G)$, given the Green's function $G(\vec{k}, \tau)$. Correlations in the imaginary time data were taken into account by considering the covariance matrix. Details about MaxEnt can be found in the comprehensive review article by J.E. Gubernatis and M. Jarrell [47].

We would like to stress fi nally, that part of the dynamical data presented below were obtained without use of MaxEnt but directly extracted from the imaginary time Green's function. This is possible due to the high statistics and stability attainable with the present algorithm, that results one the one hand, from the fact that global moves lead to very good statistics with less efforts that in single-move simulations. On the other hand, the algorithm for a single hole is free from the minus-sign problem, in contrast to previous studies [35, 36]. The slowest decaying exponential, that corresponds to the excitation with lowest energy can be extracted simply by fi tting the tail of the Green's function at large values of τ. This leads to the value of the excitation and its corresponding weight, as shown in the following subsections. Furthermore, in connection with Max-Ent, the next higher excitation can be obtained by subtracting the contribution from the lowest one from the Green's function. This procedure is used in the two-dimensional case below.

After having presented the algorithmic details, we consider next the one- and two-dimensional cases. The first case will allows us to see if the method used is accurate enough to study the vanishing of the quasi-particle weight in the thermodynamic limit, as expected for a Luttinger liquid. Once the accuracy of the method is tested, we discuss the two-dimensional case, that is relevant for high temperature superconductors.

Single hole dynamics in one dimension

One-dimensional compounds attract at present an increasing amount of interest in order to elucidate, whether signals of charge-spin separation as predicted from Luttinger-liquid theory [48, 49] can be observed experimentally. On the other hand, theoretical treatments based on Bethe-*Ansatz* led recently to a complete description of the spectral function of the Hubbard model at $U = \infty$ [50] and the low energy sector in the nearest-neighbour t-J model, where explicit results are obtained at the supersymmetric (SuSy) point [51]. We therefore consider first the one-dimensional case as a test of our algorithm [52].

The simulations were performed at temperatures $T \leq \min(J,t)/15$, such that no appreciable changes with a further decrease in temperature can be seen: the results correspond to the zero temperature limit, a limit which is in general difficult to reach in other finite-temperature fermionic algorithms.

We compare our results with the predictions of the charge-spin separation *Ansatz* (CSSA), where free holons and spinons are described by [53, 41]

$$H = -\frac{t_h}{2} \sum_{<i,j>} h_i^\dagger h_j - \frac{J_s}{2} \sum_{<i,j>} s_{i,\sigma}^\dagger s_{j,\sigma}. \tag{59}$$

Here the electron operator $c_{i,\sigma}$ is given by the product of a holon (h_i) and a spinon ($s_{i,\sigma}$) operator, $c_{i\sigma} = s_{i,\sigma} h_i^\dagger$, the holon being a boson and the spinon a spin-1/2 fermion. As a consequence of the above *Ansatz* , the dispersion relations of the free holons and spinons are given by $\varepsilon_h = -t_h \cos q_h$ and $\varepsilon_s = -J_s \cos q_s$ respectively, whereas the energy of the hole is $E(k) = \varepsilon_h - \varepsilon_s$ and by momentum conservation $k = q_h - q_s$. We take t_h and J_s as two free parameters in contrast to a mean-field approximation, where they have to be calculated self-consistently. The spectral function is then given by a convolution of the spinon and holon Green functions. The lowest attainable energy ($-t_h$) and highest one ($t_h + J_s$) define the bandwidth of the hole, $2t_h + J_s$. Since the full band-width obtained by considering the compact support of the spectral function at $J = 0$ is known to be exactly $4t$ [50], we take $t_h = 2t$. In order to determine J_s, we consider the overall bandwidth, as obtained from the simulation. For all values of J, the width of the density of states $N(\omega)$ scales approximately as $4t + J$ in the parameter range considered [52], leading to $J_s = J$.

Beyond predicting bandwidths, the CSSA in 1D describes accurately the support of the spectral function in the case $J = 0$, when compared with exact results [50, 53]. If furthermore phase string effects [53] are taken into account, the singularities of $A(k,\omega)$ related to holons and spinons can be reproduced. For finite J, the minimal (maximal) possible energy of a hole in CSSA is given by $E(k) = -F_k$ ($E(k) = F_k$) for $k < k_0$ ($k > k_0$), where $F_k \equiv \sqrt{J^2 + 4t^2 - 4tJ\cos(k)}$ contains both holon and spinon contributions, and k_0 is determined by $\cos(k_0) = J/(2t)$. The remaining parts of the compact support are given by $E(k) = \mp 2t\sin(k)$ for $k > k_0$ (lower edge) and $k < k_0$ (upper edge) respectively. Such dispersions correspond to holons with momentum $k + q_s$, and a spinon with $q_s = \mp\pi/2$ [51, 53]. As $J \to 2t$, $k_0 \to 0$ and the lower edge of the compact support is entirely determined by the dispersion of the holon.

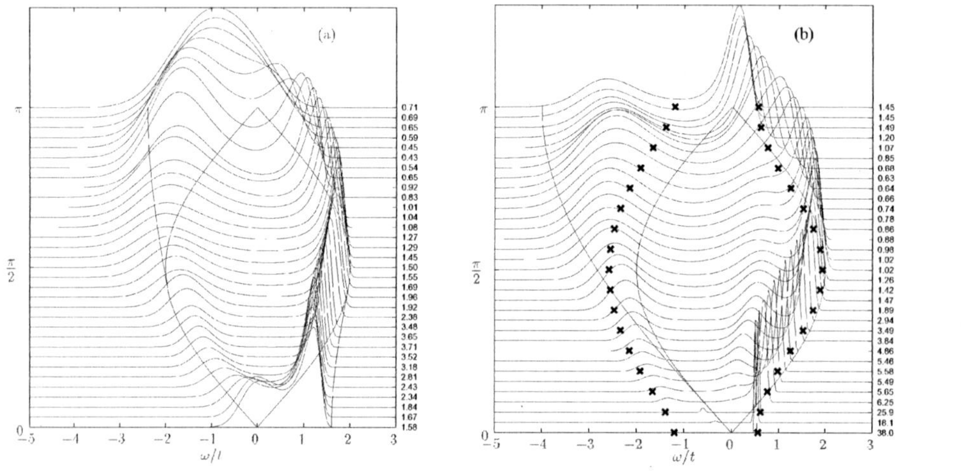

FIGURE 6. Spectral function $A(k,\omega)$ for $J = 0.4t$ (a) and $J = 2t$ (b). Here the wave vector $k \in [0,\pi]$ is given on the y-axis. For clarity, the data is rescaled by the number given at the right hand side of the plot.

We now compare the above predictions with our QMC data. Figure 6 shows $A(k,\omega)$ for $J/t = 0.4$ (a) and 2 (b). In all cases the compact support is reproduced very well by the CSSA. The *Ansatz* also predicts singularities at the lower (upper) edge for $k < k_0$ ($k > k_0$), and when phase strings are considered [53] along the edges and the holon lines ($\pm 2t\sin(k)$) for all momenta. The singularities along the lower holon line are also supported by a recent low energy theory [51]. For all parameter values we observe dominant weight along the above mentioned lines. For $J/t = 0.4$, we have checked that the results are consistent within the uncertainties of MaxEnt with a peak along the edges and a further peak along the holon lines, signaled by a broad structure between the edges and the holon lines (Fig. 6 a). We observed such a behaviour for $0.33 \leq J/t \leq 0.6$. For $J/t \geq 1.2$, the structure at the lower edge narrows considerably and the data are not any more consistent with an additional structure along the lower holon line for $k < k_0$, but only with a singularity for $k > k_0$ [52].

At $J/t = 2$ the exact holon and spinon dispersions can be obtained by Bethe-*Ansatz* [5, 6]. Figure 6 (b) shows the comparison with the CSSA, where on the one side the original dispersions are used (full line) and on the other side, with the dispersions as given by Bethe-*Ansatz* (crosses). Whereas the Bethe-*Ansatz* holon dispersion reproduces very well the lower edge, showing that as anticipated by the CSSA, at the SuSy point that edge is completely determined by the holon dispersion, the full bandwidth is better described with the original dispersions. We assign the additional weight in the region $k > \pi/2$ to processes involving one holon and more than one (Bethe-*Ansatz*) spinon. In fact, that portion resembles the difference between the supports for one-holon/one-spinon and one-holon/three-spinon processes in the so-called $1/r^2$ t-J model [54]. In this case, both the hopping matrix elements t_{ij} and the exchange couplings J_{ij} in the Hamiltonian (32) are assumed to have the form

$$t_{ij} = J_{ij}/2 = t(\pi/L)^2/\sin^2(\pi(i-j)/L) , \tag{60}$$

where L is the length of the system in units of the lattice constant. We will discuss this model in more detail when dealing with the case of fi nite doping. In our case, no limitation on the possible number of spinons exists, such that in principle all odd number of them are allowed. It is interesting to notice that using a fermionic spinon one is able to describe both the case $J = 0$ and $J = 2t$. In the fi rst case, the spinon in the exact solution is a fermion. At the SuSy point it is expected to be a semion [54, 55] and on the basis of our results, we conclude that the fermionic spinon contains all possible states with an odd number of semionic spinons.

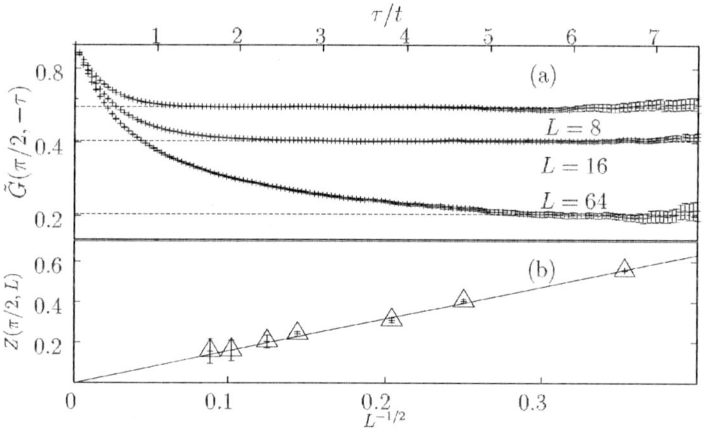

FIGURE 7. Quasiparticle weight at $k = \pi/2$. (a) $\tilde{G}(\pi/2, -\tau)$ versus τ/t. At $\tau/t \gg 1$ this quantity converges to the quasiparticle weight $Z(\pi/2)$. (b) Finite size scaling of $Z(\pi/2)$ as obtained from (a). The solid line is a least square fit to the form $L^{-1/2}$. We consider $\beta J = 30$ for $L \leq 48$ and $\beta J = 60$ for $L > 48$, to guarantee convergence in τ.

Finally, we consider the quasiparticle residue $Z_k = \left|\langle \Psi_0^{L-1} | \tilde{c}_{k\sigma} | \Psi_0^L \rangle\right|^2$ at $k = \pi/2$ for $J = 2t$. As Fig. 6 (b) shows, the lower edge is very sharp and without prior knowledge, the question may arise whether we are dealing with a quasiparticle. Z_k is related to the imaginary time Green function through:

$$\lim_{\tau \to \infty} G(k, -\tau) \propto Z_k \exp\left[\tau \left(E_0^L - E_0^{L-1}(k)\right)\right] . \tag{61}$$

Figure 7 (a) shows $\tilde{G}(\pi/2,-\tau) \equiv G(\pi/2,-\tau)\exp\left(-\tau(E_0^L - E_0^{L-1}(\pi/2))\right)$ versus τ, where the energy difference is obtained by fi tting the tail of $G(\pi/2,-\tau)$ to a single exponential form, for several sizes. The thus estimated $Z(\pi/2)$ is plotted versus system size in Fig. 7 (b). Our results are consistent with a vanishing quasiparticle weight $Z(\pi/2) \propto L^{-1/2}$ which is the scaling obtained by a combination of bosonization and conformal fi eld theory [51].

The results above show that the algorithm for a single hole discussed here is able to allow for the extraction of dynamical data with high accuracy, such that it is possible even to determine the power-law for the vanishing of the quasiparticle weight in hte thermodynamic limit. Since the CPU-times scales as $V\beta$ (V is the volume) the determination of the Z-factor may be effi ciently extended to higher dimensions, in contrast to determinantal algorithms for the Hubbard model that scale as $V^3\beta$.

Similar simulations were also performed on two- and three-leg ladders [56], where shadow bands and a fi nite quasiparticle weight could be determined in the case of a two-leg ladder, whereas the three-leg ladder shows vanishing weight in the antisymmetric channel and fi nite quasiparticle weight in the symmetric ones.

Single hole dynamics in two dimensions

An accurate description of the dynamics of a single hole in two dimensions is of central importance for theories of high temperature superconductors. In particular, to elucidate whether the hole can be consider as a quasiparticle, or quantum fluctuations of the spin-background lead to a vanishing quasiparticle weight, is decisive for theories, where departures from a Fermi liquid are proposed [41].

The Brinkman and Rice [40] treatment of a single hole in a spin background led to a fully incoherent spectrum in the so-called retraceable path approximation, for an antiferromagnetic Ising-like background, in the limit $J_z \to 0$. The retraceable path approximation, where it is assumed that a hole travels in the system on a path that does not close on itself, is exact in one [40] and in infi nite dimensions [57] but not in two dimensions since contributions of loops (Trugman paths [58]) may lead to a coherent propagation of the hole. In fact, for an Ising-like background it was shown within a Lanczos scheme [59], that a fi nite quasiparticle weight results.

For the case of physical interest, namely with a Heisenberg spin-background, a large number of numerical methods [8] led to conflicting results. Whereas exact diagonalizations found large quasiparticle peaks at the lower edge of the spectrum [60], QMC results were interpreted as leading to a vanishing quasiparticle weight [61]. Since exact diagonalizations are possible only on very small lattices, fi nite size scaling cannot be performed reliably. On the other hand, QMC simulations suffered from the minus-sign problem, such that scaling was not possible with reasonable confi dence. Further studies based on the self-consistent Born approximation (SCBA) [62-64] gave a fi nite quasiparticle weight. However, since there fluctuations of the spin-background are only taken into account in the frame of a spin-wave approximation, the results obtained are *a priori* not conclusive. Exact results for the supersymmetric point $J = 2t$ were obtained by Sorella [65], that give important benchmarks for any analytical or numerical method, but unfortunately, they cannot be rigorously extended to the physical relevant parameter range $J \sim 0.5t$.

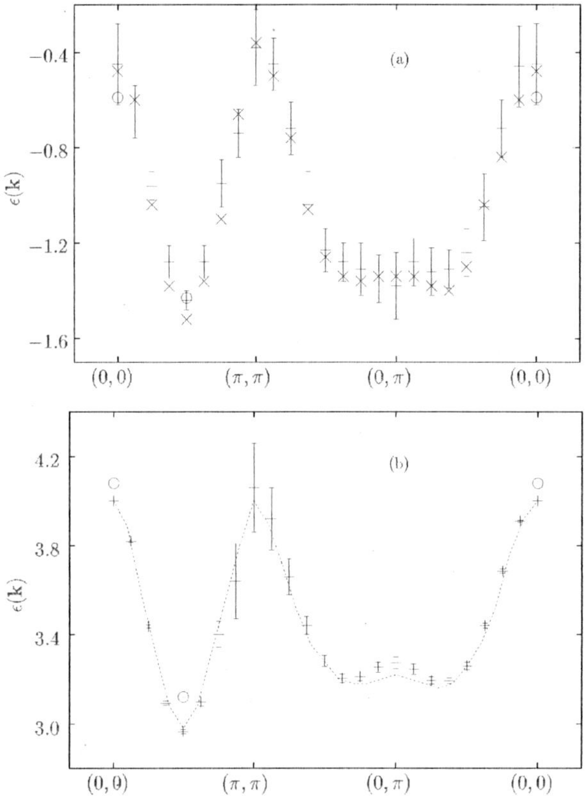

FIGURE 8. Lower edge of the spectrum along the symmetry lines of the Brillouin zone for a) $J/t = 0.4$ and b) $J/t = 2$ in a 16×16 lattice. Comparisons are made with VMC (circles), GFMC for $J/t = 0.4$ ($\times$), and series expansion [66] for $J/t = 2$ (dotted line).

The dynamics of a single hole in an antiferromagnetic background became quite recently, experimentally accessible by angle resolved photoemission spectroscopy (ARPES) in undoped materials like $Sr_2CuO_2Cl_2$ [45, 67, 68] and $Ca_2CuO_2Cl_2$ [69]. The main features observed there are a minimum of the dispersion at $\vec{k} = (\pi/2, \pi/2)$ together with a vanishing of spectral weight beyond this point along the (1,1) direction. The obtained spectra show that the very flat portion around $(\pi, 0)$, that in optimally doped materials is almost degenerate with the bottom of the spectrum at $(\pi/2, \pi/2)$ [70], is shifted upwards (in a hole representation) by approximately 300 meV. This contradicts the single hole spectra found theoretically so far, where essentially the lower edge of the spectrum at $\vec{k} = (\pi/2, \pi/2)$ and $(\pi, 0)$ are almost degenerate, such that additional second and third nearest neighbor hopping terms were suggested [68, 71], that lead to an agreement of the exact diagonalization results with experiments. Such terms were made recently responsible also for the vanishing of spectral weight close to $(\pi/2,\pi/2)$ by reducing the quasiparticle weight [71, 72]. In the following we present a summary of a detailed study of the dynamical properties of a single hole in a two-dimensional t-J model on lattices with up to 24×24 sites in the parameter range $0.4 \leq J/t \leq 4$ [46].

We consider first the lower edge of the spectrum. This is a quantity that can be obtained by several other methods, including various Monte Carlo algorithms, such that the relative accuracy of each one and the region in parameter space, where each method gives best results, can be assessed. In our case, the accuracy and stability of the data allow to obtain the lower edge of the spectrum directly from the slope of the one-particle Green's function as a function of imaginary time τ, for large values of τ.

Figure 8 shows the lower edge of the spectrum for $J/t = 0.4$ and $J/t = 2$ in a 16×16 sites system. The energies are displayed with respect to the ground-state energy of the Heisenberg antiferromagnet. The results are compared with variational Monte Carlo (VMC) [34], Green's function Monte Carlo (GFMC) [36], and series expansions [66], whenever data is available. At $J/t = 0.4$ (Fig. 8a), where our results are most affected by fluctuations, we observe good agreement with GFMC. The behavior of the statistical error is similar in both methods, with larger fluctuations around $\vec{k} = (0,0)$ and (π,π). Around $\vec{k} = (\pi,0)$ our results show somewhat larger fluctuations. For $J/t = 0.4$ VMC [34], also appears to be very accurate concerning the lower edge. When its energies are compared to our calculations and the GFMC technique, we find that their energies are within the error bars of the exact QMC calculations. At $\vec{k} = (0,0)$, the variational result is at the lower edge of the error bars of our calculation, and have the smallest statistical error of all three approaches. At this specific k-point both GFMC and our approach have large fluctuations before the state with lowest energy is clearly reached. As mentioned above, additional calculations with $\Delta\tau t = 0.2$ were performed, in order to check the results obtained, without observing significant changes.

Figure 8b shows that at $J/t = 2$, where our algorithm leads to much more accurate results, the variational results are too high in energy, but still close to our numerically exact ones. For values of $J/t \geq 1$, additional results from series expansions [66] are available. At $J/t = 2$ we observe in general a very good agreement. Only around $(\pi,0)$ we see that series expansions slightly underestimate the energy of the hole.

The resulting dispersions agree with previous results obtained within SCBA and series expansions [66] for $J/t < 1$, whereas for $J/t > 1$ only agreement with series expansions is found. In particular, a flat dispersion is obtained around $\vec{k} = (\pi,0)$ very close in value to the bottom of the band at $\vec{k} = (\pi/2,\pi/2)$, in contrast to the experiments [45, 68, 69].

As already discussed in the one-dimensional case, the asymptotics of the imaginary time Green's function delivers also the quasiparticle weight for that band. It is the weight of the exponential with the slowest decay, that is the exponential that determines the lower edge of the spectrum. Similarly to (61), it is

$$Z(\vec{k}) = \lim_{-\tau\to\infty} G(-\tau,\vec{k}) \exp\left[\left(\varepsilon_{\vec{k}} - \varepsilon_0\right)\tau\right] \tag{62}$$

In the following we focus on the thermodynamic limit of $Z(\vec{k})$ for the wave vectors $\vec{k} = (\pi, 0)$ and $\vec{k} = (\pi/2, \pi/2)$.

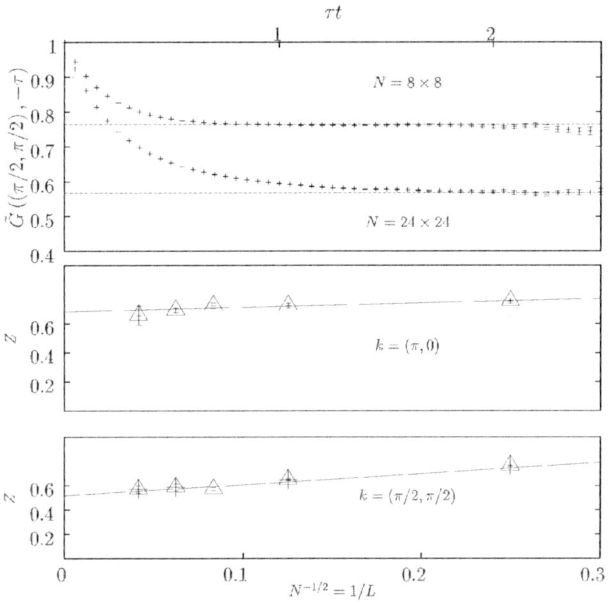

FIGURE 9. a) Extrapolation of $\tilde{G}(\vec{k}, -\tau) \equiv G(\vec{k}, -\tau)\exp[(\varepsilon_k - \varepsilon_0)\tau]$ for $N = 8 \times 8$ and 24×24 at $J/t = 2$. Finite-size scaling for b) $\vec{k} = (\pi/2, \pi/2)$ and c) $\vec{k} = (\pi, 0)$.

Figure 9 shows the finite-size scaling on these two points for $J/t = 2$. For both k-values, an appreciable quasiparticle weight is obtained, demonstrating that the lower edge of the spectrum describes the band of a coherent quasiparticle. The size dependence of $Z(\pi/2, \pi/2)$ and $Z(\pi, 0)$ is not very large and scales linearly with the inverse linear size of the system for $J/t \geq 0.6$, in agreement with SCBA [63]. The size dependence at $(\pi/2, \pi/2)$ is systematically larger than at $(\pi, 0)$. The sizes considered are $L \times L$, with $L = 16, 12, 8$, and 4. At $J/t = 2$ we use additionally a 24×24 lattice.

Figure 10 shows that the extrapolated quasiparticle weight increases with J/t both for $\vec{k} = (\pi, 0)$ and $\vec{k} = (\pi/2, \pi/2)$. At $J/t = 4$ the quasiparticle reaches about 80% of its maximal value. The changes of the quasiparticle weight with J/t are small when $J/t \geq 1$ and the slope becomes steeper for smaller values. Estimates of the quasiparticle weight were given both by VMC [34] and SCBA [63], the difference being rather small. The general trend is that VMC overestimates it at small J whereas SCBA overestimates it at large J.For definiteness we compare our results with SCBA for a 16×16 system in Fig. 10. We find a rather good agreement between both methods. As in our case $Z(\pi, 0) > Z(\pi/2, \pi/2)$ for all considered values of J/t. At small values of J ($0.01 \leq J/t \leq 0.5$) SCBA finds a scaling of $Z(\pi/2, \pi/2) = 0.31J^{2/3}$ and $Z(\pi, 0) = 0.35J^{0.7}$. For $J/t \geq 1$, the results from SCBA overestimates the quasiparticle weight at the two considered k-points, with an increasing deviation for larger values of J/t. Based on the quantitative agreement of SCBA with our results for small J, we can confidently conclude that the quasiparticle at $\vec{k} = (0, \pi)$ and $(\pi/2, \pi/2)$ should be finite for all values of J in the physically relevant region (i.e. $J/t > 0.1$).

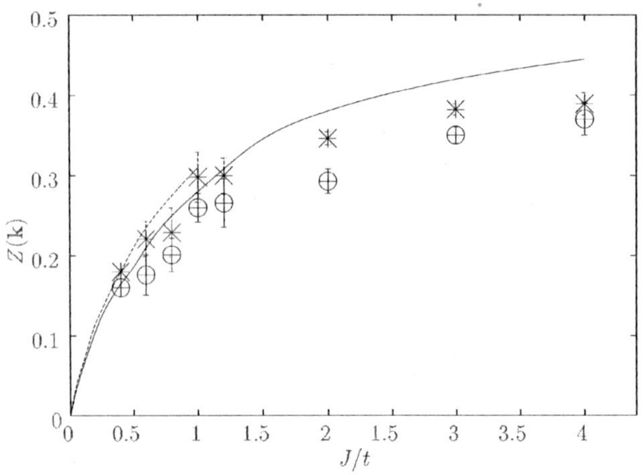

FIGURE 10. Quasiparticle weight as a function of J/t for $\vec{k} = (\pi/2, \pi/2)$ (circles) and $\vec{k} = (\pi, 0)$ ($\times$) in a 16×16 lattice. We compare our result with SCBA, where the dashed line corresponds to the quasiparticle weight for $\vec{k} = (\pi, 0)$ and the full line corresponds to $\vec{k} = (\pi/2, \pi/2)$. The SCBA data were taken from Ref. [63].

The results discussed above for the lower edge of the spectrum and for the quasiparticle weight can be recognized in the spectral function (Fig. 11) obtained by using MaxEnt. For clarity, the maximum of each curve is normalized to 1 in the plots. The small numbers on the right hand side of the figures correspond to the maximal value of $A(\vec{k}, \omega)$ when the integral $\int_{-\infty}^{\infty} d\omega A(\vec{k}, \omega)$ is properly normalized to $\pi/2$. The peaks around $(0, \pi)$ and $(\pi/2, \pi/2)$ are generally very sharp, in agreement with the fact that a finite quasiparticle weight was found. For values of the coupling in the range $J/t \leq 2$ we observe satellite peaks in the region around $\vec{k} = (\pi/2, \pi/2)$ (see vertical lines in Fig. 11) next to the lowest energy peak which is extremely sharp and corresponds to a quasiparticle. Similar structures were observed in exact diagonalization [60, 8] and in SCBA [64], and were ascribed to string excitations.

String excitations appear when quantum fluctuations of the spin-background are neglected, as in the t-J_z model, and furthermore, the motion of the hole is assumed to be restricted to paths without loops. Then, the hole is confined by a linear potential, that in the continuum limit leads to (k-independent) eigenvalues of the energy [73, 64, 74] given by

$$E_n/t = -2\sqrt{3} + a_n (J_z/t)^{2/3}, \tag{63}$$

where a_n are the eigenvalues of a dimensionless Airy equation [62]. The first three eigenvalues are given by $a_n = 2.33, 4.08, 5.52$. In Fig. 12 the results for the first three excitations are given for $\vec{k} = (\pi/2, \pi/2)$, and are compared to the predictions from

SCBA. The error bars on the second and third peak are obtained as the width of

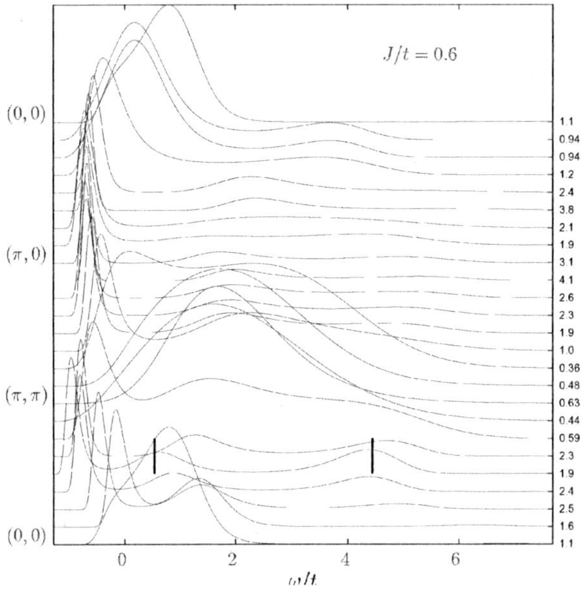

FIGURE 11. Spectral function for a 16×16 system and $J/t = 0.6$. The vertical lines indicate resonances above the quasiparticle peak at $\vec{k} = (\pi/2, \pi/2)$.

the MaxEnt peak at half intensity, the error bars of the first peak are taken directly from the determination of the lower edge of the spectrum (i.e. from the asymptotic behavior of the propagator in imaginary time, and hence independently from MaxEnt). We find, that for $J/t \leq 2$ the lowest peak can be accurately described by $\varepsilon_0(\pi/2, \pi/2) = -E_H - 3.28t + 2.33(J/t)^{2/3}t$, where E_H is the Heisenberg energy per site, and the second peak by $\varepsilon_1(\pi/2, \pi/2) = -E_H - 3.28t + 4.08(J/t)^{2/3}t$. The value of $3.28t + E_H$ is the result obtained from SCBA [64], whereas the prefactors of $(J/t)^{2/3}$ are exactly the values of the dimensionless Airy function, implying that the first two peaks behave (within our error bars) exactly as it is expected by the string picture. In contrast to this, a fit from SCBA for the first three excitations in the t-J model for values of $J/t \leq 0.4$ results in $a_n = 2.16, 5.46, 7.81$, also with the exponent $2/3$ [64], leading to a clear disagreement with our data. The third peak that can be resolved, cannot be explained by the string picture, since its distance to the lower band edge is independent of J and has a value of about $4t$.

The results above lead to the conclusion that the lowest excitations can be well described by the string picture. However, it should be kept in mind that the string picture originates in the Ising limit for $J/t \ll 1$, and that it is based on the continuum limit, that seems far away from our case with strings of lengths between two and a maximum of five lattice points, that correspond to the first two string excitations. Moreover, the string picture predicts a band without dispersion, that is clearly not the case in our simulations. A way to reconcile this paradoxical situation is given by the very good quantitative agreement between QMC and series expansions [66] for the dispersion of the quasiparticle and its bandwidth for a fairly large range in J. As shown by the expansion around the Ising limit, a coherent motion of the hole is made possible after the creation of strings due to hopping processes, by appropriate spin-flips, the shortest string being of length two. The lowest order contribution appears in third order, where the points $(\pi/2, \pi/2)$ and $(\pi, 0)$ are degenerate. Fourth and higher order processes remove this degeneracy, giving rise to a band that agrees qualitatively very well with the one obtained in QMC. Therefore on top of the coherent motion determined by $J_\perp$, string like excitations are possible and related to J_z and t. Such a possibility was already proposed by Béran, Poilblanc and Laughlin [75] on the basis of exact diagonalizations on small systems and is confi rmed unambiguously by our simulations on large systems.

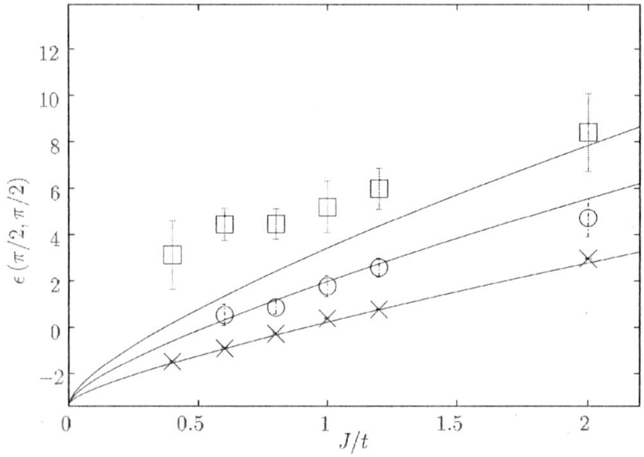

FIGURE 12. The first three excitations at $\vec{k} = (\pi/2, \pi/2)$. At $J/t = 0.4$ only two peaks were resolved. The lines represent the solutions obtained by solving the linear string potential for the hole in the $t - J_z$ model.

From the results presented above, we can conclude that antiferromagnetic fluctuations are not able to destroy the quasihole. Furthermore, the string picture with a linear potential confi ning the hole and spin-defects was shown to apply, at least for the lowest excitations of the composite quasiparticle.

THE HYBRID-LOOP ALGORITHM FOR FINITE DOPING

Certainly, the most interesting case to study with the t-J model is the one with a fi nite doping, that may lead to gain a deep insight into high temperature superconductivity. A number of such studies were performed by exact diagonalization [8], but as mentioned before, they suffer under large fi nite-size effects, so that the extrapolation to the thermodynamic limit cannot in general be performed with confi dence. Larger system sizes can be dealt with the density-matrix renormalization group (DMRG) [76], extending further the insight gained by exact diagonalization. However, the fact that such calculations are in general performed with open boundary conditions, in order to have a high accuracy, renders the calculation of correlation functions and the extraction of information relevant for the thermodynamic limit rather diffi cult.

In view of the above, it is clearly desirable to develop alternative methods capable of dealing with system sizes large enough, so that an extrapolation to the thermodynamic limit is possible, and also able to calculate various correlation functions. In this respect, QMC methods offer a possible alternative.

Quantum Monte Carlo simulations of the t-J model were performed mainly in the frame of Green's function Monte Carlo (GFMC) [77], where the ground-state is projected out of a trial wavefunction essentially by applications of H^n on it, where H is the Hamiltonian of the system under consideration. The ground-state is reached in the limit $n \longrightarrow \infty$. For fi nite doping, such an algorithm suffers under the minus-sign problem, i.e. the fact that weights during the stochastic sampling of confi gurations may become negative [78]. An approximate method to deal with this problem is the fi xed node (GFMCFN) approximation, where a variational estimate of the ground-state is obtained [79, 80]. However, it is an approximation and it is diffi cult to asses its accuracy. A recent advance on the minus-sign problem was achieved by the Green's function Monte Carlo with stochastic reconfi guration (GFMCSR) [81, 82]. Yet, this algorithm is not exact, and an accurate determination of correlation functions is diffi cult.

In the following we describe the hybrid-loop algorithm [83], which is a numerically exact algorithm, i.e. where beyond the statistical errors, the systematic error introduced by discretization in imaginary time can be controlled. The advantage of this algorithm is that a number of correlation functions and spectral functions can be obtained. However, at this point the minus-sign problem remains, so that the fi rst application is restricted to a one-dimensional case. Since a new element is needed, namely the determinantal algorithm, we fi rst introduced it in the next section.

Projection to the ground-state and determinantal algorithm for lattice fermions

In contrast to the world-line algorithms that we discussed at the beginning, we focus now on algorithms where the fermions are integrated out and the weight of confi gurations of a fi eld coupled to the fermions contains the fermionic information in the form of a determinant. Such methods can in principle be only applied to Hamiltonians bilinear in fermion operators, such as non-interacting electrons coupled to phonons (for an example of such a simulation see [84]). For interacting electrons, only those cases can be treated, where the interaction term can be decoupled by means of a Hubbard-Stratonovich transformation [85, 14], such that it becomes a bilinear form coupled to an auxiliary classical fi eld. Then, the sum over confi gurations of the auxiliary classical fi eld is accomplished by Monte Carlo methods, the weight of the confi gurations being the fermionic determinant. For a detailed discussion, see e.g. the lecture by F.F. Assaad in this same series [14].

In the present case, however, due to the canonical transformation (34), the original Hamiltonian (32), that contains a large number of nontrivial interaction terms, is brought to a bilinear form in fermions (36), without introducing auxiliary classical fi elds. Instead, quantum mechanical pseudospins that couple to the holes are subjected to a Monte Carlo simulation, with new weights, where the fermionic information enters in the form of a determinant. Our discussion is based on the Blankenbecler-Scalapino-Sugar algorithm [86] that was originally implemented for simulations at finite temperatures of the Hubbard model [87], and the projector algorithm for $T = 0$ [88, 89, 90].

We consider first the following quantity

$$< \mathcal{O} >= \lim_{\theta \to \infty} \frac{\sum_{\{\sigma_0\}} < \sigma_0| \otimes < \Psi_T | \mathscr{P} e^{-\theta \tilde{H}/2} \mathcal{O} e^{-\theta \tilde{H}/2} \mathscr{P} | \Psi_T > \otimes |\sigma_0 >}{\sum_{\{\sigma_0\}} < \sigma_0| \otimes < \Psi_T | \mathscr{P} e^{-\theta \tilde{H}} \mathscr{P} | \Psi_T > \otimes |\sigma_0 >}, \tag{64}$$

where $\{|\sigma_0 >\}$ is a complete set of spin states, $|\Psi_T >$ is a trial wavefunction for the spinless fermions and $\mathscr{P}$ is a projector to enforce the constraint (37). Furthermore, $\tilde{H}$ is the Hamiltonian in the representation given by (36). As long as the overlap of the considered state with the ground-state is finite we have

$$\lim_{\theta \to \infty} e^{-\frac{\theta}{2}\tilde{H}} \mathscr{P} |\Psi_T > \otimes |\sigma_0 >= \lim_{\theta \to \infty} e^{-\frac{\theta}{2} E_G} |\Psi_G > , \tag{65}$$

where E_G and $|\Psi_G >$ are the ground-state energy and ground-state of the t-J model respectively. This can be easily seen by using the complete set of eigenstates of $\tilde{H}$. Then,

$$< \mathcal{O} > \; = \; \lim_{\theta \to \infty} \frac{\mathscr{N} e^{-\theta E_G} < \Psi_G | \mathcal{O} | \Psi_G >}{\mathscr{N} e^{-\theta E_G} < \Psi_G | \Psi_G >} = \frac{< \Psi_G | \mathcal{O} | \Psi_G >}{< \Psi_G | \Psi_G >}, \tag{66}$$

where $\mathscr{N}$ is the number of spin-states with a finite overlap with the ground-state in the finite system.

Going back to (64), we introduce Trotter splitting and complete sets of spin states, such that

$$\begin{aligned}
< \mathcal{O} > \; = \; & \lim_{\theta \to \infty} \Bigg[\sum_{\{\sigma_i\}} < \Psi_T | \otimes < \sigma_0 | \mathscr{P} e^{-\Delta\tau \tilde{H}} | \sigma_{L-1} > \cdots < \sigma_{\frac{L}{2}+2} | e^{-\Delta\tau \tilde{H}} | \sigma_{\frac{L}{2}+1} > \\
& \times < \sigma_{\frac{L}{2}+1} | \mathcal{O} | \sigma_{\frac{L}{2}} >< \sigma_{\frac{L}{2}} | e^{-\Delta\tau \tilde{H}} | \sigma_{\frac{L}{2}-1} > \\
& \times \cdots < \sigma_1 | e^{-\Delta\tau \tilde{H}} \mathscr{P} | \sigma_0 > \otimes | \Psi_T > \Bigg] \\
& \times \Bigg[\sum_{\{\sigma_i\}} < \Psi_T | \otimes < \sigma_0 | \mathscr{P} e^{-\Delta\tau \tilde{H}} | \sigma_{L-1} > \cdots \\
& \times < \sigma_1 | e^{-\Delta\tau \tilde{H}} \mathscr{P} | \sigma_0 > \otimes | \Psi_T > \Bigg]^{-1}
\end{aligned} \tag{67}$$

where $\Delta\tau = \theta / L$. Again we split the Hamiltonian in even and odd contributions, as done in the case of a single hole.

The denominator of (67) can be considered as a "partition function" (recall that we are interested in the case $T = 0$)

$$\begin{aligned} Z &\equiv \sum_{\{\sigma_i\}} < \Psi_T | \otimes < \sigma_0 | \mathscr{P} \mathrm{e}^{-\Delta\tau\tilde{H}} | \sigma_{2L-1} > \cdots < \sigma_1 | \mathrm{e}^{-\Delta\tau\tilde{H}} \mathscr{P} | \sigma_0 > \otimes | \Psi_T > \\ &\equiv \sum_{\{\sigma_i\}} W_S[\sigma_i] \, W_F[\sigma_i] \,, \end{aligned} \tag{68}$$

where $W_S[\sigma_i]$ is the weight of a pure Heisenberg model for a spin-confi guration σ_i, as we already discussed in connection with the world-line algorithm, and

$$W_F[\sigma_i] \equiv < \Psi_T | U(\sigma_0, \sigma_{2L-1}) \cdots U(\sigma_1, \sigma_0) | \Psi_T > , \tag{69}$$

where the evolution operators $U(\sigma_\ell, \sigma_{\ell-1})$ were defi ned in eq. (54). Whereas in the case of the propagation of a single hole in an antiferromagnet, such evolution operators acted only between the imaginary times 0 and τ, they appear here over the whole extension θ. The reweighting by the matrix elements of the Heisenberg model as given by (9), leads to $W_S[\sigma_i]$ in eq. (68). Furthermore, it should be understood that the projector for the constraint is present, and we do not write it explicitly from now on.

The trial wavefunction is a Slater determinant, i.e. a product of one-particle states in a system with N sites. Then, we take a product of states like (48) and have

$$| \Psi_T > = \prod_{j=1}^{N_p} \sum_{i=1}^{N} P_{ij}^T f_i^\dagger \, | 0 > , \tag{70}$$

where N is the number of sites, N_p the number of holes, and the weights subjected to the constraint fulfi ll

$$\sum_i P_{ij}^T P_{ik}^T = \delta_{jk} \,. \tag{71}$$

In this way we have a normalized trial wavefunction:

$$\begin{aligned} < \Psi_T \, | \, \Psi_T > &= \sum_{\substack{i_1 \cdots i_{N_p} \\ j_1 \cdots j_{N_p}}} P_{i_1 1}^T \cdots P_{i_{N_p} N_p}^T P_{j_1 1}^T \cdots P_{j_{N_p} N_p}^T \\ &\qquad \times < 0 \, | \, f_{i_{N_p}} \cdots f_{i_1} f_{j_1}^\dagger \cdots f_{j_{N_p}}^\dagger \, | 0 > , \end{aligned} \tag{72}$$

where we used that

$$< 0 \, | \, f_{i_{N_p}} \cdots f_{i_1} f_{j_1}^\dagger \cdots f_{j_{N_p}}^\dagger \, | 0 > = \varepsilon^{k_1 \cdots k_{N_p}} \delta_{i_1 j_{k_1}} \cdots \delta_{i_{N_p} j_{k_{N_p}}} , \tag{73}$$

with $k_1 = 1, \ldots, N_p$ and $\varepsilon^{k_1 \cdots k_{N_p}}$ the Levi-Civita symbol in N_p dimensions. The identity above can easily be proved by induction.

From the projection we have for a given confi guration of the pseudospin fi eld,

$$\begin{aligned} | \Psi_F[\sigma_i] > &= U(\sigma_0, \sigma_{2L-1}) \cdots U(\sigma_1, \sigma_0) \, | \, \Psi_T > \\ &= \prod_{j=1}^{N_p} \sum_{i=1}^{N} P_{ij}^F f_i^\dagger \, | 0 > , \end{aligned} \tag{74}$$

where $\mid \Psi_F[\sigma_i] >$ is the wavefunction reached after an evolution for a given configuration σ_i.

As can be seen from the calculation of the matrix elements (46), the evolution operators $U(\sigma_\ell, \sigma_{\ell-1})$ consist of exponentials of bilinear forms in fermionic operators, such that its action on a Slater determinant produces another one. In order to see this, consider the action of an exponential of a bilinear form on a Slater determinant

$$\exp\left(f_i^\dagger A_{ij} f_j\right) \prod_{j=1}^{N_p} \sum_{i=1}^{N} P_{ij} f_i^\dagger \mid 0 > \; = \; \prod_{j=1}^{N_p} \sum_{i=1}^{N} \left(e^A P\right)_{ij} f_i^\dagger \mid 0 > , \qquad (75)$$

where we assumed A to be a real symmetric $N \times N$ matrix, such that it can be diagonalized by a similarity transformation. Hence, going back to (74), we have that P_{ij}^F is an $N \times N_p$ matrix with the amplitudes for each fermion j on a site i at the end of the evolution. This generalizes the discussion performed in the previous section for one hole. In particular, this shows that the same matrices as defined in Table 1 can be used.

On considering the weight (69), we have

$$\begin{aligned} W_F[\{\sigma_i\}] &= <\Psi_T \mid U(\sigma_0, \sigma_{2L-1}) \cdots U(\sigma_1, \sigma_0) \mid \Psi_T > \\ &= <\Psi_T \mid \Psi_F> = <0 \mid \prod_{k=1}^{N_p} \sum_{i}^{N} P_{ij}^T f_i \prod_{m=1}^{N_p} \sum_{j}^{N} P_{jm}^F f_j^\dagger \mid 0 > \\ &= \sum_{\substack{i_1,\ldots,i_{N_p} \\ j_1,\ldots,j_{N_p}}}^{N} P_{i_1 1}^T \cdots P_{i_{N_p} N_p}^T P_{j_1 1}^F \cdots P_{j_{N_p} N_p}^F < 0 \mid f_{i_1} \cdots f_{i_{N_p}} f_{j_{N_p}}^\dagger \cdots f_{j_1}^\dagger \mid 0 > \\ &= \det P^T P^F , \end{aligned} \qquad (76)$$

where the identity (73) was used again. Thus, starting with Slater determinants, we have at the end a weight related to the fermions in the form of a determinant. This has the same form as in the cases where an auxiliary field is introduced [14]. However, in our case, the spin-field is a quantum mechanical one.

Once we discussed the evolution of the fermionic wavefunction, we consider the updating of pseudospin configurations. As shown in (68), the weight of a configuration of pseudospins is given by

$$W[\sigma_i] = W_S[\sigma_i] \; W_F[\sigma_i] \; , \qquad (77)$$

such that the ratio

$$r \equiv \frac{W\left[\sigma_i^{\text{new}}\right]}{W\left[\sigma_i^{\text{old}}\right]} \, , \qquad (78)$$

that determines the probability for a change, is given by the product of the ratio of weights for pseudospins and of the ratio of determinants. For the pseudospins we implement the loop-algorithm in order to propose a new pseudospin configuration, whereas the contribution of the holes given the pseudospin configuration can be obtained as described above.

Stable evolution of the fermionic wavefunction and measurements

One diffi culty of the determinantal method is, that the evolution operator in imaginary time is not unitary, and therefore, although P^T was initially chosen as a set of orthogonal vectors, after a time τ this will be in general not true any more. Due to the fi nite precision of computers, this could have serious consequences for the simulation, since if one of the vectors becomes linearly dependent of any one of the others within machine precision, the determinant would become zero. Such a problem was recognized early on in the algorithms for the Hubbard model, and a stabilization procedure was suggested [90, 91], that is particularly effective in projector algorithms.

We start the discussion by introducing the notation

$$| \Psi[\sigma_0], \tau > \equiv \exp\left(-\tau\tilde{H}\right) \mathscr{P}|\sigma_0 > \otimes|\Psi_T > , \tag{79}$$

describing a state reached at time τ after the evolution by $\tilde{H}$ starting from a state $\mathscr{P}|\sigma_0 > \otimes|\Psi_T >$. Then, the expectation value (64) can be written as follows:

$$\begin{aligned} < \mathscr{O} > &= \lim_{\theta\to\infty} \frac{\sum_{\{\sigma_0\}} < \Psi[\sigma_0], \theta/2 \mid \mathscr{O} \mid \Psi[\sigma_0], \theta/2 >}{\sum_{\{\sigma_0\}} < \Psi[\sigma_0], \theta/2 \mid \Psi[\sigma_0], \theta/2 >} \\ &= \lim_{\theta\to\infty} \sum_{\{\sigma_0\}} \frac{< \Psi[\sigma_0], \theta/2 \mid \mathscr{O} \mid \Psi[\sigma_0], \theta/2 >}{< \Psi[\sigma_0], \theta/2 \mid \Psi[\sigma_0], \theta/2 >} \\ &\qquad \times \frac{< \Psi[\sigma_0], \theta/2 \mid \Psi[\sigma_0], \theta/2 >}{\sum_{\{\sigma_0\}} < \Psi[\sigma_0], \theta/2 \mid \Psi[\sigma_0], \theta/2 >} \\ &= \lim_{\theta\to\infty} \sum_{\{\sigma_0\}} < \bar{\Psi}[\sigma_0], \theta/2 \mid \mathscr{O} \mid \bar{\Psi}[\sigma_0], \theta/2 > \frac{W_S[\sigma_0]\, W_F[\sigma_0]}{Z} , \end{aligned} \tag{80}$$

where we defi ned a normalized wavefunction

$$| \bar{\Psi}[\sigma_0], \theta/2 > \equiv \frac{| \Psi[\sigma_0], \theta/2 >}{\sqrt{< \Psi[\sigma_0], \theta/2 \mid \Psi[\sigma_0], \theta/2 >}} . \tag{81}$$

Next, let a consider a wavefunction $| \Psi(\sigma_0), \tau >$, for some imaginary time τ. As discussed in connection with (74), this wavefunction is directly mapped into an $N \times N_p$ matrix, which we call P_τ. This matrix can be splitted by a Gramm-Schmidt orthogonalization as follows,

$$P_\tau = U_\tau D_\tau V_\tau , \tag{82}$$

where $U_\tau^\dagger U_\tau = \vec{1}_{N_p\times N_p}$, D is a diagonal matrix with (exponentially) large and small energy scales containing the norm of the one particle states, and V is an upper triangular matrix with unity in the diagonal, and hence, with determinant one. The norm of the wavefunction is given by

$$\begin{aligned} < \Psi[\sigma_0], \tau \mid \Psi[\sigma_0], \tau > &= \det\left(V_\tau^\dagger D_\tau U_\tau^\dagger U_\tau D_\tau V_\tau\right) \\ &= \det\left(V_\tau^\dagger D_\tau D_\tau V_\tau\right) = (\det D)^2 , \end{aligned} \tag{83}$$

such that the normalized wavefunction $\mid \bar{\Psi}(\sigma_0), \tau >$ is given by U_τ. Therefore, in the calculation of the expectation value of $\mathscr{O}$ for a given world-line confi guration, only the information contained in U is needed. Since the (exponentially) large and small scales contained in D do not enter the calculation of expectation values for a given pseudospin confi guration, the evaluation can be made in a numerically stable way by applying the Gramm-Schmidt orthogonalization procedure after a given number of time slices.

The stabilization discussed above can be made explicit for the observables related to fermionic quantities. In order to see this, we recall that, since the holes do not interact with each other in the formulation (36), the expectation values of operators related with the hole can be decomposed in a product of single-particle Green's functions using Wick's theorem. Next we show that these quantities can be calculated entirely based on the matrices U obtained from a Gramm-Schmidt orthogonalization.

The one-particle Green's function can be calculated explicitly by introducing an appropriate source term in the generating functional or partition function (68)

$$\begin{aligned} Z[J] &= \sum_{\{\sigma_i\}} <\Psi_T|U(\sigma_0,\sigma_{2L-1})\cdots U\left(\sigma_{L/2+2},\sigma_{L/2+1}\right) e^{Jc_i^\dagger c_j} \\ &\qquad \times U\left(\sigma_{L/2},\sigma_{L/2-1}\right)\cdots U(\sigma_1,\sigma_0)|\Psi_T> , \end{aligned} \tag{84}$$

such that

$$\begin{aligned} <c_i^\dagger c_j> &= \left.\frac{\partial \ln Z}{\partial J}\right|_{J=0} \\ &= \left\{ \sum_{\{\sigma_0\}} <\Psi_T \mid U(\sigma_0,\sigma_{2L-1})\cdots U(\sigma_{L+2},\sigma_{L+1}) c_i^\dagger c_j \right. \\ &\qquad \left. \times U(\sigma_L,\sigma_{L-1})\cdots U(\sigma_1,\sigma_0) \mid \Psi_T > \right\} \\ &\quad \times \left\{ \sum_{\{\sigma_0\}} <\Psi_T \mid U(\sigma_0,\sigma_{2L-1})\cdots U(\sigma_1,\sigma_0) \mid \Psi_T > \right\}^{-1} . \end{aligned} \tag{85}$$

On the other hand, the source term introduces just another exponential of a bilinear form, such that again one can write for a given confi guration of the pseudospin fi elds

$$\begin{aligned} <c_i^\dagger c_j> &= \left.\frac{\partial \ln Z}{\partial J}\right|_{J=0} \\ &= \frac{\partial}{\partial J} \ln\det \left[P^T B_{2L-1}\cdots B_{L+1} e^{JC} B_L \cdots B_0 P^T\right]\Big|_{J=0} \\ &= \frac{\partial}{\partial J} \mathrm{Tr}\ln \left(P^T B_{2L-1}\cdots B_{L+1} e^{JC} B_L \cdots B_0 P^T\right)\Big|_{J=0} \\ &= \mathrm{Tr}(LR)^{-1} LCR \\ &= \mathrm{Tr} R(LR)^{-1} LC = \left[R(LR)^{-1} L\right]_{ij} , \end{aligned} \tag{86}$$

where $C_{kl} = \delta_{kj}\delta_{li}$, and we used the notation introduce in (55). Furthermore, following the notation in [91], we introduced the $N_p \times N$ and $N \times N_p$ matrices

$$L = P^T B_{2L-1} \cdots B_{L+1} \ , \tag{87}$$

and

$$R = B_L \cdots B_0 P^T \ . \tag{88}$$

From the above, the Green's function is given by

$$G = 1 - R(LR)^{-1} L \ . \tag{89}$$

We can now apply the decomposition (82) to the matrices L and R

$$L = V_L D_L U_L \ , \qquad R = U_R D_R V_R \ , \tag{90}$$

leading to

$$\begin{aligned} G &= 1 - U_R D_R V_R \left(V_L D_L U_L U_R D_R V_R \right)^{-1} V_L D_L U_L \\ &= 1 - U_R \left(U_L U_R \right)^{-1} U_L \ , \end{aligned} \tag{91}$$

such that we have only to deal with orthogonal matrices and no large scales are generated.

The discussion above shows that in the case of the t-J model, it is possible to combine the advantages of the loop-algorithm for global updates of the spin-background and those of the determinantal algorithm for the holes. This is a consequence of the canonical transformation (34) that leads to a charge-spin separated formulation. Since this canonical transformation is generally valid for fermionic operators in a lattice model, it remains as a task for the future to explore its advantges for the Hubbard model, and other strongly correlated systems.

After the description of the algorithm, we discuss some results about its performance. Figure 13 shows a comparison of ground-state energies from QMC and exact diagonalization on a system with $L = 20$ sites for various values of J at a density $n = 0.9$. The correct value is reached for values of the projection parameter $\Theta \sim 10/t - 20/t$, demonstrating that the algorithm leads to the correct ground-state with high accuracy (statistical errors are smaller than the size of the symbols). In spite of the possibility of having minus-signs, the average sign approaches one very fast by increasing the system size. For $L \sim 40$, the average sign is about 0.97. In the following section we give a brief account of results obtained for the one-dimensional t-J model at fi nite doping, where the algorithm proved to be accurate enough to allow for the calculation of spectral functions.

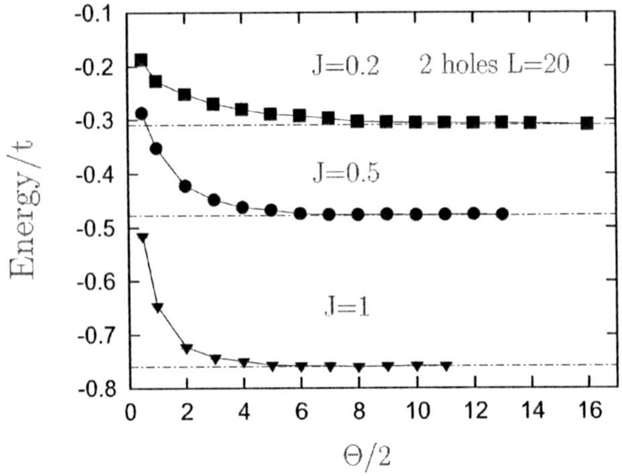

FIGURE 13. Ground-state energies *vs.* exact diagonalization results for $L = 20$ with two holes as a function of the projection parameter Θ.

One dimensional t-J model with finite doping

We focus first on the spectral function for one-particle excitations in the one-dimensional (1D) nearest neighbor t-J model [83], since in this case the present algorithm attains a high accuracy being free of the minus-sign problem. On the other hand, the 1D nearest neighbor t-J model has a phase with dominating superconducting correlation, a spin-gap phase, and also phase-separation [92], resembling thus many of the phases found in high temperature superconductors.

It is well established that electrons in 1D metals generally lead to a Luttinger liquid [93, 49], where charge-spin separation (CSS) takes place. The most direct evidence of CSS was predicted for the spectral function of such systems [94, 95]. While experimental evidence of CSS has accumulated in recent years [42, 43, 96], an exact theoretical evaluation of the one-particle spectral function $A(k,\omega)$ could until now only be fully accomplished for the Hubbard model at $U = \infty$ for arbitrary doping on the basis of the Bethe-*Ansatz* solution [97, 98]. However, recent progress was made for spectral properties of the supersymmetric (SuSy) t-J model with $1/r^2$ interaction (see eq. (60) for a description of the coupling constants), where beyond the exact ground state [99], the thermodynamics [100], the compact support of $A(k,\omega)$ [101], the single-hole dynamics [54], and the electron addition spectrum [102] could be calculated analytically. In addition to spinons and holons, the inverse squared (IS) SuSy t-J model was shown to contain antiholons with charge $Q = 2e$, spin $S = 0$, and twice the mass of the holons, i.e. they are not merely charge conjugate to the holons.

We present in the following one-particle spectral functions for the 1D nearest neighbor t-J model with finite doping, obtained by QMC simulations based on the hybrid-loop algorithm described above. We show that the electron addition part of $A(k,\omega)$ presents a clear structure following the antiholon dispersion found in $A(k,\omega)$ of the IS SuSy t-J model for the same doping. Our results show moreover, that also away from that point, a corresponding feature is present, strongly indicating that antiholons originally identifi ed in the IS SuSy t-J model are generically present in the nearest neighbor model.

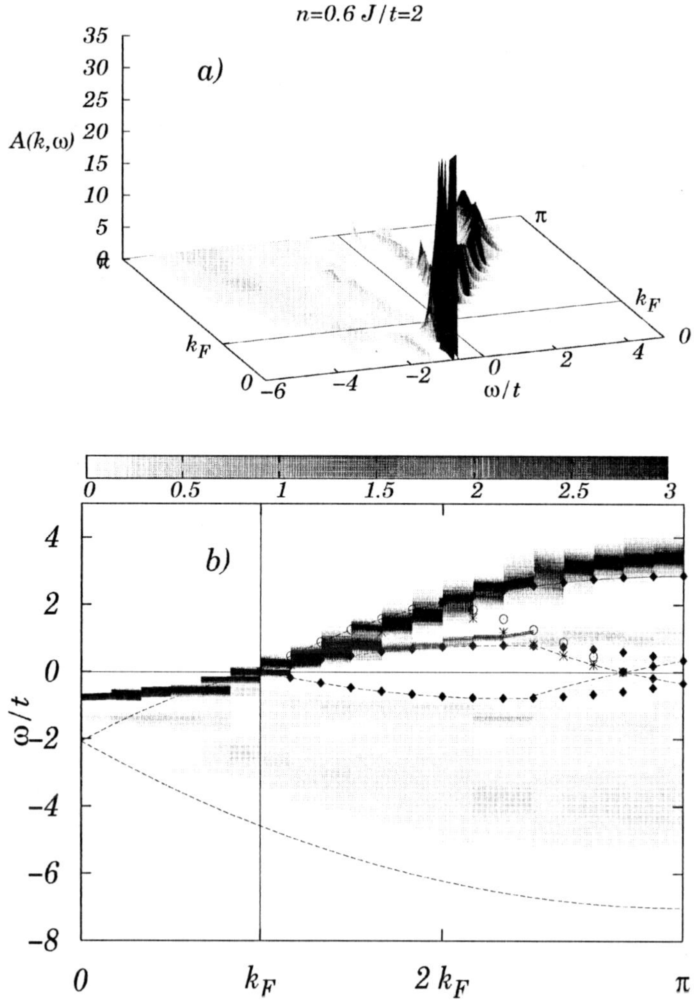

FIGURE 14. a) $A(k,\omega)$ for $J=2t$ at a density $n=0.6$. b) Projection of intensities on the (ω,k) plane. Solid lines: compact support of the IS SuSy t-J model. Red crosses: spinons, blue asterisks: holons, magenta diamonds: antiholons. See text for the dispersions.

We consider $A(k,\omega)$ both for electron-removal (ER) and electron-addition (EA) processes. igure 14 a) shows $A(k,\omega)$ obtained from the QMC simulations for $J=2t$ ($L=40$, $\Theta=24/t$), at a density $n=0.6$. The Fermi energy is taken as the zero of the energy scale. A splitting of the spectral weight into two branches can be readily seen on the EA side, in contradiction with what is expected for a single band. Figure 14 b) shows the projection of $A(k,\omega)$ on the (ω,k) plane, revealing the dispersion of the main features in the spectrum, together with the compact support for the IS SuSy t-J model at the same density. Furthermore, the dispersions of spinon (s), holon (h), and antiholon ($\bar{h}$) branches that determine the compact support for EA processes in the IS SuSy t-J model are also shown. The dispersions for right (R) and left (L) going spinons and holons are given by $\varepsilon_{\mathrm{sR(L)}}(q)/t = q(\pm v_{\mathrm{s}}^0 - q)$, and $\varepsilon_{\mathrm{hR(L)}}(q)/t = q(q \pm v_{\mathrm{c}}^0)$, respectively, where $v_{\mathrm{c}}^0 = \pi(1-n)$ and $v_{\mathrm{s}}^0 = \pi$. The antiholon dispersion is $\varepsilon_{\bar{\mathrm{h}}}(q)/t = q\left(2v_{\mathrm{c}}^0 - q\right)/2$. The accessible range of momenta is for $\varepsilon_{sR(L)}$ and $\varepsilon_{hR(L)}$, $0 \le q \le k_F$ $(-k_F \le q \le 0)$, and for $\varepsilon_{\bar{h}}$, $0 \le q \le 2\pi - 4k_F$ [100-102]. The compact support is obtained by assuming that the energy and momenta of the particle (EA) or hole (ER) are given by the addition of energy and momenta of s, h, and $\bar{h}$ with the dispersions above [101].

In the ER part of the spectrum, only the corresponding part of the compact support and the dispersion of an antiholon branch along the support is shown for clarity, since in contrast to the EA processes, where only one spinon, one holon, and one antiholon are present, in the ER part three spinons, three holons and one antiholon contributions [101] are possible. Therefore, a large number of features would appear, whose intensity is at the moment unknown, and hence, their importance is diffi cult to assess. A sharp feature is visible on the ER side that escapes from the compact support of the IS SuSy model. It is due to a holon branch, and as already discussed in the limit of a single hole [52], the actual dispersion of the holon is needed, in order to describe this feature correctly. Also a deviation from the IS SuSy compact support is observed on the EA side at high energies, where the differences in the models is expected to become noticeable.

There are however, several features that are well described by the excitations of the IS model. The strongest feature on the EA side is followed closely by the spinon and holon branches between k_F and $2k_F$, and for $k > 2k_F$ by a spinon at k_F together with a dispersing antiholon. The analytic results for EA processes in the IS SuSy model [102] show that the largest portion of spectral weight is along this line. More striking is a second, weaker, but clearly visible branch that follows very closely the dispersion of an antiholon between k_F and $2\pi - 3k_F$. The analytic results of $A(k,\omega)$ for the IS SuSy model [102] predict a stepwise discontinuity at this edge and, in fact, the explicit evaluation of the weight shows for the present range of doping a higher value than in the interior of the support. Also the upper edge of the compact support on the ER part, that in the IS model corresponds to an antiholon, is well reproduced, with spectral weight down to very low energies around $3k_F$, as predicted by the IS SuSy model. Therefore, at the SuSy point, the clearest signal of CSS in $A(k,\omega)$ for the nearest neighbor t-J model are present in the EA part of the spectrum and through the comparison with the IS t-J model, it is clear that a sizeable part of the spectral weight goes to the antiholon excitation.

Since the exact solution of the IS model is restricted to the SuSy point, it is of much interest to see whether the features discussed above correspond to a generic behavior of the nearest neighbor t-J model or whether it is better described, e.g. by the solution of the nearest neighbor model at $J = 0$ [97]. Figure 15 a) shows $A(k,\omega)$ for $n = 0.6$ and

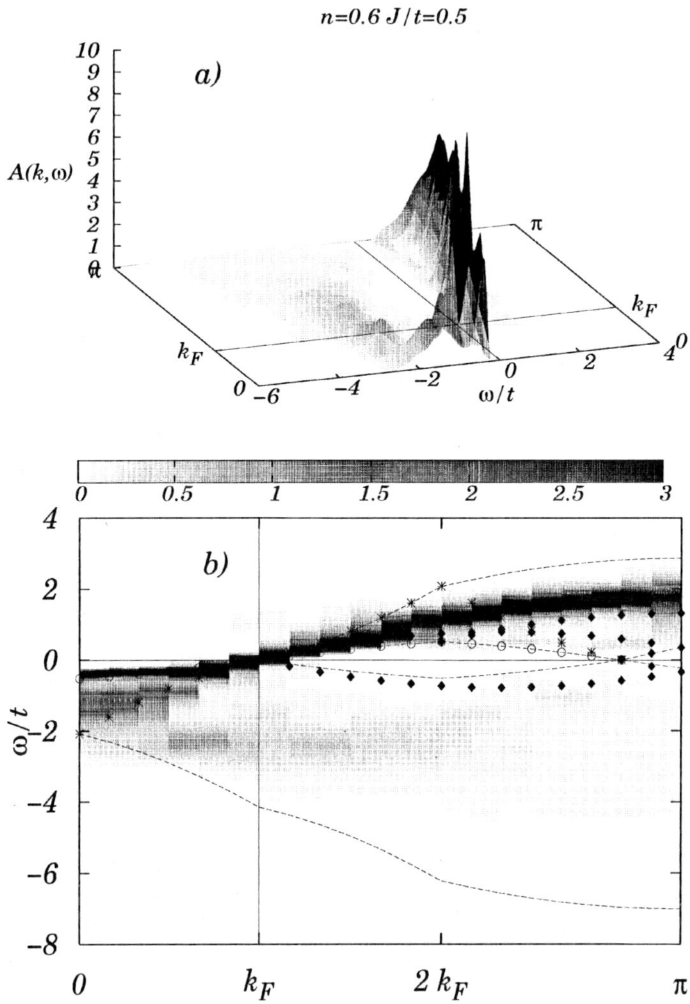

FIGURE 15. a) $A(k,\omega)$ for $J = 0.5t$ at $n = 0.6$. b) Projection of intensities on the (ω,k) plane. Symbols coded as in Fig. 14

$J = 0.5t$ ($L = 40$, $\Theta = 56/t$), i.e. very far away from the SuSy point and at a value of J/t of experimental relevance for cuprate compounds. A perspective was chosen, so that it is already visible that as in the SuSy case, a structure splits off the main feature for k between $2k_F$ and π. Figure 15 b) shows the projection of $A(k,\omega)$ on the (ω,k) plane. As a model for free spinons, holons, and antiholons, we use the same dispersions as for the IS SuSy model, but with $\varepsilon_{\mathrm{sR(L)}}(q) = (J/2)q(\pm v_s^0 - q)$, i.e. assuming that away from the SuSy point, only the energy scale of spinons is changed. The corresponding compact support, spinon, holon, and antiholon dispersions are encoded as in Fig. 14. In the present case, the compact support encloses rather well all the spectral weight. Moreover, on the ER part, the strongest feature is very accurately followed by a spinon, whereas a second structure is also closely followed by a holon. A more detailed view of these structures is given below in Fig. 16. They correspond to the generally expected signal in photoemission for CSS, that were also found in previous numerical studies of the Hubbard model [103]. However, as shown in Fig. 16, the present algorithm seems to lead to results accurate enough, so that after application of maximum entropy, CSS is seen below E_F in a wider range in k-space than previously. On the EA side, the feature with largest intensity is followed close to k_F by a holon, a spinon, and an antiholon. However, further away from k_F, the dispersion of the maximum is, up to $k \sim 2k_F$, closer to an antiholon going from k_F to $2\pi - 3k_F$ and beyond $2k_F$ by a curve corresponding to a spinon at k_F and a dispersing antiholon. Moreover, a second maximum develops beyond $2k_F$ that follows the antiholon dispersing from k_F to $2\pi - 3k_F$, in a similar way as for $J = 2t$ but with a smaller gap between both curves. In particular, the results from the simulations show appreciable weight between $3k_F$ and the zone boundary, where only antiholons are present.

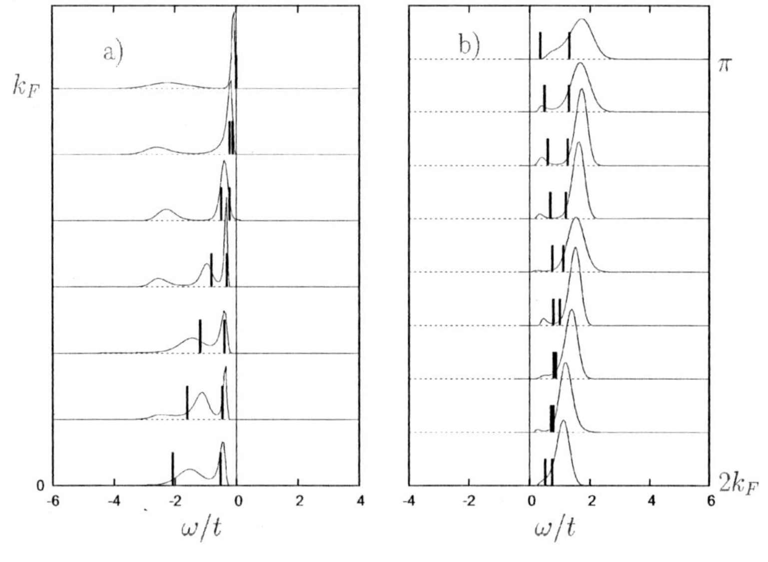

FIGURE 16. Detailed view of $A(k,\omega)$ at $J = 0.5t$ and $n = 0.6$. a) ER side for $0 \le k \le k_F$ at $J = 0.5t$ and $n = 0.6$. Vertical bars denote the positions for free spinon (closer to $\omega = 0$) and holon excitations. b) EA side for $2k_F \le k \le \pi$. Vertical bars denote here antiholon dispersions.

A closer look to both features signaling CSS is given in Fig. 16. Figure 16 a) shows $A(k,\omega)$ on the ER side and the location of the excitation energies for one spinon and one holon. Whereas the spinon dispersion follows the QMC data very closely, a deviation is seen for the holon for the farthest points from k_F, as can be expected, since at higher energies, details of the dispersion matter in general. Yet, the agreement is good enough to enable an identifi cation of the excitation content of the spectrum. The details of the splitted maxima for $2k_F \le k \le \pi$ on the EA side are shown in Fig. 16 b), where both an antiholon dispersing from k_F to $2\pi - 3k_F$ (closer to $\omega = 0$) and an antiholon dispersing from $2k_F$ to $2\pi - 2k_F$ on top of a spinon at k_F are shown. Whereas the latter follows the larger maximum, the former can be associated with the second maximum. As at the SuSy point, there seems to be almost no weight associated with the left propagating spinon and holon that give rise to the contributions between $2k_F$ and $3k_F$. This is consistent with the analytic results obtained for the IS SuSy model [102]. Results for other values of doping ($0.6 \leq n \leq 0.9$) and J ($0.5 \leq J/t \leq 3$) not presented here, show the same qualitative behavior, in particular the presence of a branch on the EA side below the main dispersing structure, that is closely followed by an antiholon branch under the assumption of free spinons, holons, and antiholons.

SUMMARY

We have presented in these lectures an account of a new algorithm that combines recent advances in the simulation of strongly correlated systems. Although the lectures were focused on the hybrid-loop algorithm, we hope that this presentation was also useful to get acquainted with the world-line, the loop-, and the determinantal ($T = 0$) algorithms.

We expect also that not only algorithmic aspects but the physical content were useful as well. From the discussion of the single hole problem it should become clear that some physical quantities like the quasiparticle weight, that is fundamental in order to decide whether Fermi-liquid like or alternative theories are appropriate, can only be accessed numerically, since it would be extremely diffi cult to obtain a clear answer from experiments.

Also the question about charge-spin separation is up to now rather diffi cult to clarify experimentally. Quantum Monte Carlo simulations able to deliver highly accurate results can help to unravel such phenomena. In our case we have seen that the nearest neighbor t-J model presents a specifi c form of charge-spin separation, very much like in the inverse squared supersymmetric t-J model.

There are certainly still many possible further developments of the hybrid-loop algorithm like the extension to fi nite temperatures, that is currently under way, or the application to the Hubbard model or other strongly correlated systems.

ACKNOWLEDGMENTS

I would like to thank M. Brunner and C. Lavalle, who decisively contributed to the development to the single-hole and hybrid-loop algorithms. I am very grateful to M. Arikawa for teaching us about the inverse squared supersymmetric t-J model and also to F. Assaad and S. Capponi for collaborations on this subject. Financial support by the DFG under SFB 382 is gratefully acknowledge. We are thankful to HLR-Stuttgart and NIC-Jülich for helpfull allocation of computer time on massive parallel machines.

REFERENCES

1. G. J. Bednorz and K. A. Müller, Z. Phys. B **64**, 188 (1986).
2. Science **288**, 461 (2000).
3. M. Imada, A. Fujimori, and Y. Tokura, Rev. Mod. Phys. **70**, 1039 (1998).

4. E. H. Lieb and F. Y. Wu, Phys. Rev. Lett. **20**, 1445 (1968).
5. P.-A. Bares and G. Blatter, Phys. Rev. Lett. **64**, 2567 (1990).
6. P.-A. Bares, G. Blatter, and M. Ogata, Phys. Rev. B **44**, 130 (1991).
7. A. Georges, G. Kotliar, W. Krauth, and M. J. Rozenberg, Rev. Mod. Phys. **68**, 13 (1996).
8. E. Dagotto, Rev. Mod. Phys. **66**, 763 (1994).
9. K. Binder and D. W. Heermann, *Monte Carlo Simulations in Statistical Physics: An Introduction 4th ed.* (Springer, Berlin, 2002).
10. H. G. Evertz, M. Marcu, and G. Lana, Phys. Rev. Lett **70**, 875 (1993).
11. H. G. Evertz, Adv. Phys. **52**, 1 (2003).
12. J. E. Hirsch, R. L. Sugar, D. J. Scalapino, and R. Blankenbecler, Phys. Rev. B **26**, 5033 (1982).
13. F. Assaad, in *Quantum Simulations of Complex Many-Body Systems: from Theory to Algorithms*, edited by J. Grotendorst, D. Marx, and A. Muramatsu (NIC Series, Vol. 10, FZ-Jülich, 2002).
14. F. Assaad, in *Lectures on the Physics of Highly Correlated Electron Systems VII*, edited by A. Avella and F. Mancini (AIP Conference Proceedings, Melville, New York, 2003), Vol. 678.
15. P. Jordan and E. Wigner, Z. Phys. **47**, 631 (1928).
16. H. F. Trotter, Proc. Am. Math. Soc. **10**, 545 (1959).
17. M. Suzuki, in *Quantum Monte Carlo Methods*, edited by M. Suzuki (Springer-Verlag, Heidelberg, 1986).
18. F. F. Assaad and D. Würtz, Phys. Rev. B **50**, 136 (1991).
19. N. Kawashima, J. E. Gubernatis, and H. G. Evertz, Phys. Rev. B **50**, 136 (1994).
20. R. Swendsen and J.-S. Wang, Phy. Rev. Lett. **58**, 86 (1987).
21. N. Kawashima and J. E. Gubernatis, Phys. Rev. Lett. **73**, 1295 (1994).
22. N. Kawashima, J. Stat. Physics **82**, 131 (1996).
23. U.-J. Wiese and H.-P. Ying, Z. Phys. B **93**, 147 (1994).
24. B. Ammon *et al.*, Phys. Rev. B **58**, 4304 (1998).
25. M. Brunner and A. Muramatsu, Phys. Rev. B **58**, R10100 (1998).
26. M. Troyer, H. Tsunetsugu, and T. M. Rice, Phy. Rev. B **53**, 251 (1996).
27. B. B. Beard and U.-J. Wiese, Phy. Rev. Lett. **77**, 5130 (1996).
28. N. V. Prokof'ev, B. V. Svistunov, and I. S. Tupitsyn, JETP **87**, 310 (1998).
29. V. A. Kashurnikov, N. V. Prokof?ev, B. V. Svistunov, and M. Troyer, Phys. Rev. B **59**, 1162 (1999).
30. H. G. Evertz and W. von der Linden, Phys. Rev. Lett **86**, 5164 (2001).
31. S. Chandrasekharan and U. J. Wiese, Phys. Rev. Lett. **83**, 3116 (1999).
32. O. F. Syljuåsen and A. W. Sandvik, Phys. Rev. E **66**, 046701 (2002).
33. X. Y. Zhang, E. Abrahams, and G. Kotliar, Phys. Rev. Lett. **66**, 1236 (1991).
34. M. Boninsegni and E. Manousakis, Phys. Rev. B **45**, 4877 (1992).
35. M. Boninsegni and E. Manousakis, Phys. Rev. B **46**, 560 (1992).
36. M. Boninsegni, Phys. Lett. A **188**, 330 (1994).
37. G. Kotliar and A. E. Ruckenstein, Phys. Rev. Lett **57**, 1362 (1986).
38. G. Khaliullin, JETP Lett. **52**, 389 (1990).
39. A. Angelucci, Phys. Rev. B **51**, 11580 (1995).
40. W. F. Brinkman and T. M. Rice, Phys. Rev. B **2**, 1324 (1970).
41. P. W. Anderson, Science **235**, 1196 (1987).
42. C. Kim *et al.*, Phys. Rev. Lett **77**, 4054 (1996).
43. K. Kobayashi *et al.*, Phys. Rev. Lett **82**, 803 (1999).
44. T. Takahashi *et al.*, Phy. Rev. B **56**, 7870 (1997).
45. B. O. Wells *et al.*, Phys. Rev. Lett. **74**, 964 (1995).
46. M. Brunner, F. F. Assaad, and A. Muramatsu, Phys. Rev. B **62**, 15480 (2000).
47. M. Jarrell and J. Gubernatis, Phys. Rep. **269**, 133 (1996).
48. K. Schönhammer and V. Meden, Phys. Rev. B **47**, 16205 (1993).
49. J. Voit, Rep. Prog. Phys. **58**, 977 (1995).
50. S. Sorella and A. Parola, J. Phys. Condens. Matter **4**, 3589 (1992).
51. S. Sorella and A. Parola, Phys. Rev. Lett. **76**, 4604 (1996).
52. M. Brunner, F. F. Assaad, and A. Muramatsu, Eur. Phys. J. B **16**, 209 (2000).
53. H. Suzuura and N. Nagaosa, Phys. Rev. B **56**, 3548 (1997).
54. Y. Kato, Phys. Rev. Lett **81**, 5402 (1998).

55. F. D. M. Haldane, in *Correlation Effects in Low-Dimensional Electron Systems*, edited by A. Okiji and N. Kawakami (Springer Verlag, Berlin, 1994).
56. M. Brunner, S. Capponi, F. F. Assaad, and A. Muramatsu, Phys. Rev. B **63**, 180511(R) (2001).
57. W. Metzner, P. Schmit, and D. Vollhardt, Phys. Rev. B **45**, 2237 (1992).
58. S. A. Trugman, Phys. Rev. B **35**, 1597 (1988).
59. Q. F. Zhong and S. Sorella, Phys. Rev. B **51**, 16135 (1995).
60. D. Poilblanc, H. Schulz, and T. Ziman, Phys. Rev. B **47**, 3268 (1993).
61. S. Sorella, Phys. Rev. B **46**, 11670 (1992).
62. C. L. Kane, P. A. Lee, and N. Read, Phys. Rev. B **39**, 6880 (1989).
63. G. Martinez and P. Horsch, Phys. Rev. B **44**, 317 (1991).
64. Z. Liu and E. Manousakis, Phys. Rev. B **45**, 2425 (1992).
65. S. Sorella, Phys. Rev. B **53**, 15119 (1996).
66. C. J. Hamer, Z. Weihong, and J. Otimaa, Phys. Rev. B **58**, 15508 (1998).
67. S. LaRosa *et al.*, Phys. Rev. B **56**, R525 (1997).
68. C. Kim *et al.*, Phys. Rev. Lett. **80**, 4245 (1998).
69. F. Ronning *et al.*, Science **282**, 2067 (1998).
70. D. Marshall *et al.*, Phys. Rev. Lett. **76**, 4841 (1996).
71. T. Tohyama *et al.*, cond-mat/9904231 (unpublished).
72. G. B. Martins, C. Gazza, and E. Dagotto, Phys. Rev. B **59**, 13596 (1999).
73. L. N. Bulaevskii, E. L. Nagaev, and D. I. Khomskii, JETP **27**, 836 (1968).
74. V. Elser, D. Huse, B. Shraiman, and E. Siggia, Phys. Rev. B **41**, 6715 (1990).
75. P. Béran, D. Poilblanc, and R. B. Laughlin, Nuc. Phys. B **473**, 707 (1996).
76. S. White and D. Scalapino (unpublished).
77. C. S. Hellberg and E. Manousakis, Phys. Rev. B **61**, 11787 (2000).
78. E. Y. Loh *et al.*, Phys. Rev. B **41**, 9301 (1990).
79. H. J. M. van Bemmel *et al.*, Phy. Rev. Lett. **72**, 2442 (1994).
80. D. F. B. ten Haaf *et al.*, Phys. Rev. B **51**, 13039 (1995).
81. M. Calandra, F. Becca, and S. Sorella, Phys. Rev. Lett. **81**, 5185 (1998).
82. S. Sorella and L. Capriotti, Phys. Rev. B **61**, 2599 (2000).
83. C. Lavalle *et al.*, Phys. Rev. Lett. **90**, 216401 (2003).
84. R. T. Scalettar, N. E. Bickers, and D. J. Scalapino, Phys. Rev. B **40**, 197 (1989).
85. J. E. Hirsch, Phys. Rev. B **29**, 4059 (1983).
86. R. Blankenbecler, D. J. Scalapino, and R. L. Sugar, Phys. Rev. D **24**, 2278 (1981).
87. J. E. Hirsch, Phys. Rev. Lett. **54**, 1317 (1985).
88. G. Sugiyama and S. E. Koonin, Annals of Phys. **168**, 1 (1986).
89. S. Sorella, S. Baroni, R. Car, and M. Parinello, Europhys. Lett. **8**, 663 (1989).
90. S. Sorella *et al.*, Int. J. Mod. Phys B **1**, 993 (1989).
91. E. Y. Loh and J. E. Gubernatis, in *Modern Problems of Condensed Matter Physics*, edited by W. Hanke and Y. Kopaev (North Holland, Amsterdam, 1992).
92. M. Ogata, M. Lucchini, S. Sorella, and F. Assaad, Phys. Rev. Lett. **66**, 2388 (1991).
93. F. D. M. Haldane, J. Phys. C **14**, 2585 (1981).
94. V. Meden and K. Schönhammer, Phys. Rev. B **46**, 15753 (1992).
95. J. Voit, Phys. Rev. B **47**, 6740 (1993).
96. R. Claessen *et al.*, Phys. Rev. Lett. **88**, 096402 (2002).
97. K. Penc, K. Hallberg, F. Mila, and H. Shiba, Phys. Rev. Lett **77**, 1390 (1996).
98. J. Favand *et al.*, Phys. Rev. B **55**, R4859 (1997).
99. Y. Kuramoto and H. Yokoyama, Phys. Rev. Lett. **67**, 1338 (1991).
100. Y. Kuramoto and Y. Kato, J. Phys. Soc. Jpn. **64**, 4518 (1995).
101. Z. N. C. Ha and F. D. M. Haldane, Phys. Rev. Lett **73**, 2887 (1994).
102. M. Arikawa, Y. Saiga, and Y. Kuramoto, Phys. Rev. Lett **86**, 3096 (2001).
103. M. Zacher, E. Arrigoni, W. Hanke, and J. Schrieffer, Phys. Rev. B **57**, 6370 (1998).

PART II

PARTICIPANT CONTRIBUTIONS

Beyond RPA: dynamical exchange effects and the two-dimensional electron gas

K.J. Hameeuw*, F. Brosens* and J.T. Devreese*

*TFVS, Departement Natuurkunde, Universiteit Antwerpen, Universiteitsplein 1, B-2610 Antwerpen, Belgium

Abstract. Dynamic exchange interactions can be introduced in the dielectric function via a dynamic local field factor. We study the effects of this inclusion on the energy-loss function of a two-dimensional electron gas, using the dynamic local field factor that we derived recently via the dynamical exchange decoupling method. The results are compared with the dielectric function in the Random Phase Approximation, showing a drastic influence of the dynamic exchange interactions.

INTRODUCTION

The electron gas is an ideal system of interacting electrons in a uniform positive background in the presence of an external electric field. At zero temperature, it is determined by one single parameter, its density. The electron gas forms a theoretical model for the study of electron-electron interactions in systems with nearly-free-electrons, where the ion-core lattice is replaced by a homogeneous positive background, thereby ignoring all lattice influence.

Knowledge of the linear-response functions of this model provides us with a lot of information on its properties. The dielectric function $\varepsilon(\mathbf{q},\omega)$, which gives a linear correlation between an external field and the induced charge density fluctuations in a system in this field, is a main ingredient in the description of the electron gas. It allows for the calculation of many fundamental properties like the pair correlation function, the static and dynamic structure factor, the ground state energy, plasmon dispersion relations etc.

In its most simple form, $\varepsilon(\mathbf{q},\omega)$ can be written as $\varepsilon(\mathbf{q},\omega) = 1 + Q_0(\mathbf{q},\omega)$, where $Q_0(\mathbf{q},\omega)$ is the Lindhard polarizability. This approximation, known as the Random Phase Approximation (RPA), only takes into account long range Hartree interactions. To include the effects of short range interactions, one can introduce the so called local-field factor $\mathcal{G}(\mathbf{q},\omega)$ by writing $\varepsilon(\mathbf{q},\omega) = 1 + Q_0(\mathbf{q},\omega) / \left(1 - \mathcal{G}(\mathbf{q},\omega) Q_0(\mathbf{q},\omega)\right)$. In an exact theory of the electron gas, $\mathcal{G}(\mathbf{q},\omega)$ accounts for all correlation and exchange effects. This concept was first introduced by Hubbard [1], who made a static approximation $\mathcal{G}(\mathbf{q})$. However, a consistent theory can only be achieved if the full frequency dependence of $\mathcal{G}(\mathbf{q},\omega)$ is taken into account, since causality arguments [2] and internal consistency requirements in the theory of the electron gas [3] imply the need for a frequency dependent local field factor $\mathcal{G}(\mathbf{q},\omega)$.

The interest in the properties of the two-dimensional electron gas (2DEG) has been driven by the discovery of peculiar behavior that was not found in its three-dimensional

CP715, *Lectures on the Physics of Highly Correlated Electron Systems VIII: Eighth Training Course*, edited by A. Avella and F. Mancini

(3D) counterpart. The 2DEG has thus shown itself to be a fascinating system to study, revealing behavior that is quite different from that of the well known 3DEG. Already in the seventies, Jonson [4] showed that in a 2DEG, correlations are much more important then in the 3D case. In the present paper, an expression for the frequency dependent local field factor of a 2DEG is derived. We follow the dynamic exchange decoupling (DED) method, as proposed by Brosens, Devreese and Lemmens [5] for the 3D case. By applying this approach, we extend the perturbative calculation of $\mathcal{G}(\mathbf{q},\omega)$ by Czachor *et al.* [6] to a variational description [7].

CONCEPTS AND FRAMEWORK

We consider a gas of interacting electrons, whose motion is confined to two dimensions. The fact that we are dealing with charged particles, implies the need for a neutralizing background to prevent the electrons from being pushed apart due to the Coulomb repulsion. Their fermionic nature ensures us that they obey Pauli's principle and points to the possibility of important exchange effects. In experimental situations, an external perturbing electromagnetic field is applied to probe the system. To handle the 2DEG mathematically, the jellium model is applied, assuming that the electrons are moving in a homogeneous neutralizing background so that the system in equilibrium is neutral. The only parameter left is the density, often expressed via the Winger-Seitz radius r_s, defined as the radius of a sphere that contains one electron. If an external electrical field $\varphi(\mathbf{r},t)$ is considered, the model Hamiltonian in second quantization looks like

$$\begin{aligned} H = {} & \frac{\hbar^2}{2m}\sum_\sigma \int d^3r \nabla \psi_\sigma^\dagger(\mathbf{r}) \cdot \nabla \psi_\sigma(\mathbf{r}) + \frac{ie\hbar}{mc}\sum_\sigma \int d^3r \psi_\sigma^\dagger(\mathbf{r})\left(\mathbf{A}(\mathbf{r},t)\cdot\nabla\right)\psi_\sigma(\mathbf{r}) \\ & + e\sum_\sigma \int d^3r \psi_\sigma^\dagger(\mathbf{r})\varphi(\mathbf{r},t)\psi_\sigma(\mathbf{r}) \\ & + \frac{1}{2}\sum_\sigma\sum_{\sigma'} \int d^3r \int d^3r' \psi_\sigma^\dagger(\mathbf{r})\psi_{\sigma'}^\dagger(\mathbf{r}')\, v(\mathbf{r}-\mathbf{r}')\,\psi_{\sigma'}(\mathbf{r}')\psi_\sigma(\mathbf{r}), \end{aligned} \quad (1)$$

where $\psi_\sigma^\dagger(\mathbf{r})$ and $\psi_\sigma(\mathbf{r})$ are the Fermi field operators and $v(\mathbf{r})$ is the Coulomb potential.

The 2DEG as described above, can be experimentally realized in different ways. The most pure system is formed by electrons moving on the surface of liquid Helium. A totally different realization that is most widely used in industry is the Si-MOSFET, whose conduction relies on electrons moving in two dimensions underneath the insulating surface of silicium oxide. Furthermore, one can study the 2DEG in different types of semiconductor heterostructures such as GaAs-AlGaAs. More recently, it was discovered that the electrons in high-T_c superconducting cuprates move in 2D planes formed by the copper oxides.

The testcharge-testcharge dielectric function $\varepsilon(q,\omega)$ is defined as

$$V_{q,\omega}^{tc} = \varphi_{q,\omega}/\varepsilon(q,\omega), \quad (2)$$

where $V_{q,\omega}^{tc}$ is the potential felt by a test charge in the presence of the medium and $\varphi_{q,\omega}$ is the Fourier transform of the Coulomb potential. The potential felt by a test charge in the presence of the medium can also be written as

$$V^{tc}_{q,\omega} = \varphi_{q,\omega} + v_q n_{q,\omega}, \tag{3}$$

if $n_{q,\omega}$ represents the density induced in the medium due to $\varphi_{q,\omega}$ and the factor v_q is the Fourier transform of the Coulomb potential. As such, the only quantity that has to be calculated is the induced density in the medium, since combining equations (2) and (3) then gives an expression for $\varepsilon(q,\omega)$.

One of the first attempts was performed by Lindhard for the 3DEG. Starting from a gas of free electrons and assuming a perturbation $W^{tot}_{q,\omega}$ as the total potential felt by the electrons, first order perturbation theory tells us that $v_q n_{q,\omega} = -Q_0(q,\omega) W^{tot}_{q,\omega}$, where the so called Lindhard polarizability is given by

$$Q_0(q,\omega) = -\frac{4\pi e^2}{q^2} \sum_{k<k_F} \left(\frac{1}{\hbar\omega + \frac{\hbar^2 k^2}{2m} - \frac{\hbar^2(\mathbf{k}+\mathbf{q})^2}{2m}} - \frac{1}{\hbar\omega - \frac{\hbar^2 k^2}{2m} + \frac{\hbar^2(\mathbf{k}+\mathbf{q})^2}{2m}} \right). \tag{4}$$

The roots of the denominators define the boundaries of the single particle excitation spectrum, also known as the Landau continuum. In the resonant structure of this function, one clearly distinguishes two possible processes, which are related to absorption and emission respectively.

If we assume that the total potential felt by the electrons $W^{tot}_{q,\omega}$ can be written as $W^{tot}_{q,\omega} = \varphi_{q,\omega} + v_q n_{q,\omega}$, i.e. the electrons are assumed to be mere charged particles, the well known result of the Random Phase Approximation (RPA) is found, where $\varepsilon(q,\omega) = 1 + Q_0(q,\omega)$. However, since the electrons are fermions, exchange and correlation effects should be taken into account. In a formal way, we can do that by writing

$$W^{tot}_{q,\omega} = \varphi_{q,\omega} + v_q \left(1 - \mathscr{G}(\mathbf{q},\omega)\right) n_{q,\omega}, \tag{5}$$

where the so-called local field factor $\mathscr{G}(\mathbf{q},\omega)$ is introduced to describe the exchange and correlations hole that surrounds the electrons. The wavevector dependence accounts for the spatial extension of this hole, while the frequency dependence originates from the retarded nature of the exchange and correlation effects, i.e. there is a time scale for screening. As mentioned earlier, this is merely a formal way of introducing $\mathscr{G}(\mathbf{q},\omega)$, because the exchange and correlation effects are many-body effects by definition and can thus never be properly taken into account in the above single particle picture.

To find an expression for the local field factor $\mathscr{G}(\mathbf{q},\omega)$, one has to follow a many-body approach when deriving an expression for the induced density. In the following section, we will introduce the dynamical-exchange decoupling (DED) approach that was developed for the 3DEG by Brosens, Devreese and Lemmens [5, 8] and adapt it to the 2D case. Since we are interested in the frequency dependence of the local field factor, we start from the Liouville equation for the density matrix $n(\mathbf{r},t)$

$$i\hbar \frac{d}{dt} n(\mathbf{r},t) = \left\langle \left[\psi^\dagger_\sigma(\mathbf{r}') \, \psi_\sigma(\mathbf{r}), H \right] \right\rangle_t,$$

where $\psi^\dagger_\sigma(\mathbf{r})$ and $\psi_\sigma(\mathbf{r})$ are the Fermi field operators and H is the model Hamiltonian (1) introduced before. Elaboration of the commutator leads to a BBGKY- type of hierarchy for the density matrix, which is broken by applying the Hartree-Fock decoupling scheme to the four-field operator terms $\langle \psi_1^\dagger \psi_2^\dagger \psi_3 \psi_4 \rangle \approx \langle \psi_1^\dagger \psi_4 \rangle \langle \psi_2^\dagger \psi_3 \rangle - \langle \psi_1^\dagger \psi_3 \rangle \langle \psi_2^\dagger \psi_4 \rangle$. This decoupling dynamically preserves the full exchange contribution. The resulting Time-Dependent Hartree-Fock equation is linearized in the external field. For technical convenience, this equation of motion is expressed in terms of the Wigner distribution function $\widetilde{f}(\mathbf{p},\mathbf{q},\omega)$, where $\widetilde{f}(\mathbf{p},\mathbf{q},\omega)$ is the Fourier transform in space and time of $f(\mathbf{p},\mathbf{R},t) - f^0(\mathbf{p})$ and

$$f(\mathbf{p},\mathbf{R},t) = \frac{1}{(2\pi\hbar)^2} \int d^2 r e^{-i\mathbf{p}\cdot\mathbf{r}/\hbar} \left\langle \psi^\dagger \left(\mathbf{R} - \frac{\mathbf{r}}{2}\right) \psi \left(\mathbf{R} + \frac{\mathbf{r}}{2}\right) \right\rangle_t . \tag{6}$$

The equilibrium distribution function $f^0(\mathbf{p})$ is assumed to be the paramagnetic Fermi-Dirac distribution at zero temperature. Under these assumptions, the equation of motion for $\widetilde{f}(\mathbf{p},\mathbf{q},\omega)$ becomes

$$\widetilde{f}(\mathbf{p},\mathbf{q},\omega) = -\frac{1}{2} \frac{U(\mathbf{q},\omega) N_{\mathbf{q}}(\mathbf{p})}{\omega + i\varepsilon - \frac{\mathbf{p}\cdot\mathbf{q}}{m}} + \widetilde{X}(\mathbf{p},\mathbf{q},\omega), \tag{7}$$

where $U(\mathbf{q},\omega)$ is the sum of the induced Hartree potential and the external potential $e\varphi_{\mathbf{q},\omega}$ and $\widetilde{X}(\mathbf{p},\mathbf{q},\omega)$ accounts for the exchange effects

$$\widetilde{X}(\mathbf{p},\mathbf{q},\omega) = 2\pi^2 \int d^2 p' v_{|\mathbf{p}-\mathbf{p}'|/\hbar} \frac{\widetilde{f}(\mathbf{p}',\mathbf{q},\omega) N_{\mathbf{q}}(\mathbf{p}) - \widetilde{f}(\mathbf{p},\mathbf{q},\omega) N_{\mathbf{q}}(\mathbf{p}')}{\omega + i\varepsilon - \frac{\mathbf{p}\cdot\mathbf{q}}{m}}. \tag{8}$$

The factor $v_{|\mathbf{p}-\mathbf{p}'|/\hbar}$ denotes the Fourier transform of the two dimensional Coulomb potential, $v_q = e^2/2\pi q$. The function $N_{\mathbf{q}}(\mathbf{p})$ is a geometrical factor, determined by the equilibrium distribution function $N_{\mathbf{q}}(\mathbf{p}) = 2\left(f^0(\mathbf{p}+\hbar\mathbf{q}/2) - f^0(\mathbf{p}-\hbar\mathbf{q}/2)\right)/\hbar$.

If one neglects the exchange contribution (8), equation (7) can be solved exactly. The solution $f_\sigma(\mathbf{p},\mathbf{q},\omega)_L$ is known as the Lindhard distribution. To include the exchange interactions, we adopt a variational scheme [9]. It seems natural to reformulate the problem in terms of the Lindhard solution, by introducing a new function $\gamma_{\mathbf{q}\omega}(\mathbf{p})$:

$$\widetilde{f}_\sigma(\mathbf{p},\mathbf{q},\omega) = f_\sigma(\mathbf{p},\mathbf{q},\omega)_L \, \gamma_{\mathbf{q}\omega}(\mathbf{p}). \tag{9}$$

One can then introduce a functional $F_{\mathbf{q}\omega}[\gamma]$ so that by performing the first variation of this functional with respect to $\gamma_{\mathbf{q}\omega}(\mathbf{p})$, and by taking into account (9), one finds the integral equation (7). The simplest possible trial function $\gamma_{\mathbf{q}\omega}(\mathbf{p})$ is independent of momentum $\mathbf{p}$. In this approximation, the first variation of the functional $F_{\mathbf{q}\omega}[\gamma]$ with respect to $\gamma_{\mathbf{q}\omega}$ gives

$$\gamma_{\mathbf{q}\omega} = \left(1 + Q_0(\mathbf{q},\omega)\left(1 - \mathscr{G}(\mathbf{q},\omega)\right)\right)^{-1}, \tag{10}$$

leading to a dielectric function

$$\varepsilon(\mathbf{q},\omega) = 1 + Q_0(\mathbf{q},\omega) / \left(1 - \mathscr{G}(\mathbf{q},\omega) Q_0(\mathbf{q},\omega)\right), \tag{11}$$

where $\mathcal{G}(\mathbf{q},\omega)$ is again the local field factor. This frequency dependent local field correction $\mathcal{G}(\mathbf{q},\omega)$ accounts for the dynamic exchange effects. The function $Q_0(\mathbf{q},\omega)$ is known in closed form [10]. Although the momentum dependence of the trial function $\gamma_{\mathbf{q}\omega}$ is neglected, the above solution conserves the equation of motion for the charge and current density, thereby satisfying an important consistency requirement.

Expressed in dimensionless Fermi units $\mathbf{k} = \mathbf{q}/k_F$ and $\nu = \hbar\omega/2E_F$, where k_F is the Fermi wavevector and $E_F = \hbar^2 k_F^2/2m$ is the Fermi energy, the local field factor obtained via the DED approach is a universal function of the density

$$G(k,\nu) = \mathcal{G}(kk_F, \nu 2E_F/\hbar) = \lim_{\delta\to 0} \frac{I_G(k,\nu)}{4k\pi^2 Q_d^2(k,\nu)}, \tag{12}$$

where

$$\begin{aligned} I_G(k,\nu) &= \int d^2r \int d^2r' \frac{1}{|\mathbf{r}-\mathbf{r}'|} \frac{1}{\nu+i\delta-\mathbf{r}\cdot\mathbf{k}} \left(\frac{1}{\nu+i\delta-\mathbf{r}'\cdot\mathbf{k}} - \frac{1}{\nu+i\delta-\mathbf{r}\cdot\mathbf{k}} \right) \\ &\times \left(\mathcal{N}(\mathbf{r}+\mathbf{k}/2) - \mathcal{N}(\mathbf{r}-\mathbf{k}/2) \right) \left(\mathcal{N}(\mathbf{r}'+\mathbf{k}/2) - \mathcal{N}(\mathbf{r}'-\mathbf{k}/2) \right), \end{aligned}$$

and the dimensionless variant $Q_d(k,\nu) = Q_0(\mathbf{q},\omega)/r_s$ of the Lindhard distribution is introduced. The function $\mathcal{N}(\mathbf{r}) = \Theta(1-|\mathbf{r}|)$, i.e. $\mathcal{N}(\mathbf{r}) = 1$ if $|\mathbf{r}| \leq 1$ and $\mathcal{N}(\mathbf{r}) = 0$ if $|\mathbf{r}| > 1$. The fourfold integral for $I_G(k,\nu)$ can analytically be reduced two a double integral [11], leaving two integrations to be done numerically.

Figure 1 shows the real and imaginary parts of both $I_G(k,\nu)$ and $Q_d(k,\nu)$ for different values of the wave vector k. The regions around the boundaries $\nu = |k \pm k^2/2|$ of the Landau continuum are left out of the graphs, since the singular behavior that is displayed here by the real and imaginary parts of $I_G(k,\nu)$ is an artefact of the choice of the trial function as a product of $f_\sigma(\mathbf{p},\mathbf{q},\omega)_L$ and $\gamma_{\mathbf{q}\omega}$ [12]. Since we choose $\gamma_{\mathbf{q}\omega}$ to be independent of the momentum $\mathbf{p}$, the many body effects are squeezed into the single particle excitation spectrum, producing nonphysical behavior at these boundaries.

Figure 2 shows the real and imaginary parts of $G(k,\nu)$, again for several values of the wave vector k. The inset shows the cut in the (k,ν)−plane along which the graphs are produced. Both $I_G(k,\nu)$ and $G(k,\nu)$ display a strong frequency dependence that is nontrivial. Furthermore, the real and imaginary parts of the functions $I_G(k,\nu)$ and $Q_d(k,\nu)$ are of the same order of magnitude. As a consequence, the local field factor itself $G(k,\nu)$ is substantially different from zero. Therefor any perturbative treatment of $\mathcal{G}(\mathbf{q},\omega)$, as e.g. performed by Czachor *et al.* [6], can be expected to be valid only for small values of r_s.The expression for $G(k,\nu)$ furthermore contains no parametrizations and fulfills important consistency requirements, including the correct long and short wavelength behavior [13], both in the static and the frequency dependent case. The local field factor $G(k,\nu)$ is not directly measurable in experiments. Nevertheless, Larson *et al.* managed to obtain experimental results for the static local field factor $G(k,0)$ of aluminum in 3D [14], showing that the local field factor derived in the DED approach leads to a significantly better agreement with the experimental results. For a more detailed discussion of $G(k,0)$ in the DED approach in 2D, we refer to Ref. [11].

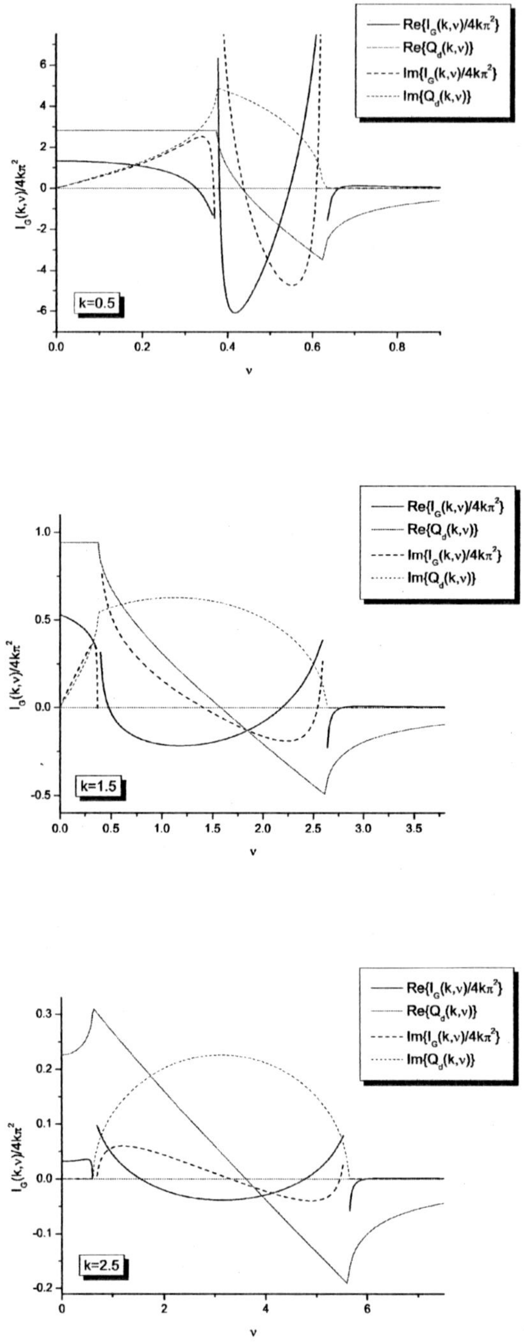

FIGURE 1. Real (full curves) and imaginary (dashed curves) part of $I_G(k,v)$ and $Q_d(k,v)$ as a function of the frequency v for $k = 0.5$, $k = 1.5$ and $k = 2.5$.

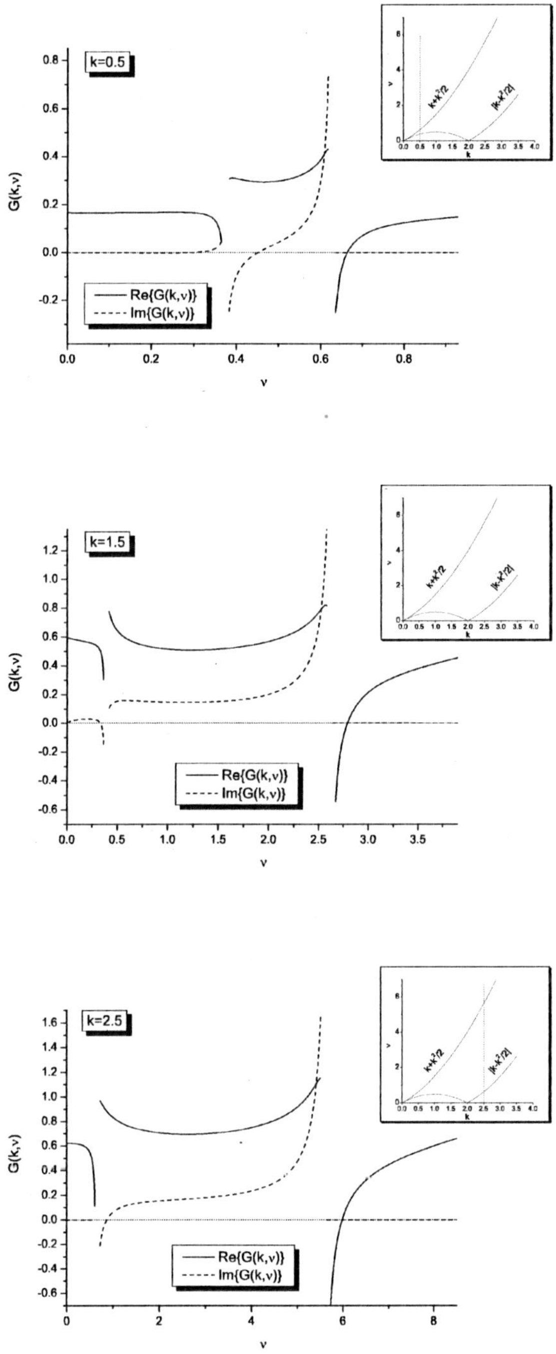

FIGURE 2. Real (full curve) and imaginary (dashed curve) part of $G(k,\nu)$ as a function of the frequency ν for $k = 0.5$, $k = 1.5$ and $k = 2.5$.

RESULTS

To be able to confront the theoretical results with experimental values, we should look for experimentally measurable quantities. We will therefor discuss the results for the energy-loss function $S(k,\nu)$, which describes the probability density for an incoming particle with wavevector $\mathbf{q}$ and frequency ω to be scattered with a change $\mathbf{k}$ and ν in wavevector and frequency respectively. Figures 3-5 show the contribution of the Landau continuum to the energy-loss function as a function of frequency for different values of the wavevector k and the Wigner-Seitz radius r_s. The results were obtained via the following equation that relates $S(k,\nu)$ to the imaginary part of the inverse dielectric function $\varepsilon(k,\nu)$

$$S(k,\nu) \propto -Im\left\{\varepsilon^{-1}(k,\nu)\right\}, \tag{13}$$

combined with the expressions (12) and (11). For results of the real and imaginary part of $\varepsilon(k,\nu)$ as well as a discussion of the static dielectric function $\varepsilon(k,0)$, we again refer to Ref. [11]. Comparison with the RPA approach shows a pronounced effect of the inclusion of dynamic exchange effects on the frequency dependence of $S(k,\nu)$. The discrepancy between both approaches increases with increasing r_s, as could be expected. Unfortunately, up till now experimental results are unavailable for the energy-loss function of the 2DEG. However, absolute IXS-measurements made on Aluminum for $q = k_F$ and $q = 1.5k_F$ in 3D performed by Tischler *et al.* [15] again show a good agreement between the DED approach and the experimental results. This suggests that also in the 2DEG, our results could lead to better agreement with future experimental results.

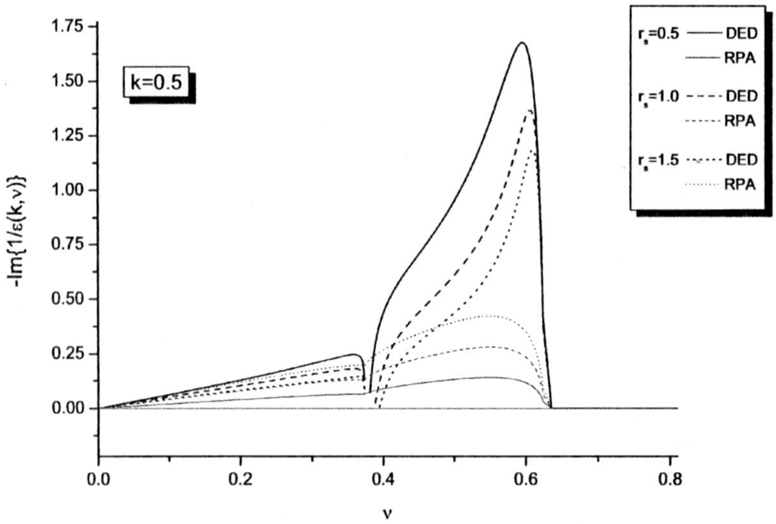

FIGURE 3. The imaginary part of the inverse dielectric function $-Im\{1/\varepsilon(k,\nu)\}$ in the DED approach and the RPA approach for different values of r_s, i.e. $r_s = 0.5$ (full curves), $r_s = 1.0$ (dashed curves) and $r_s = 1.5$ (dotted curves), as a function of the frequency ν for $k = 0.5$.

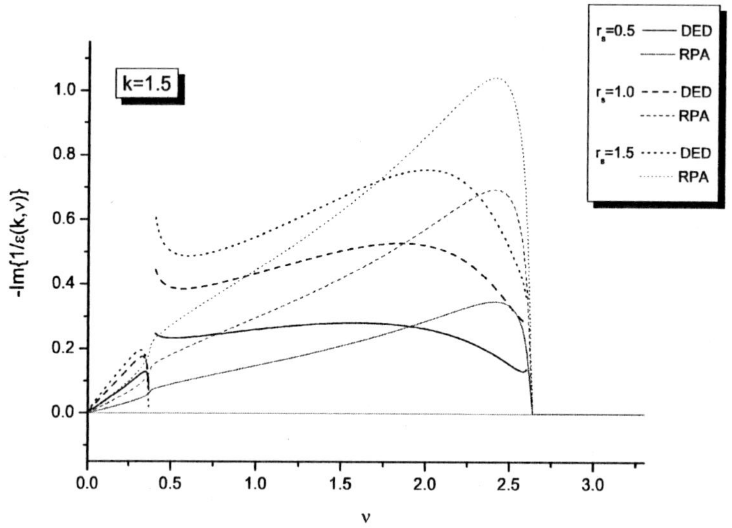

FIGURE 4. The imaginary part of the inverse dielectric function $-Im\{1/\varepsilon(k,\nu)\}$ in the DED approach and the RPA approach for different values of r_s, i.e. $r_s = 0.5$ (full curves), $r_s = 1.0$ (dashed curves) and $r_s = 1.5$ (dotted curves), as a function of the frequency ν for $k = 1.5$.

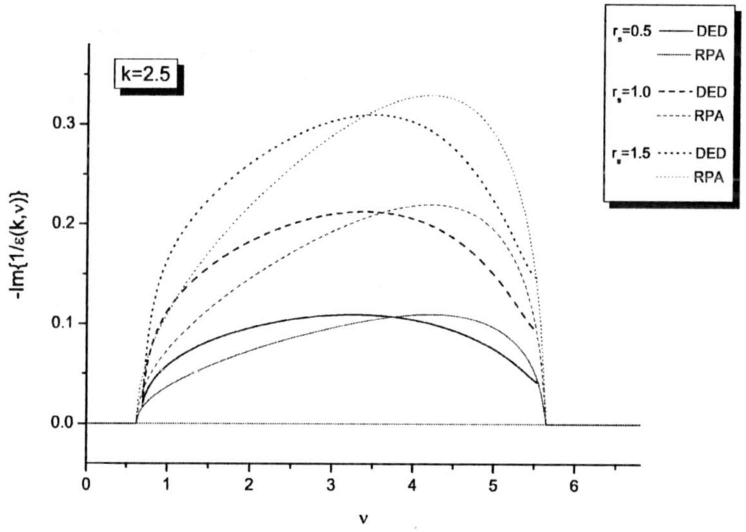

FIGURE 5. The imaginary part of the inverse dielectric function $-Im\{1/\varepsilon(k,\nu)\}$ in the DED approach and the RPA approach for different values of r_s, i.e. $r_s = 0.5$ (full curves), $r_s = 1.0$ (dashed curves) and $r_s = 1.5$ (dotted curves), as a function of the frequency ν for $k = 2.5$.

CONCLUSIONS

We derived an expression for the frequency dependent local field factor $G(k, \nu)$ of a two-dimensional electron gas via a variational solution of the Time-Dependent Hartree-Fock equation for the density matrix, following the dynamical exchange decoupling method that was developed in 3D by Brosens, Devreese and Lemmens [5, 8]. This local field factor $G(k, \nu)$ takes into account the full wave vector and frequency dependence, without relying on parametrizations and is shown to be independent of the density. The dependence on the frequency was shown to be nontrivial, while $G(k, \nu)$ was substantially different from zero indicating the limitations of a perturbative treatment of the exchange effects. Results for the energy-loss function $S(k, \nu)$ are given for different values of the wave vector k and the Wigner-Seitz radius r_s. These results clearly show the pronounced effect of the inclusion of dynamical exchange effects on the energy-loss function.

ACKNOWLEDGMENTS

This work has been supported by the GOA BOF UA 2000, IUAP, FWO-V project No. G.0274.01N and the WOG WO.025.99N. K.J.H. (Aspirant bij het Fonds voor Wetenschappelijk Onderzoek -Vlaanderen), acknowledges the FWO - Vlaanderen for financial support.

REFERENCES

1. J. Hubbard, *Proc. Roy. Soc.* **A243** (1957) 336.
2. A. Kugler, J. Stat. Phys. **8**, 107 (1973); *ibid.* **12**, 35 (1975).
3. J.C. Kimball, Phys. Rev. A **7**, 1648 (1973); J.C. Kimball, Phys. Rev. B **14**, 2371 (1976).
4. M. Jonson, J. Phys. C **9**, 3055 (1976).
5. F. Brosens, L. F. Lemmens, and J. T. Devreese, phys. stat. sol. (b) **74,** 45 (1976).
6. A. Czachor, A. Holas, S.R. Sharma, and K.S. Singwi, Phys. Rev. B **25**, 2144 (1982).
7. F. Brosens, L.F. Lemmens, and J.T. Devreese, phys. stat. sol. (a) **59**, 447 (1980).
8. J. T. Devreese, F. Brosens, and L. F. Lemmens, Phys. Rev. B **21,** 1349 (1980); F. Brosens, J. T. Devreese, and L. F. Lemmens, *ibid.* **21**, 1363 (1980).
9. A.K. Rajagopal, Phys. Rev. **142**, 152 (1965).
10. F. Stern, Phys. Rev. Lett. **18**, 546 (1967).
11. K. J. Hameeuw, F. Brosens and J. T. Devreese, Eur. Phys. J. **B35**, 93 (2003).
12. K. J. Hameeuw, F. Brosens and J. T. Devreese, Solid State Commun. **126**, 695 (2003).
13. N. Iwamoto, Phys. Rev. A **30**, 3289 (1984).
14. B.C. Larson, J.Z. Tischler, E.D. Isaacs, P. Zschack, A. Fleszar, and A.G. Eguiluz, Phys. Rev. Lett. **77**, 1346 (1996).
15. J. Z. Tischler, B. C. Larson, P. Zschack, A. Fleszar en A. G. Eguiluz, phys. stat. sol. (b) **237**, 280 (2003).

Spectral functions and their applications

Valery N. Marachevsky [1]

V. A. Fock Institute of Physics, St. Petersburg State University, 198504 St. Petersburg, Russia

Abstract. We give an introduction to the heat kernel technique and ζ-function. Two applications are considered. First we derive the high temperature asymptotics of the free energy for boson fields in terms of the heat kernel expansion and ζ-function. Another application is chiral anomaly for local (MIT bag) boundary conditions.

1. INTRODUCTION

In this paper we give an introduction to the technique of the heat-kernel expansion and ζ-function regularization. The heat kernel became a standard tool in calculations of the vacuum polarization, the Casimir effect and quantum anomalies. In the case of non-trivial background fields, and especially in the presence of boundaries or singularities, the heat kernel technique seems to be the most adequate one for the analysis of the one-loop effects (see [1] for a recent review).

A possible application of heat kernel methods to the evaluation of high temperature asymptotics of the free energy in the presense of boundaries or singularities has not received due attention yet. At high temperatures the expansion of the free energy can be determined by the heat kernel expansion and ζ-function [2].

Chiral anomaly, which was discovered more than 30 years ago [3], still plays an important role in physics. On smooth manifolds without boundaries many successful approaches to the anomalies exist [4]. The heat kernel approach to the anomalies is essentially equivalent to the Fujikawa approach [5] and to the calculations based on the finite-mode regularization [6], but it can be more easily extended to complicated geometries. The chiral anomaly in the case of non-trivial boundary conditions (MIT bag boundary conditions) has been calculated only recently [7].

The paper is organized as follows. In Sec. 2 we give an introduction to the formalism of the heat kernel and heat kernel expansion. Also we introduce a zeta function and calculate the one-loop effective action in terms of the zeta function. In Sec. 3 we consider two examples. First we derive the high temperature expansion of the free energy for boson fields in terms of the heat kernel expansion and ζ-function. Then we discuss a chiral anomaly in four dimensions for an euclidean version of the MIT bag boundary conditions [8].

---

[1] email: maraval@mail.ru, root@VM1485.spb.edu

CP715, *Lectures on the Physics of Highly Correlated Electron Systems VIII: Eighth Training Course*, edited by A. Avella and F. Mancini

2. SPECTRAL FUNCTIONS

2.1. Heat kernel

Consider a second order elliptic partial differential operator L of Laplace type on an n-dimensional Riemannian manifold. Any operator of this type can be expanded locally as

$$L = -(g^{\mu\nu}\partial_\mu\partial_\nu + a^\sigma\partial_\sigma + b), \tag{1}$$

where a and b are some matrix valued functions and $g^{\mu\nu}$ is the inverse metric tensor on the manifold. For a flat space $g^{\mu\nu} = \delta^{\mu\nu}$.

The heat kernel can be defined as follows:

$$K(t;x;y;L) = \langle x|\exp(-tL)|y\rangle = \sum_\lambda \phi^\dagger_\lambda(x)\phi_\lambda(y)\exp(-t\lambda), \tag{2}$$

where ϕ_λ is an eigenfunction of the operator L with the eigenvalue λ.

It satisfies the heat equation

$$(\partial_t + L_x)K(t;x;y;L) = 0 \tag{3}$$

with an initial condition

$$K(0;x;y;L) = \delta(x,y). \tag{4}$$

If we consider the fields in a finite volume then it is necessary to specify boundary conditions. Different choices are possible. In section 3.1 we will consider the case of periodic boundary conditions on imaginary time coordinate, which are specific for boson fields. In section 3.2 we will study bag boundary conditions imposed on fermion fields. If the normal to the boundary component of the fermion current $\psi^\dagger\gamma_n\psi$ vanishes at the boundary, one can impose bag boundary conditions, a particular case of mixed boundary conditions. We assume given two complementary projectors $\Pi_\pm$, $\Pi_- + \Pi_+ = I$ acting on a multi component field (the eigenfunction of the operator L) at each point of the boundary and define mixed boundary conditions by the relations

$$\Pi_-\psi|_{\partial M} = 0, \quad (\nabla_n + S)\Pi_+\psi|_{\partial M} = 0, \tag{5}$$

where S is a matrix valued function on the boundary. In other words, the components $\Pi_-\psi$ satisfy Dirichlet boundary conditions, and $\Pi_+\psi$ satisfy Robin (modified Neumann) ones.

It is convenient to define

$$\chi = \Pi_+ - \Pi_- . \tag{6}$$

Let $\{e_j\}$, $j = 1,\dots,n$ be a local orthonormal frame for the tangent space to the manifold and let on the boundary e_n be an inward pointing normal vector.

The extrinsic curvature is defined by the equation

$$L_{ab} = \Gamma^n_{ab}, \tag{7}$$

where Γ is the Christoffel symbol. For example, on the unit sphere S^{n-1} which bounds the unit ball in R^n the extrinsic curvature is $L_{ab} = \delta_{ab}$.

Curved space offers no complications in our approach compared to the flat case. Let $R_{\mu\nu\rho\sigma}$ be the Riemann tensor, and let $R_{\mu\nu} = R^{\sigma}{}_{\mu\nu\sigma}$ be the Ricci tensor. With our sign convention the scalar curvature $R = R^{\mu}_{\mu}$ is $+2$ on the unit sphere S^2. In flat space the Riemann and Ricci tensors are equal to zero.

One can always introduce a connection ω_μ and another matrix valued function E so that L takes the form:

$$L = -(g^{\mu\nu}\nabla_\mu\nabla_\nu + E) \tag{8}$$

Here ∇_μ is a sum of covariant Riemannian derivative with respect to metric $g_{\mu\nu}$ and connection ω_μ. One can, of course, express E and ω in terms of a^μ, b and $g_{\mu\nu}$:

$$\omega_\mu = \frac{1}{2} g_{\mu\nu}(a^\nu + g^{\rho\sigma}\Gamma^\nu_{\rho\sigma}), \tag{9}$$

$$E = b - g^{\mu\nu}(\partial_\nu\omega_\mu + \omega_\mu\omega_\nu - \omega_\rho\Gamma^\rho_{\mu\nu}) \tag{10}$$

For the future use we introduce also the field strength for ω:

$$\Omega_{\mu\nu} = \partial_\mu\omega_\nu - \partial_\nu\omega_\mu + [\omega_\mu, \omega_\nu]\,. \tag{11}$$

The connection ω_μ will be used to construct covariant derivatives. The subscript $;\mu\ldots\nu\sigma$ will be used to denote repeated covariant derivatives with the connection ω and the Christoffel connection on M. The subscript $:a\ldots bc$ will denote repeated covariant derivatives containing ω and the Christoffel connection on the boundary. Difference between these two covariant derivatives is measured by the extrinsic curvature (7). For example, $E_{;ab} = E_{:ab} - L_{ab}E_{;n}$.

Let us define an integrated heat kernel for a hermitian operator L by the equation:

$$K(Q,L,t) := \mathrm{Tr}\,(Q\exp(-tL)) = \int_M d^n x\sqrt{g}\,\mathrm{tr}\,(Q(x)K(t;x;x;L))\,, \tag{12}$$

where $Q(x)$ is an hermitian matrix valued function, tr here is over matrix indices. For the boundary conditions we consider in this paper there exists an asymptotic expansion [9] as $t \to 0$:

$$K(Q,L,t) \simeq \sum_{k=0}^{\infty} a_k(Q,L)t^{(k-n)/2}\,. \tag{13}$$

According to the general theory [9] the coefficients $a_k(Q,L)$ are locally computable. This means that each $a_k(Q,L)$ can be represented as a sum of volume and boundary integrals of local invariants constructed from Q, Ω, E, the curvature tensor, and their derivatives. Boundary invariants may also include S, L_{ab} and χ. Total mass dimension of such invariants should be k for the volume terms and $k-1$ for the boundary ones.

At the moment several coefficients of the expansion (13) are known for the case of mixed boundary conditions (5) and matrix valued function Q (see [7] for details of derivation; the formula (49) for a_4 was derived in [7] with additional restrictions $L_{ab} = 0$

and $S=0$) :

$$a_0(Q,L) = (4\pi)^{-n/2}\int_M d^n x\sqrt{g}\,\mathrm{tr}\,(Q). \tag{14}$$

$$a_1(Q,L) = \frac{1}{4}(4\pi)^{-(n-1)/2}\int_{\partial M} d^{n-1}x\sqrt{h}\,\mathrm{tr}\,(\chi Q). \tag{15}$$

$$a_2(Q,L) = \frac{1}{6}(4\pi)^{-n/2}\Big\{\int_M d^n x\sqrt{g}\,\mathrm{tr}\,(6QE+QR)$$

$$+\int_{\partial M} d^{n-1}x\sqrt{h}\,\mathrm{tr}\,(2QL_{aa}+12QS+3\chi Q_{;n})\Big\}. \tag{16}$$

$$a_3(Q,L) = \frac{1}{384}(4\pi)^{-(n-1)/2}\int_{\partial M} d^{n-1}x\sqrt{h}\,\mathrm{tr}\{Q(-24E+24\chi E\chi$$
$$+48\chi E+48E\chi-12\chi_{:a}\chi_{:a}+12\chi_{:aa}-6\chi_{:a}\chi_{:a}\chi+16\chi R$$
$$+8\chi R_{anan}+192S^2+96L_{aa}S+(3+10\chi)L_{aa}L_{bb}$$
$$+(6-4\chi)L_{ab}L_{ab})+Q_{;n}(96S+192S^2)+24\chi Q_{;nn}\}. \tag{17}$$

For a scalar function Q and mixed boundary conditions the coefficients a_4 and a_5 were already derived [10].

2.2. ζ-function

Zeta function of an operator L is defined by

$$\zeta_L(s) = \sum_\lambda \frac{1}{\lambda^s}, \tag{18}$$

where the sum is over all eigenvalues of the operator L. The zeta function is related to the heat kernel by the transformation

$$\zeta_L(s) = \frac{1}{\Gamma(s)}\int_0^{+\infty} dt\, t^{s-1}K(I,L,t). \tag{19}$$

Residues at the poles of the zeta function are related to the coefficients of the heat kernel expansion:

$$a_k(I,L) = \mathrm{Res}_{s=(n-k)/2}(\Gamma(s)\zeta_L(s)). \tag{20}$$

Here I is a unit matrix with a dimension of the matrix functions a^μ, b in (1).From (20) it follows that

$$a_n(I,L) = \zeta_L(0). \tag{21}$$

In Euclidean four dimensional space the zero temperature one-loop path integral over the boson fields $\phi = \sum_\lambda C_\lambda \phi_\lambda$ can be evaluated as follows (up to a normalization factor):

$$Z = \int d\phi e^{-\int d^4x\phi L\phi} \simeq \prod_\lambda \int \mu dC_\lambda e^{-\lambda C_\lambda^2} \simeq \mu^{\zeta_L(0)}\mathrm{det}L^{-1/2}. \tag{22}$$

Here we introduced the constant μ with a dimension of mass in order to keep a proper dimension of the measure in the functional integral. $\zeta_L(0)$ can be thought of as a number of eigenvalues of the operator L. For the operator L in the form (1) the number of eigenvalues is infinite, so $\zeta_L(0)$ yields a regularized value for this number.

The zero temperature one-loop effective action is defined then by

$$W = -\ln Z = -\frac{1}{2}\ln \det L + \frac{1}{2}\zeta_L(0)\ln\mu^2 = \frac{1}{2}\zeta_L'(0) + \frac{1}{2}\zeta_L(0)\ln\mu^2 =$$
$$= \frac{1}{2}\frac{\partial}{\partial s}(\mu^{2s}\zeta_L(s))|_{s=0} \tag{23}$$

The term $\zeta_L(0)\ln\mu^2 = a_4(I,L)\ln\mu^2$ in the effective action W determines the one-loop beta function, this term describes renormalization of the one-loop logarithmic divergences appearing in the theory.

3. APPLICATIONS

3.1. Free energy for boson fields

A finite temperature field theory is defined in Euclidean space, since for boson fields one has to impose periodic boundary conditions on imaginary time coordinate (antiperiodic boundary conditions for fermion fields respectively). A partition function is defined by

$$Z(\beta) = \mathrm{Tr}\, e^{-\beta H}, \tag{24}$$

where H is a hamiltonian of the problem and $\beta = \hbar/T$. Let us choose the lagrangian density ρ in the form

$$\rho = -\frac{\partial^2}{\partial\tau^2} + L, \tag{25}$$

where τ is an imaginary time coordinate and L is a three dimensional spatial part of the density in the form (1). The free energy of the system is defined by

$$F(\beta) = -\frac{\hbar}{\beta}\ln Z(\beta) = -\frac{\hbar}{\beta}\ln\Big(N_\beta\int D\phi \exp\big(-\int_0^\beta d\tau\int d^3x\,\phi\rho\phi\big)\Big), \tag{26}$$

the integration is over all periodic fields satisfying $\phi(\tau+\beta) = \phi(\tau)$ (N_β is a normalization coefficient). As a result the eigenfunctions of ρ have the form $\exp(i\tau\omega_n)\phi_\lambda$, where $\omega_n = 2\pi n/\beta$ and $L\phi_\lambda = \lambda\phi_\lambda$. The free energy is thus equal to [11]

$$F = \frac{\hbar}{2\beta}\sum_{n=-\infty}^{+\infty}\sum_\lambda \ln\frac{(\omega_n^2+\lambda)}{\mu^2} = -\frac{\hbar}{2\beta}\frac{\partial}{\partial s}(\mu^{2s}\zeta(s))|_{s=0}, \tag{27}$$

where we introduced ζ-function

$$\zeta(s) = \sum_{n=-\infty}^{+\infty}\sum_\lambda(\omega_n^2+\lambda)^{-s} \tag{28}$$

and the parameter μ with a mass dimensionality in order to make the argument of the logarithm dimensionless (also see a previous section).

Then it is convenient to use the formula

$$\zeta(s) = \frac{1}{\Gamma(s)} \int_0^{+\infty} dt\, t^{s-1} \sum_{n=-\infty}^{+\infty} \sum_{\lambda} e^{-t(\omega_n^2+\lambda)}, \tag{29}$$

and separate $n = 0$ and other terms in the sum. For $n \neq 0$ terms we substitute the heat kernel expansion for the operator L at small t

$$\sum_{\lambda} e^{-\lambda t} = K(I;L;t) \simeq \sum_{k=0}^{\infty} a_k(I,L) t^{(k-3)/2} \tag{30}$$

and perform t integration, then we arrive at the high temperature expansion ($\beta \to 0$) for the free energy F:

$$\begin{aligned} F/\hbar = & -\frac{1}{2\beta}\zeta_L'(0) - \frac{1}{2\beta}\zeta_L(0)\ln(\mu^2) + (4\pi)^{3/2}\left[-\frac{a_0}{\beta^4}\frac{\pi^2}{90} - \frac{a_1}{\beta^3}\frac{\zeta_R(3)}{4\pi^{3/2}} - \frac{a_2}{\beta^2}\frac{1}{24} \right. \\ & + \frac{a_3}{\beta}\frac{1}{(4\pi)^{3/2}}\ln\left(\frac{\beta\mu}{2\pi}\right) - \frac{a_4}{16\pi^2}\left(\gamma + \ln\frac{\beta\mu}{2\pi}\right) \\ & \left. - \sum_{n\geq 5} \frac{a_n}{\beta^{4-n}} \frac{(2\pi)^{3/2-n}}{2\sqrt{2}} \Gamma\left(\frac{n-3}{2}\right)\zeta_R(n-3)\right]. \end{aligned} \tag{31}$$

Here $a_k \equiv a_k(I,L)$, $\zeta_R(s) = \sum_{n=1}^{+\infty} n^{-s}$ is a Riemann zeta function, $\zeta_L(s) = \sum_\lambda \lambda^{-s}$ is a zeta function of an operator L, γ is the Euler constant. The first two terms on the r.h.s. of (31) follow from the $n = 0$ term.

The term

$$-\frac{(4\pi)^{3/2}\hbar a_0}{\beta^4}\frac{\pi^2}{90} = -V\frac{\mathrm{tr} I}{\hbar^3}\frac{\pi^2}{90}T^4 \tag{32}$$

is the leading high temperature contribution to the free energy.

The classical limit terms due to the equality $\zeta_L(0) = a_3$ can be rewritten as follows:

$$T\left(-\frac{1}{2}\zeta_L'(0) + \zeta_L(0)\ln\frac{\hbar}{2\pi T}\right) = T\sum_{\lambda}\ln\frac{\hbar\sqrt{\lambda}}{2\pi T}. \tag{33}$$

The terms on the l.h.s. of (33) yield a renormalized value of the terms on the r.h.s. of (33), since the sum on the righthandsight is generally divergent when the number of modes is infinite.

The term with a_4 determines the part of the free energy that appears due to one-loop logarithmic divergences and thus it depends on the dimensional parameter μ as in the zero temperature case.

3.2. Chiral anomaly in four dimensions for MIT bag boundary conditions

Consider the Dirac operator on an n-dimensional Riemannian manifold

$$\widehat{D} = \gamma^{\mu}\left(\partial_{\mu} + V_{\mu} + iA_{\mu}\gamma^{5} - \frac{1}{8}[\gamma_{\rho},\gamma_{\sigma}]\sigma_{\mu}^{[\rho\sigma]}\right) \tag{34}$$

in external vector V_{μ} and axial vector A_{μ} fields. We suppose that V_{μ} and A_{μ} are anti-hermitian matrices in the space of some representation of the gauge group. $\sigma_{\mu}^{[\rho\sigma]}$ is the spin-connection[2].

The Dirac operator transforms covariantly under infinitesimal local gauge transformations (the local gauge transformation is $\widehat{D} \to \exp(-\lambda)\widehat{D}\exp(\lambda)$):

$$\begin{aligned}
&\delta_{\lambda}A_{\mu} = [A_{\mu},\lambda] \\
&\delta_{\lambda}V_{\mu} = \partial_{\mu}\lambda + [V_{\mu},\lambda] \\
&\widehat{D} \to \widehat{D} + [\widehat{D},\lambda]
\end{aligned} \tag{35}$$

and under infinitesimal local chiral transformations (the local chiral transformation is $\widehat{D} \to \exp(i\varphi\gamma_{5})\widehat{D}\exp(i\varphi\gamma_{5})$):

$$\begin{aligned}
&\tilde{\delta}_{\varphi}A_{\mu} = \partial_{\mu}\varphi + [V_{\mu},\varphi], \\
&\tilde{\delta}_{\varphi}V_{\mu} = -[A_{\mu},\varphi], \\
&\widehat{D} \to \widehat{D} + i\{\widehat{D},\gamma^{5}\varphi\}.
\end{aligned} \tag{36}$$

The parameters λ and φ are anti-hermitian matrices.

We adopt the zeta-function regularization and write the one-loop effective action for Dirac fermions at zero temperature as [3]

$$W = -\ln\det\widehat{D} = -\frac{1}{2}\ln\det\widehat{D}^{2} = \frac{1}{2}\zeta'_{\widehat{D}^{2}}(0) + \frac{1}{2}\ln(\mu^{2})\zeta_{\widehat{D}^{2}}(0), \tag{37}$$

where

$$\zeta_{\widehat{D}^{2}}(s) = \mathrm{Tr}(\widehat{D}^{-2s}), \tag{38}$$

prime denotes differentiation with respect to s, and Tr is the functional trace.

The following identity holds:

$$\zeta_{A}(s) = \mathrm{Tr}A^{-s} \Rightarrow \delta\zeta_{A}(s) = -s\mathrm{Tr}((\delta A)A^{-s-1}). \tag{39}$$

Due to the identity (39)

$$\delta_{\lambda}\zeta_{\widehat{D}^{2}}(s) = -\left(2s\mathrm{Tr}([\widehat{D},\lambda]\widehat{D}^{-2s-1})\right) = -2s\left(\mathrm{Tr}([\widehat{D}^{-2s},\lambda])\right) = 0, \tag{40}$$

---

[2] The spin-connection must be included even on a flat manifold if the coordinates are not Cartesian.

[3] The one-loop effective action is proportional to Planck constant $\hbar$, in what following we put $\hbar = 1$.

so the effective action (37) is gauge invariant, $\delta_\lambda W = 0$.

The chiral anomaly is by definition equal to the variation of W under an infinitesimal chiral transformation. Using (39) we obtain:

$$\tilde{\delta}_\varphi \zeta_{\widehat{D}^2}(s) = -\left(2is\mathrm{Tr}(\{\widehat{D},\gamma^5\varphi\}\widehat{D}^{-2s-1})\right) = -4is\left(\mathrm{Tr}(\gamma^5\varphi\widehat{D}^{-2s})\right), \qquad (41)$$

and the anomaly reads

$$\mathscr{A} := \tilde{\delta}_\varphi W = \frac{1}{2}\tilde{\delta}_\varphi \zeta'_{\widehat{D}^2}(0) = -2\mathrm{Tr}(i\gamma^5\varphi\widehat{D}^{-2s})|_{s=0}. \qquad (42)$$

The heat kernel is related to the zeta function by the Mellin transformation:

$$\mathrm{Tr}(i\gamma^5\varphi\widehat{D}^{-2s}) = \Gamma(s)^{-1}\int_0^\infty dt\, t^{s-1}K(i\gamma^5\varphi,\widehat{D}^2,t). \qquad (43)$$

In particular, after the substitution of the heat kernel expansion (13) into the formula (43) we obtain

$$\mathscr{A} = -2a_n(i\gamma^5\varphi,\widehat{D}^2). \qquad (44)$$

The same expression for the anomaly follows also from the Fujikawa approach [5].

We impose local boundary conditions:

$$\Pi_-\psi|_{\partial M} = 0, \qquad \Pi_- = \frac{1}{2}\left(1-\gamma^5\gamma_n\right), \qquad (45)$$

which are nothing else than a Euclidean version of the MIT bag boundary conditions [8]. For these boundary conditions $\Pi_-^\dagger = \Pi_-$, and the normal component of the fermion current $\psi^\dagger\gamma_n\psi$ vanishes on the boundary.

Since $\widehat{D}$ is a first order differential operator it was enough to fix the boundary conditions (45) on a half of the components. To proceed with a second order operator $L = \widehat{D}^2$ we need boundary conditions on the remaining components as well. They are defined by the consistency condition [12]:

$$\Pi_-\widehat{D}\psi|_{\partial M} = 0, \qquad (46)$$

which is equivalent to the Robin boundary condition

$$(\nabla_n + S)\Pi_+\psi|_{\partial M} = 0, \qquad \Pi_+ = \frac{1}{2}\left(1+\gamma^5\gamma_n\right) \qquad (47)$$

with

$$S = -\frac{1}{2}\Pi_+ L_{aa}. \qquad (48)$$

In the paper [7] the following expression for a coefficient $a_4(Q,L)$ with an hermitian matrix valued function Q and conditions (5), $L_{ab} = 0$ (flat boundaries), $S = 0$ was obtained:

$$a_4(Q,L)=\frac{1}{360}(4\pi)^{-n/2}\Big\{\int_M d^nx\sqrt{g}\,\mathrm{tr}\,\{Q(60E_{;\mu}{}^{\mu}+60RE+180E^2$$
$$+30\Omega_{\mu\nu}\Omega^{\mu\nu}+12R_{;\mu}{}^{\mu}+5R^2-2R_{\mu\nu}R^{\mu\nu}+2R_{\mu\nu\rho\sigma}R^{\mu\nu\rho\sigma})\}$$
$$+\int_{\partial M}d^{n-1}x\sqrt{h}\,\mathrm{tr}\,\{Q\{30E_{;n}+30\chi E_{;n}\chi+90\chi E_{;n}+90E_{;n}\chi$$
$$+18\chi\chi_{:a}\Omega_{an}+12\chi_{:a}\Omega_{an}\chi+18\Omega_{an}\chi\chi_{:a}-12\chi\Omega_{an}\chi_{:a}$$
$$+6[\chi\Omega_{an}\chi,\chi_{:a}]+54[\chi_{:a},\Omega_{an}]+30[\chi,\Omega_{an:a}]+12R_{;n}+30\chi R_{;n}\}+$$
$$+Q_{;n}(-30E+30\chi E\chi+90\chi E+90E\chi-$$
$$-18\chi_{:a}\chi_{:a}+30\chi_{:aa}-6\chi_{:a}\chi_{:a}\chi+30\chi R)+30\chi Q_{;\mu}{}^{\mu n}\}\Big\}. \tag{49}$$

To obtain the chiral anomaly in four dimensions[4] with MIT bag boundary conditions one has to calculate the coefficient $a_4(Q,L)$ (49) with $L=\widehat{D}^2$, $Q=i\gamma^5\phi$ and substitute it into (44). We define $V_{\mu\nu}=\partial_\mu V_\nu-\partial_\nu V_\mu+[V_\mu,V_\nu]$, $A_{\mu\nu}=D_\mu A_\nu-D_\nu A_\mu$, $D_\mu A_\nu=\partial_\mu A_\nu-\Gamma^\rho_{\mu\nu}A_\rho+[V_\mu,A_\nu]$. The anomaly contains two contributions:

$$\mathcal{A}=\mathcal{A}_V+\mathcal{A}_b. \tag{50}$$

In the volume part

$$\mathcal{A}_V=\frac{-1}{180\,(2\pi)^2}\int_M d^4x\sqrt{g}\,\mathrm{tr}\varphi\Big(-120\,[D_\mu V^{\mu\nu},A_\nu]$$
$$+60\,[D_\mu A_\nu,V^{\mu\nu}]-60\,D_\mu D^\mu D_\nu A^\nu+120\,\{\{D_\mu A_\nu,A^\nu\},A^\mu\}$$
$$+60\,\{D_\mu A^\mu,A_\nu A^\nu\}+120\,A_\mu D_\nu A^\nu A^\mu+30\,[[A_\mu,A_\nu],A^{\mu\nu}]$$
$$+\varepsilon_{\mu\nu\rho\sigma}\,\{-45\,i\,V^{\mu\nu}V^{\rho\sigma}+15\,i\,A^{\mu\nu}A^{\rho\sigma}-30\,i\,(V^{\mu\nu}A^\rho A^\sigma+A^\mu A^\nu V^{\rho\sigma})$$
$$-120\,i\,A^\mu V^{\nu\rho}A^\sigma+60\,i\,A^\mu A^\nu A^\rho A^\sigma\}-60\,(D_\sigma A_\nu)R^{\nu\sigma}+30\,(D_\mu A^\mu)R$$
$$-\frac{15i}{8}\varepsilon_{\mu\nu\rho\sigma}R^{\mu\nu}{}_{\eta\theta}R^{\rho\sigma\eta\theta}\Big) \tag{51}$$

only the $DA-R$ terms seem to be new [7] (for flat space it can be found e.g. in [6]).

The boundary part

$$\mathcal{A}_b=\frac{-1}{180\,(2\pi)^2}\int_{\partial M}d^3x\sqrt{h}\,\mathrm{tr}\Big(12\,i\,\varepsilon^{abc}\,\{A_b,\varphi\}D_aA_c$$
$$+24\{\varphi,A^a\}\{A_a,A_n\}-60\,[A^a,\varphi](V_{na}-[A_n,A_a])$$
$$+60(D_n\varphi)D_\mu A^\mu\Big) \tag{52}$$

is new [7]. It has been derived under the two restrictions: $S=0$ and $L_{ab}=0$. Note, that in the present context, the first condition ($S=0$) actually follows from the second one ($L_{ab}=0$) due to (48).

---

[4] In two dimensions ($n=2$) the boundary part of the chiral anomaly with MIT bag boundary conditions is equal to zero [7].

ACKNOWLEDGMENTS

V.M. thanks the organizers of the VIII Training Course in the Physics of Correlated Electron Systems and High-Tc Superconductors for partial support and hospitality in Vietri sul Mare (Salerno). V.M. is grateful to Dr. Adolfo Avella for his help and efforts that made the visit to Vietri possible. V.M. thanks Dmitri Vassilevich for his suggestions during the preparation of the paper.

REFERENCES

1. D. V. Vassilevich, *Phys.Rept.* **388**, 279 (2003) [arXiv:hep-th/0306138].
2. J.S.Dowker and G.Kennedy, *J.Phys.A* **11**, 895 (1978).
3. S.L.Adler, *Phys. Rev.* **177**, 2496 (1969); J.S.Bell and R.Jackiw, *Nuovo Cim. A* **60**, 47 (1969).
4. R. A. Bertlmann, "Anomalies In Quantum Field Theory", Clarendon Press, Oxford, 1996.
5. K. Fujikawa, *Phys. Rev. Lett.* **42**, 1195 (1979).
6. A. A. Andrianov and L. Bonora, *Nucl. Phys. B* **233**, 232 (1984).
7. V.N.Marachevsky and D.V.Vassilevich, "Chiral anomaly for local boundary conditions", [arXiv: hep-th/0309019] , to appear in *Nucl.Phys.B*.
8. A. Chodos, R. L. Jaffe, K. Johnson, C. B. Thorn and V. F. Weisskopf, *Phys. Rev. D* **9**, 3471 (1974) ; A. Chodos, R. L. Jaffe, K. Johnson and C. B. Thorn, *Phys. Rev. D* **10**, 2599 (1974) ; T. DeGrand, R. L. Jaffe, K. Johnson and J. E. Kiskis, *Phys. Rev. D* **12**, 2060 (1975) .
9. P. B. Gilkey, "Invariance theory, the heat equation, and the Atiyah-Singer index theorem", CRC Press, 1994.
10. T. P. Branson and P. B. Gilkey, *Commun. Part. Diff. Equat.* **15**, 245 (1990); D. V. Vassilevich, *J. Math. Phys.* **36**, 3174 (1995) [arXiv:gr-qc/9404052]; T. P. Branson, P. B. Gilkey, K. Kirsten and D. V. Vassilevich, *Nucl. Phys. B* **563**, 603 (1999) [arXiv:hep-th/9906144].
11. M.Bordag, U.Mohideen, V.M.Mostepanenko, *Phys.Rept.* **353**, 1 (2001) (pp.113 - 117).
12. T. Branson and P. Gilkey, *J. Funct. Anal.* **108**, 47 (1992); *Differential Geom. Appl.* **2**, 249 (1992).

Local Mott metal-insulator transition in confined fermions on optical lattices

M. Rigol*, A. Muramatsu*, G. G. Batrouni† and R. T. Scalettar**

*Institut für Theoretische Physik III, Universität Stuttgart, Pfaffenwaldring 57, D-70550 Stuttgart, Germany
†Institut Non-Linéaire de Nice, Université de Nice–Sophia Antipolis, 1361 route des Lucioles, 06560 Valbonne, France
**Physics Department, University of California, Davis, California 95616, USA

Abstract. Using quantum Monte Carlo (QMC) simulations we study the ground state properties of the one-dimensional fermionic Hubbard model in traps with an underlying lattice. We show that this model displays quantum critical behavior at the boundaries of the local Mott-insulating regions. A local compressibility defined to characterize the Mott-insulating phase has a non-trivial critical exponent. Both the local compressibility and the variance of the local density show universality with respect to the confining potential. We study the momentum distribution function and we find that it is not appropriate to characterize the transitions in the system.

INTRODUCTION

The realization of Bose-Einstein condensation (BEC) of trapped atomic gases [1, 2, 3] has generated in the last years a lot of experimental and theoretical research [4, 5]. BEC was achieved by confining atoms in magnetic traps and lowering the temperature of the system via the evaporative cooling technique. In this technique, the hottest atoms are selectively removed from the system and the remaining ones rethermalize via two-body collisions. A common feature of these experiments is that the trapped gases are dilute; mean-field theory then provides a useful framework to study the role of the interaction between particles.

Recently, a new feature has been added to the experiments: the magnetically trapped condensate is transferred into an optical lattice generated by interfering laser beams [6]. With this new experimental set-up the Mott metal-insulator transition (MMIT), a paradigm of strong correlations, was recently realized in ultracold atoms confined on an optical lattice [6]. Due to the fact that the atoms interact only via a contact potential, this system constitutes the most direct experimental realization of the Hubbard model which is the prototype generally used to study the MMIT. Whereas the experiments with optical lattices were realized with bosonic atoms, recent progress in cooling techniques allow fermionic systems to go well below the degeneracy temperature [7, 8], such that even superfluidity appears within reach [9]. It is therefore, to be expected that soon a fermionic MMIT will be realized on an optical lattice, offering the possibility to confront in a controlled way our knowledge of the MMIT in solid-state systems, without extrinsic effects always present there. This possibility is especially important since the MMIT is not only a long standing problem in condensed matter physics, but has also received

CP715, *Lectures on the Physics of Highly Correlated Electron Systems VIII: Eighth Training Course*, edited by A. Avella and F. Mancini

renewed attention in recent years due to its different manifestations in a number of transition metal oxides, the most prominent being high temperature superconductors [10].

Motivated by the possibility of such cross-fertilization, we performed in Refs. [11, 12] QMC simulations for the ground-state of the one-dimensional Hubbard model with a harmonic potential, as in experiments with ultracold atoms. In the present work we review and summarize the results presented in these references. The one-dimensional case was chosen since in one dimension, the quantum critical properties for the unconfined system are well characterized by the *Bethe Ansatz* solution. It is well known that there is a MMIT phase transition at half filling and for any value of the on-site repulsive interaction. The global compressibility $\kappa \sim \partial n/\partial \mu$ (that is zero in the Mott insulating phase) diverges when the system approaches to the MMIT as δ^{-1}, where $\delta = 1 - n$, and n is the expectation value of the density [13]. However, as shown theoretically [14] and numerically [15], in the presence of a confining potential the Mott-insulating phase is restricted to a domain that coexists with a compressible phase, and the global compressibility never vanishes [15]. This is in contrast to the global character typical of solid state systems.

We found that a properly defined local order parameter (local compressibility) displays critical behavior on approaching the edges of the Mott-insulating phase, revealing a new critical exponent. Furthermore, it was shown that both the variance of the local density ($\Delta_i \equiv < n_i^2 > - < n_i >^2$) and the local compressibility as functions of the local density n_i show universal behavior for $n_i \to 1$ independently of the power and strength of trapping potential and of the strength of the on-site repulsion. These results are reviewed in the next section. In the case of global quantities like the momentum distribution function we found that they are not appropriate to characterize the MMIT in the trapped case, as it will be shown in the third section. Conclusions are given in the last section.

LOCAL QUANTUM CRITICALITY AND UNIVERSALITY

The Hamiltonian of the fermionic Hubbard model with an arbitrary confining potential has the form

$$H = -t \sum_{i,\sigma} \left(c^{\dagger}_{i\sigma} c_{i+1\sigma} + h.c. \right) + U \sum_{i} n_{i\uparrow} n_{i\downarrow} + V_{\alpha} \sum_{i\sigma} \left(i - \frac{N}{2} \right)^{\alpha} n_{i\sigma},$$

where t is the hopping parameter, U is the on-site interaction that in the present work will be considered repulsive ($U > 0$), V_{α} characterizes the strength of the arbitrary confining potential with power α, and N the number of lattice sites considered. The QMC simulations were performed using a projector algorithm [16, 17, 18, 19], which applies $\exp(-\theta H)$ to a trial wavefunction (in our case the solution for $U = 0$). A projector parameter $\theta \simeq 20/t$ suffices to reach well converged values of the observables discussed here. A time slice of $\Delta\tau = 0.05/t$ was used in general.

Following the experiments done so far for bosons confined on optical lattices [6], where the superfluid-Mott insulating phase transition is achieved by increasing the ratio between the on-site repulsive interaction and the hopping parameter, we did a similar study for the fermionic case. The results are shown in Fig. 1 (a) for a trapped system with $N = 100$, $N_f = 70$ and a harmonic confining potential $V_2 = 0.0025t$. The ratio U/t is increased from 2 to 8 (for more details of these density profiles see Fig. 5). It can be seen that for small values of U/t ($U = 2t$) the density profile is approximately parabolic, indicating that the whole system is in a metallic phase. As the value of U/t ($U = 4t$) is increased, a Mott phase tries to develop at $n = 1$ while a metallic phase with $n > 1$ is present in the center of the system. As the on-site repulsion is increased even further ($U = 6t,\ 8t$), a Mott domain (plateau with $n = 1$) appears in the middle of the trap suppressing the metallic phase that was present there. At this point, the identification of the insulating regions is only based on the occupation number, using our knowledge from unconfined periodic systems, however a local physical quantity should be defined to characterize them. A first quantity that can be used is the variance of the density Δ_i [15], since on entering the Mott-insulating region a suppression of double occupancy should occur, leading to a decrease of the variance.

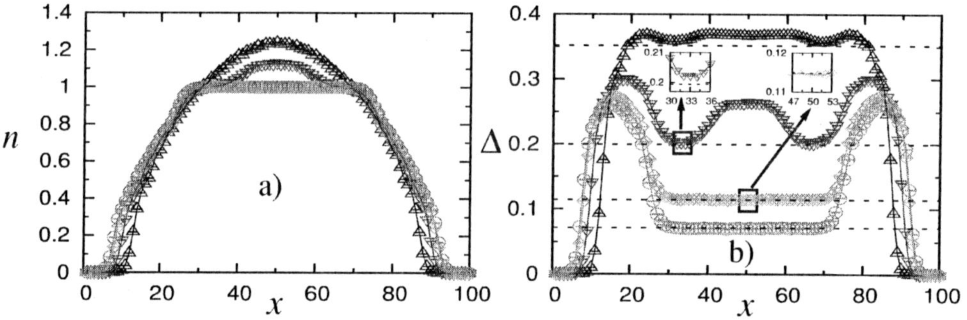

FIGURE 1. Profiles for a trap with $V_2 = 0.0025t$ and $N_f = 70$, the on-site repulsions are U =2t (△), 4t (▽), 6t (◇), and 8t (○). (a) Local density (b) Variance of the local density. The dashed lines in (b) are the values of the variance in the $n = 1$ homogeneous system for $U = 2t,\ 4t,\ 6t,\ 8t$ (from top to bottom).

In Fig. 1 (b) we show the variance of the density for the profiles in Fig. 1 (a) (from top to bottom, the values presented are for $U = 2t,\ 4t,\ 6t,\ 8t$). As expected, the variance decreases in both the metallic and Mott phases when the on-site repulsion is increased. When the Mott plateau is formed in the density profile, a plateau with constant variance appears in the variance profile with a value that will vanish only in the limit $U/t \to \infty$. We found that whenever a Mott domain is formed in the trap, the value of the variance in it is exactly the same as the one for the Mott phase in the homogeneous system for the same value of U/t. This is shown by the insets in Fig. 1 (b), where the values of the variance in the $n = 1$ homogeneous system are shown as horizontal dashed lines. This fact supports further our working assumption up to this point that the plateaus correspond to a Mott insulating region. Moreover, a very careful study shows that for $U = 4t$ (inset in Fig. 1 (b)) the value of the variance in the Mott phase of the homogeneous system is still not reached in the trap, which is an indication that there is still not a Mott insulating phase although a density $n = 1$ is reached. Then, in contrast to the homogeneous case, a Mott insulating region is not only determined by the filling. In the cases of $U = 6t$ (inset in Fig. 1 (b) for a closer look) and $U = 8t$ the value of the variance in the homogeneous system is reached and then we can say that Mott phases are formed there.

Therefore, the variance helps to characterize the system when a local Mott insulator appears or is within reach. This because it decreases on approaching the Mott insulating phase and when the variance at $n = 1$ is equal to the variance of the homogeneous system, a Mott insulator is formed. But there are also other characteristics of the variance that imply that a clearer order parameter is still needed to characterize the local Mott insulator in the trap: a) in the Mott domain the variance does not attain the smallest possible value b) The value of the variance in the local Mott phase changes when the value of the on site repulsion is varied, and finally c) metallic regions with densities very close to $n = 0$ and $n = 2$ can have even smaller variances than the ones of the Mott insulating phases.

The proper local order parameter we devised to characterize the Mott insulating phase is a local compressibility that is defined as

$$\kappa_i^l = \sum_{|j| \leq l(U)} \chi_{i,i+j} \,, \tag{1}$$

where $\chi_{i,j} = \langle n_i n_j \rangle - \langle n_i \rangle \langle n_j \rangle$ is the density-density correlation function. We take the length scale $l(U) \simeq b\xi(U)$, where $\xi(U)$ is the correlation length given by $\chi_{i,j}$ in the *unconfined* system at half filling for the given value of U. The value of b is such that κ^l becomes insensitive to the value chosen. In general we have $b \sim 10$, with $\xi(U) \sim a$ (a is the lattice constant) for the values of U used here. The local compressibility thus gives the response to a constant shift of the potential over a finite range but over distances larger than $\xi(U)$ in the periodic, unconfined system. If a region is in a Mott-insulating phase, and hence incompressible, no density response over distances larger than ξ is expected, leading to $\kappa_i^l = 0$. Figure 2 shows the profile along the trap of κ_i^l, the compressibility becomes zero in the outlying regions, where no particles are present and also where a Mott plateau is present. For $U = 4t$, it can be seen that in the

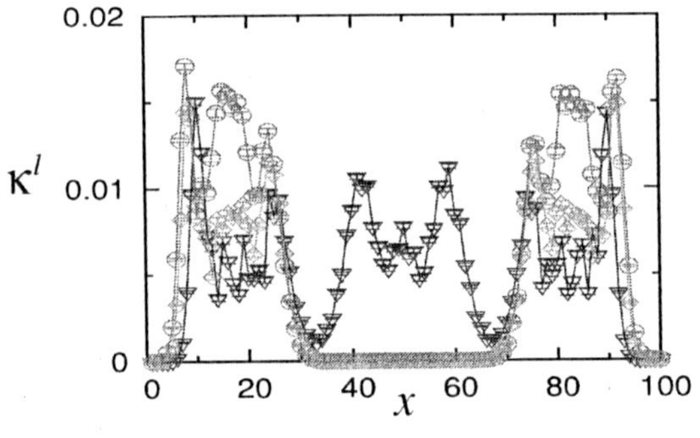

FIGURE 2. Local compressibility for a trap with $V_2 = 0.0025t$ and $N_f = 70$, the on-site repulsions are U =2t (△), 4t (▽), 6t (◇), and 8t (○). The density and variance profiles were shown in Fig. 1.

region with $n \sim 1$ the local compressibility, although small, does not vanish. This is compatible with the fact that the variance is not equal to the value in the homogeneous system there, so that although there is a shoulder in the density profile this region is not a Mott-insulator. Therefore, the local compressibility defined here serves as a genuine local order parameter to describe the insulating regions that coexist, in general, with a surrounding metallic zone or even with metallic intrusions, beyond the intuitive pictures on the basis of the density profiles.

Once the phases were characterized quantitatively, we concentrated on the regions where the system goes from one phase to another. Criticality can arise, despite the microscopic spatial size, due to the extension in imaginary time that reaches a thermodynamic limit at $T = 0$, very much like the case of the single impurity Kondo problem [20], where long-range interactions in imaginary time appear for the local degree of freedom as a result of the interaction with the rest of the system. Recent experiments leading to a MMIT [6] consider a system with linear dimension $\sim 65a$, i.e. still in this microscopic range. An intriguing future question, for both theory and experiment, will be the role of spatial dimension in the critical behavior of systems in the thermodynamic limit.

Figure 3 shows the local compressibility *vs.* δ for $\delta \to 0$ in a double logarithmic plot. A power law $\kappa^l \sim \delta^{\varpi}$ is obtained, with $\varpi < 1$, such that a divergence results in its derivative with respect to n, showing that critical fluctuations are present in this region. Since the QMC simulation is affected by systematic errors due to discretization

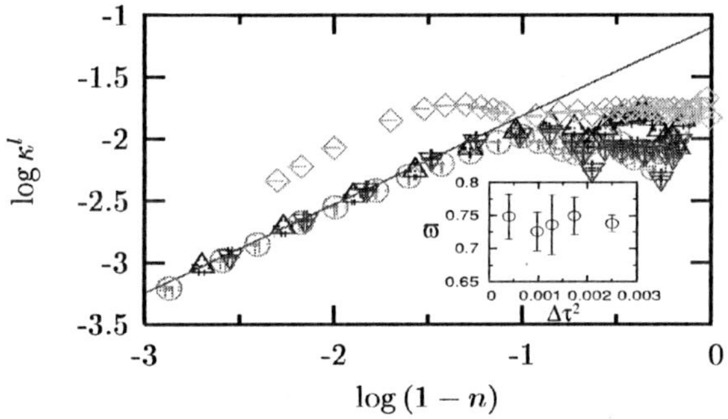

FIGURE 3. The local compressibility κ^l *vs.* $\delta = 1 - n$ at $\delta \to 0$ for ($\triangle$) $N_f = 70$, $U = 8t$ and $V_2 = 0.0025t$; ($\bigtriangledown$) $N_f = 70$, $U = 6t$ and $V_2 = 0.0025t$; ($\bigcirc$) $N_f = 72$, $U = 6t$ and a quartic potential with $V_4 = 1 \times 10^{-6}t$; ($\diamond$) unconfined periodic system with $U = 6t$. Inset: Dependence of the critical exponent ϖ on $\Delta\tau^2$.

in imaginary time, it is important to consider the limit $\Delta\tau \to 0$ in determining the critical exponent. The inset in Fig. 3 shows such an extrapolation leading to $\varpi \simeq 0.68 - 0.78$. In addition to the power law behavior, Fig. 3 shows that for $\delta \to 0$, the local compressibility of systems with a harmonic potential but different strengths of the interaction or even with a quartic confining potential, collapse on the same curve. Hence, universal behavior as expected for critical phenomena is observed also in this case. However, Fig. 3 shows also that the unconfined case departs from all the others. Up to the largest systems we simulated (400 sites), we observe an increasing slope rather than the power law of the confined systems.

Having shown that the local compressibility displays universality on approaching a Mott-insulating region, we consider the variance Δ as a function of the density n for various values of U and different confining potentials. Figure 4 shows Δ *vs.* n for a variety of systems, where not only the number of particles and the size of the system are changed, but also different forms of the confining potential were used. It appears at first glance that the data can only be distinguished by the strength of the interaction U, showing that the variance is rather insensitive to the form of the potential. The different insets, however, show that a close examination leads to the conclusion that only near $n = 1$ and only in the situations where at $n = 1$ a Mott-insulator exists, universality sets in. The inset for n around 0.6 and $U = 8t$, shows that the unconfined system has different variance from the others albeit very close on a raw scale. This difference is well beyond the error bars. Also the inset around $n = 1$ and for $U = 4t$, shows that systems that do not form a Mott insulating phase in spite of reaching a density $n = 1$, have a different variance from those having a Mott-insulator. The features above show that even a very local quantity like the variance cannot be accurately described using a local density approximation (Thomas-Fermi) [21], and can lead to even qualitatively wrong results, as for $U = 4t$ and $n = 1$, where such an approximation would predict a Mott-insulator instead of a metal as in our simulations. Only the case where all systems have a Mott-insulating phase at $n = 1$ ($U = 8t$), shows universal behavior independent of the potential, a universality that encompasses also the unconfined systems. For the unconfined system, the behavior of the variance can be examined with Bethe-*Ansatz* [22] in the limit $\delta \to 0$. In this limit and to leading order in δ, the ground state energy is given by [23] $E_0(\delta)/N - E_0(\delta = 0)/N \propto \delta$, such that the double occupancy, which can be obtained as the derivative of the ground-state energy with respect to U, will also converge as δ towards its value at half-filling. Such behavior is also obtained in our case as shown by the inset at $n = 1$ ($U = 8t$) in Fig. 4.

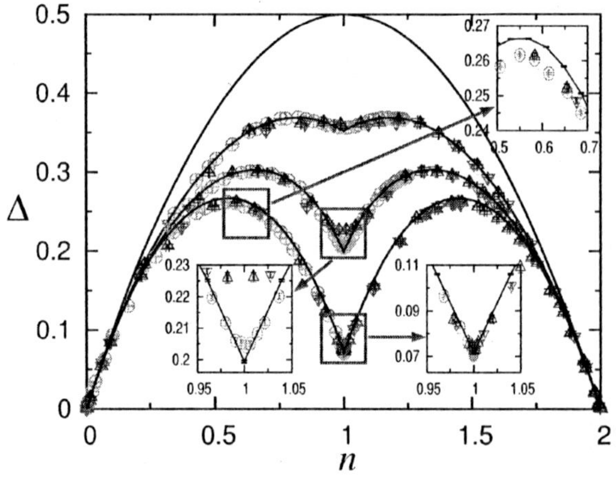

FIGURE 4. a) Variance Δ *vs.* n for (○) harmonic potential $V_2 = 0.0025t$ with $N = 100$; (△) quartic potential $V_4 = 5 \times 10^{-7}t$ with $N = 150$; (▽) harmonic potential $V_2 = 0.016t$ + cubic $V_3 = 1.6 \times 10^{-4}t$ + quartic $V_4 = 1.92 \times 10^{-5}t$ with $N = 50$; and (full line) unconfined periodic potential with $N = 102$ sites. The curves correspond from top to bottom to $U/t = 0, 2, 4, 8$. For a discussion of the insets, see text.

MOMENTUM DISTRIBUTION FUNCTION

There is, an important global quantity related to all the experiments realized so far for bosons and fermions. This quantity is the momentum distribution function, which is determined in time-of-flight measurements and which allowed the study of the superfluid-Mott insulator transition [6] in the bosonic case. A QMC study relating this quantity to the density profiles and proposing how to determine the point at which the superfluid-Mott insulator transition occurs was presented in Ref. [24].

In Fig. 5 we show the normalized momentum distribution function (bottom) for density profiles (top) in which the on site repulsion was increased from $U = 2t$ to $8t$. For the trapped systems, we normalized the momentum distribution to be unity at $k = 0$. We first notice that n_k for the pure metallic phase in the harmonic trap (Fig. 5 (b)) does

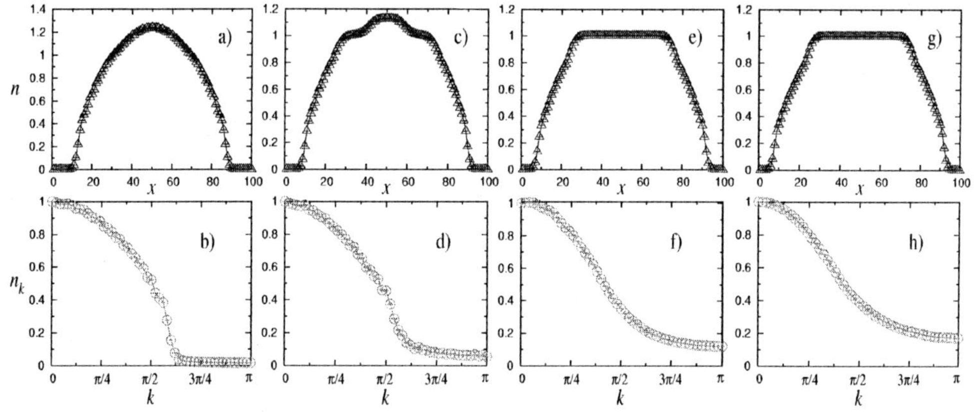

FIGURE 5. Density profiles (top) and their normalized momentum distribution functions (bottom) for $U = 2t$ (a,b), $4t$ (c,d), $6t$ (e,f), $6t$ (h,i) and $N = 100$, $N_f = 70$, $V = 0.0025t$.

not display any sharp feature corresponding to a Fermi surface, in clear contrast to the homogeneous case. The lack of a sharp feature for the Fermi surface is independent of the presence of the interaction and is also independent of the size of the system. In the non-interacting case, this can be easily understood: the spatial density and the momentum distribution will have the same functional form because the Hamiltonian is quadratic in both coordinate and momentum. When the interaction is present, it could be expected that the formation of local Mott domains generates a qualitatively and quantitatively different behavior of the momentum distribution, like in the homogeneous case where in the Mott insulating phase there is no Fermi surface (*i.e.* the singularity at k_f disappears). In Figs. 5 (b), (d), (f), (h) it can be seen that there is no qualitative change of the momentum distribution when the Mott-insulating phase appears in the middle of the trap. The only effect seen is that n_k becomes smoother and clear quantitative changes appear only when the on-site repulsion goes to the strong coupling regime, but this is long after the Mott insulating phase has appeared in the system.

At this point one might think that in order to study the MMIT using the momentum distribution function it is necessary to avoid the inhomogeneous trapping potential and use instead a kind of magnetic box with infinitely high potential on the boundaries. However, in that case one of the most important achievements of the inhomogeneous system is lost, namely the possibility of creating Mott insulating phases for a continuous range of fillings. In the perfect magnetic box, the Mott insulating phase would only be possible at half filling, which would be extremely hard (if possible at all) to adjust experimentally. The other possibility is to create traps which are almost homogenous in the middle and which have an appreciable trapping potential only close to the boundaries. This can be studied theoretically by considering traps with higher powers of the trapping potentials. We will approach the physics of these systems by considering trapped particles that do not interact, so that it is possible to exactly diagonalize the Hamiltonian for big systems.

In Fig. 6 (a) we show the results for the density profile of a system with 1000 sites, $N_f = 840$, and a trapping potential of the form $V_{10}(i-N/2)^{10}$ with $V_{10} = 7\times10^{-27}t$. It

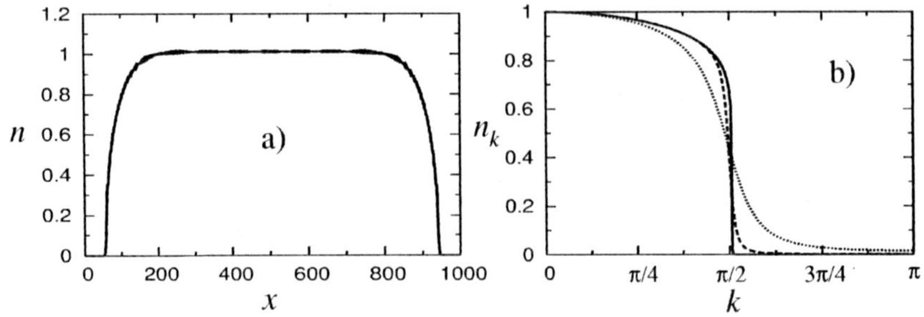

FIGURE 6. Exact results for $N_f = 840$ non-interacting trapped fermions in a lattice with 1000 sites and a confining potential $V_{10} = 7\times10^{-27}t$. Density profile (a) and the normalized momentum distribution function (b): the continuous line corresponds to a), the dashed line is the result when an alternating potential $V_a = 0.1t$ is superposed on the system, and the dotted line corresponds to $V_a = 0.5t$.

can be seen that the density profile is very homogeneous all over the trap with a density of the order of one particle per site. Only a small part of the system at the borders has the variation of the density required for the particles to be trapped. In Fig. 6 (b) (continuous line) we show the corresponding normalized momentum distribution. It can be seen that a kind of Fermi surface develops in the system but for smaller values of k, n_k is always smooth and its value starts decreasing at $k = 0$. At this point, the question that arises is the possibility of detecting the formation of a local incompressible region in the system measuring n_k. In order to answer that question, we introduced an additional alternating potential, so that in this case the new Hamiltonian has the form

$$H = -t\sum_{i,\sigma}\left(c^{\dagger}_{i\sigma}c_{i+1\sigma}+h.c.\right)+V_{10}\sum_{i\sigma}\left(i-\frac{N}{2}\right)^{10}n_{i\sigma}+V_a\sum_{i\sigma}(-1)^i n_{i\sigma}, \tag{2}$$

where V_a is the strength of the alternating potential. For the parameters presented in Fig. 6 (a), we find that a small value of V_a ($V_a = 0.1t$) generates a band insulator in the trap which extends over the region with $n\sim1$ (when $V_a = 0$). However, the formation of this band insulator is barely reflected in n_k, as can be seen in Fig. 6 (b) (dashed line). Only when the value of V_a is increased and the system departs from the phase transition ($V_a = 0.5t$, dotted line in Fig. 6 (c)), does a quantitatively appreciable change in n_k appear.

Another question that is important to answer is how the picture described above changes when the size of the system is increased. If one thinks in terms of open boundary conditions, an improvement in the behavior of the momentum distribution function is expected when the size of the system is increased, so that in the thermodynamic limit the momentum distribution function will be equal to the one in the homogeneous system. However, this reasoning does not apply to a system trapped by an inhomogeneous potential; there one has to define properly how to reach the thermodynamic limit and then study how quantities (like the momentum and densities) scale in this case. In the homogeneous case, this is done in such a way that the density of the system is kept constant, but this does not work for an inhomogeneous system. In the case of fermions confined in optical lattices when there is a confining potential with a power α, a characteristic length of the system (ζ) is given by $(V_\alpha/t)^{-1/\alpha}$, so that a characteristic density ($\tilde{\rho}$) can be defined as $N_f(V_\alpha/t)^{1/\alpha}$. We find that this characteristic density is the one meaningful in the thermodynamic limit. In Fig. 7 we show three systems in which the total number of particles and the curvature of the confining potential V_{10} were changed, keeping $\tilde{\rho}$ constant. We centered the systems at $x = 0$ and measured the positions in units of the characteristic length ζ. These systems have occupied regions

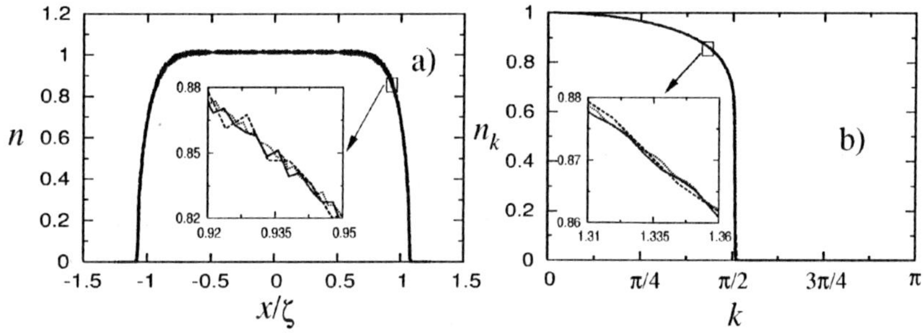

FIGURE 7. Exact results for non-interacting trapped fermions in a confining potential V_{10}. a) Density profiles b) Normalized momentum distribution function. Dashed line corresponds to $V_{10} = 6\times10^{-24}t$, $N_f = 428$ ($N \sim 500$), continuous line corresponds to $V_{10} = 7\times10^{-27}t$, $N_f = 840$ ($N \sim 1000$) and dotted line corresponds to $V_{10} = 1\times10^{-29}t$, $N_f = 1620$ ($N \sim 2000$). In the density profile the position is given in units of the characteristic length (ζ) defined in the text.

($n > 0$) with very different sizes, of the order of 500 lattice sites for the dashed line, of the order of 1000 sites for the continuous line, and of the order of 2000 lattice sites for the dotted line. Fig. 7 (a) shows that the density profiles scale perfectly when the curvature of the confining potential and the number of fermions are changed in the system, so that in order to show the changes in the density profile, we introduced an inset that expands a region around a density 0.85. In the case of the momentum distribution function (Fig. 7 (b)), it is possible to see that this quantity also scales very well so that almost no changes occur in the momentum distribution function when the occupied system size is increased, implying that in the thermodynamic limit the behavior of n_k is different from the one in the homogeneous system. An expanded view of the region with n_k around 0.87 is introduced to better see the scale of the differences in this region.

CONCLUSIONS

In summary, on the basis of QMC simulations of the Hubbard model, we found a number of new and unexpected features for the MMIT. (*i*) A local compressibility κ^l that appropriately characterizes Mott-insulating regions, shows critical behavior on entering those regions. Due to the microscopic nature of the phases, spatial correlations appear not to contribute to the critical behavior discussed here. This is a new form of MMIT, not observed so far in simple periodic systems, that might be realized in fermionic gases trapped on optical lattices. Therefore, our observation adds a new aspect to this long-standing problem in condensed matter physics. We expect that a similar local critical behavior will arise in higher dimensions as long as the spatial extent of the Mott domain remains finite. (*ii*) Universal behavior is found for κ^l for $n \to 1$, independent of the confining potential and/or strength of the interaction, excluding, however, the unconfined case. Also universal behavior is found for the variance Δ when $n \to 1$. In this case, this behavior is shared by the unconfined model.

In the case of the momentum distribution function we can say that it does not seem to be an adequate quantity to study the MMIT in the trapped system. Increasing the power of the confining potential brings about a quantitative improvement of the behavior of n_k, but maybe not enough to resolve small local gaps in the system. We have also shown that the increase of the system size, with the proper thermodynamic limit definition, does not change at all the behavior of the density and momentum distribution functions in the trapped case so that they will always be different from the homogeneous case. We would like to point out that for the case of attractive interactions, where superfluidity is expected to appear, we think the situation will be even worse because the superfluid gap is very small there and its effects on n_k could be less evident because of having an inhomogeneous system (not to speak about temperature effects).

REFERENCES

1. M. H. Anderson, J. R. Ensher, M. R. Matthews, C. E. Wieman and E. A. Cornell, Science **269**, 198 (1995).
2. C. C. Bradley, C. A. Sackett, J. J. Tollett and R. G. Hulet, Phys. Rev. Lett. **75**, 1687 (1995).
3. K. B. Davis, M.-O. Mewes, M. R. Andrews, N. J. van Druten, D. S. Durfee, D. M. Kurn and W. Ketterle, Phys. Rev. Lett. **75**, 3969 (1995).
4. *Bose-Einstein Condensation in Atomic Gases*, Proceedings of the International School of Physics "Enrico Fermi", edited by M. Inguscio, S. Stringari and C. E. Wieman (IOS Press, Amsterdam 1999).
5. F. Dalfovo, S. Giorgini, L. P. Pitaevskii and S. Stringari, Rev. Mod. Phys. **71**, 463 (1999).
6. M. Greiner, O. Mandel, T. Esslinger, T. W. Hänsch and I. Bloch, Nature (London) **415**, 39 (2002).
7. Z. Hadzibabic *et al.*, Phys. Rev. Lett. **88**, 160401 (2002).
8. G. Roati, F. Riboli, G. Modugno, and M. Inguscio, Phy. Rev. Lett. **89**, 150403 (2002).
9. K. M. O'Hara, S. L. Hemmer, M. E. Gehm, S. R. Granade, and J. E. Thomas, Science **298**, 2179 (2002).
10. M. Imada, A. Fujimori, and Y. Tokura, Rev. Mod. Phys. **70**, 1039 (1998).
11. M. Rigol, A. Muramatsu, G. G. Batrouni, and R. T. Scalettar, Phys. Rev. Lett. **91**, 130403 (2003).
12. M. Rigol and A. Muramatsu, cond-mat/0309670 (2003).
13. T. Usuki, N. Kawakami, and A. Okiji, Phys. Lett. A **135**, 476 (1989).
14. D. Jaksch, C. Bruder, J. I. Cirac, C. W. Gardiner and P. Zoller, Phys. Rev. Lett. **81**, 3108 (1998).
15. G. G. Batrouni, V. Rousseau, R. T. Scalettar, M. Rigol, A. Muramatsu, P. J. H. Denteneer and M. Troyer, Phys. Rev. Lett. **89**, 117203 (2002).
16. G. Sugiyama and S. E. Koonin, Annals of Phys. **168**, 1 (1986).
17. S. Sorella, S. Baroni, R. Car, and M. Parinello, Europhys. Lett. **8**, 663 (1989).
18. E. Y. Loh and J. E. Gubernatis, in *Modern Problems of Condensed Matter Physics*, edited by W. Hanke and Y. Kopaev (North Holland, Amsterdam, 1992).
19. M. Imada, *Quantum Monte Carlo Methods in Condensed Matter Physics* (World Scientific, Singapore, 1993).
20. G. Yuval and P. Anderson, Phys. Rev. B **1**, 1522 (1970).
21. D. Butts and D. Rokhsar, Phys. Rev. A **55**, 4346 (1997).
22. E. H. Lieb and F. Y. Wu, Phys. Rev. Lett. **20**, 1445 (1968).
23. A. Schadschneider and J. Zittartz, Z. Phys. B **82**, 387 (1991).
24. V. A. Kaskurnikov, N. V. Prokof'ev and B. V. Svistunov, Phys. Rev. A **66**, 031601(R) (2002).

Physical properties of correlated electrons in nanochains from EDABI method

Adam Rycerz* and Jozef Spałek*

*Marian Smoluchowski Institute of Physics, Jagiellonian University, Reymonta 4, 30-059 Kraków, Poland

Abstract. We implement the recently proposed approach combining *Exact Diagonalization* in the Fock space with an *Ab Initio* method (EDABI) to obtain a fairly complete description of correlated nanochains. In particular, the microscopic parameters are determined and the evolution of the system properties is traced in a systematic manner as a function of the interatomic distance. Both ground–state and dynamical correlation functions are discussed within a single scheme. The principal physical results are: (i) the appearance of mixed metallic and insulating features for a *one–dimensional* nanochain in the *half–filled* band case, and (ii) the transformation from a *highly–conducting* nanometal to the charge–ordered nanoinsulator away from the half–filling. The analysis is performed using the Gaussian $1s$–like basis and includes *long–range* Coulomb interactions.

INTRODUCTION

Recent development in computational, as well as analytical, methods has lead to a successful determination of electronic properties of semiconductors and metals starting from LDA [1], LDA+U [2], and related [3] approaches. Even strongly correlated systems, such as V_2O_3 (undergoing the Mott transition) and high–temperature superconductors, have been treated in that manner [4]. However, the discussion of the metal–insulator transition of the Mott–Hubbard type is not as yet possible in a systematic manner, particularly for low–dimensional systems. These difficulties are caused by the circumstance that the electron–electron interaction is comparable, if not stronger than the single–particle energy. In effect, the procedure starting from the single–particle picture (band structure) and including subsequently the interaction via a *local* potential might not be appropriate then. In this situation, one resorts to parametrized models of correlated electrons, where the single–particle and the interaction-induced aspects of the electronic states are treated on equal footing [5]. The single–particle wave–functions are contained in the formal expressions for model parameters. We propose to combine the two efforts in an exact manner, at least for model systems.

In our method of approach (EDABI), we determine *first* rigorously the energy of interacting particles in terms of the microscopic parameters and then optimize this energy with respect to the wave–functions contained in those parameters by deriving the *self–adjusted wave equation* for this state. Physically, the last step amounts to allowing the single–particle wave functions relax in the correlated state. The EDABI method has been overviewed in number of papers [6, 7, 8], so we concentrate here on its application to one–dimensional (1D) nanochains of $N \leqslant 16$ atoms, close to the metal–insulator

CP715, *Lectures on the Physics of Highly Correlated Electron Systems VIII: Eighth Training Course*, edited by A. Avella and F. Mancini

transition (MIT). The paper complements our recent study of such systems [6, 7] with the systematic analysis of both *half–* and *quarter–filled* band cases, as well as with the analysis of its transport properties. Throughout the paper we are using adjustable Wannier functions composed of Gaussians, which are determined explicitly from the minimization of the system ground–state energy as a function of interatomic distance.

THE CORRELATED NANOCHAIN

We consider the system of N_e electrons on N lattice sites, each containing a single valence orbital and an infinite–mass ion (i.e. we start from hydrogenic–like atoms). Including *all* long–range Coulomb interaction and neglecting other terms, one can write down the system Hamiltonian in the form

$$H = \varepsilon_a^{\rm eff} \sum_i n_i + t \sum_{i\sigma} \left(a_{i\sigma}^\dagger a_{i+1\sigma} + \mathrm{HC} \right) + U \sum_i n_{i\uparrow} n_{i\downarrow} + \sum_{i<j} K_{ij} \delta n_i \delta n_j, \qquad (1)$$

where $\delta n_i \equiv n_i - 1$, $\varepsilon_a^{\rm eff} = \varepsilon_a + N^{-1} \sum_{i<j} (2/R_{ij} + K_{ij})$ (in Ry) is the effective atomic level, R_{ij} is the distance between the i–th and j–th atoms, t is the nearest–neighbor hopping, U and K_{ij} are the intra– and inter–site Coulomb repulsions, respectively. Thus all the *mean–field* Coulomb terms are collected in $\varepsilon_a^{\rm eff}$, whereas the last term in the above Hamiltonian represent the *correlated* part of the long–range interactions.

The Hamiltonian (1) is diagonalized in the Fock space with the help of Lanczos technique. As the microscopic parameters $\varepsilon_a^{\rm eff}$, t, U, and K_{ij} are calculated numerically in the Gaussian basis, the orbital size of the $1s$–like state is subsequently adjusted to obtain the minimal ground state energy E_G as a function of the interatomic distance R. Earlier, have shown [7] that such variational procedure converges rapidly with the latice size N, so one can extrapolate the optimal orbital parameters for larger N using those obtained for small systems (i.e. for $N = 6 \div 10$) and speed up the computation.

The Tomonaga–Luttinger scaling

We discuss now the electron momentum distribution of 1D chain of $N = 6 \div 16$ atoms to address the question whether the system compose either the *Luttinger–liquid* state or forms the insulating (*Mott–Hubbard*) state. But first we summarize, following Voit [9], the salient properties of 1D conductors, which include the two principal features:

(i) a continuous momentum distribution function, showing the singularity near the Fermi level $k \approx k_F$ in the form (Solyom, Ref. [9])

$$n_{k\sigma} = n_F + A \left| k_F - k \right|^\theta \mathrm{sgn}(k_F - k), \qquad (2)$$

where θ is a non–universal (*interaction–dependent*) exponent; in consequence, it yields the non–existence of fermionic quasi–particles (the quasi–particle residue in vanishes as $z_k \sim |k_F - k|^\theta$ when $k \to k_F$);

(*ii*) similar power–law behavior of all physical properties, particularly of the single–particle density of states $\mathcal{N}(\omega) \sim |\omega - \mu|^{\theta}$ (*pseudogap*), that imply a finite Drude weight $D > 0$ for $\theta < 1$.

In the case of lattice models, such as (*extended*) Hubbard, the Luttinger liquid behavior is predicted by the renormalization group (RG) mapping onto Tomonaga–Luttinger model [9]. Through such mapping, one can also expect the convergence of the momentum distribution $n_{k\sigma}$ (discrete for finite N) to the continuous power–law form (2) with increasing N. This hypothesis was checked numerically for the Hubbard model [10].

Here we present a similar approach to a finite 1D chain with long–range Coulomb interaction, as described by the Hamiltonian (1). The corresponding electron–momentum distribution for the half–filling ($N_e = N$) is depicted in Fig. 1a in the linear, and in Fig. 1b in the *log–log* scale. We use the boundary conditions that minimize the ground–state energy for a given N (namely, the *periodic* for $N = 4n+2$ and the *antiperiodic* for $N = 4n$ at the half–filling). In order to extract the exponent θ accurately from the data for finite N, it was necessary to include also the higher scaling corrections. They can be obtained from the Tomonaga mapping in the form of an expansion in power of $\ln(\pi/|k_F - k|a)$,

$$\ln|n_F - n_{k\sigma}| = -\theta \ln z + b \ln\ln z + c + O(1/\ln z), \tag{3}$$

where $z \equiv \pi/|k_F - k|$. This singular form is required by the especially slow approach to the RG fixed point (Solyom, Ref. [9]); neglecting logarithmic corrections one can obtain the asymptotic form (2) for $k \approx k_F$. Solid lines in Figs. 1a and 1b represent the formula (3), the best fitted values of the parameters θ, b and c are listed in Tab. 1. The exponent θ is also plotted in Fig. 1c as a function of the lattice parameter R showing, that it crosses the critical value $\theta = 1$ (corresponding to the metal–insulator boundary in 1D) for $R_{\text{crit}} = 2.60a_0$ (a_0 is the Bohr radius). We also supply the *residual sum of squares* (*cf.* inset in Fig. 1c) to show, that a quality of the fit (3) becomes worst for $R \approx R_{\text{crit}}$, where the system is close to the metal–insulator transition.

The electron momentum distribution for the *quarter–filled* (QF) band case ($N_e = N/2$) is shown in Figure 2. The available number of datapoints was too small to fit the singular formula (3) to a reasonable accuracy, so the lines on the plot are a guide to the eye only. However, the smooth behavior of the Luttinger–liquid type is evident for $R \lesssim 4a_0$, whereas it changes dramatically for the larger values of R. Such behavior corresponds to the onset of the charge–density wave ordering, as illustrated in Fig. 3. In QF chain

TABLE 1. The parameters of the expansion (3) for the *half-filled* 1D chain with *long–range* Coulomb interactions. The standard deviation σ is also specified in each case.

R/a_0	θ	$\sigma(\theta)$	b	$\sigma(b)$	c	$\sigma(c)$
1.5	0.138	0.015	0.147	0.024	-0.567	0.015
2.0	0.387	0.055	0.425	0.089	-0.346	0.053
2.5	0.893	0.122	0.971	0.196	0.084	0.118
3.0	1.307	0.128	1.315	0.207	0.357	0.125
4.0	1.455	0.186	1.113	0.299	-0.032	0.180
5.0	1.413	0.133	0.943	0.214	-0.823	0.129

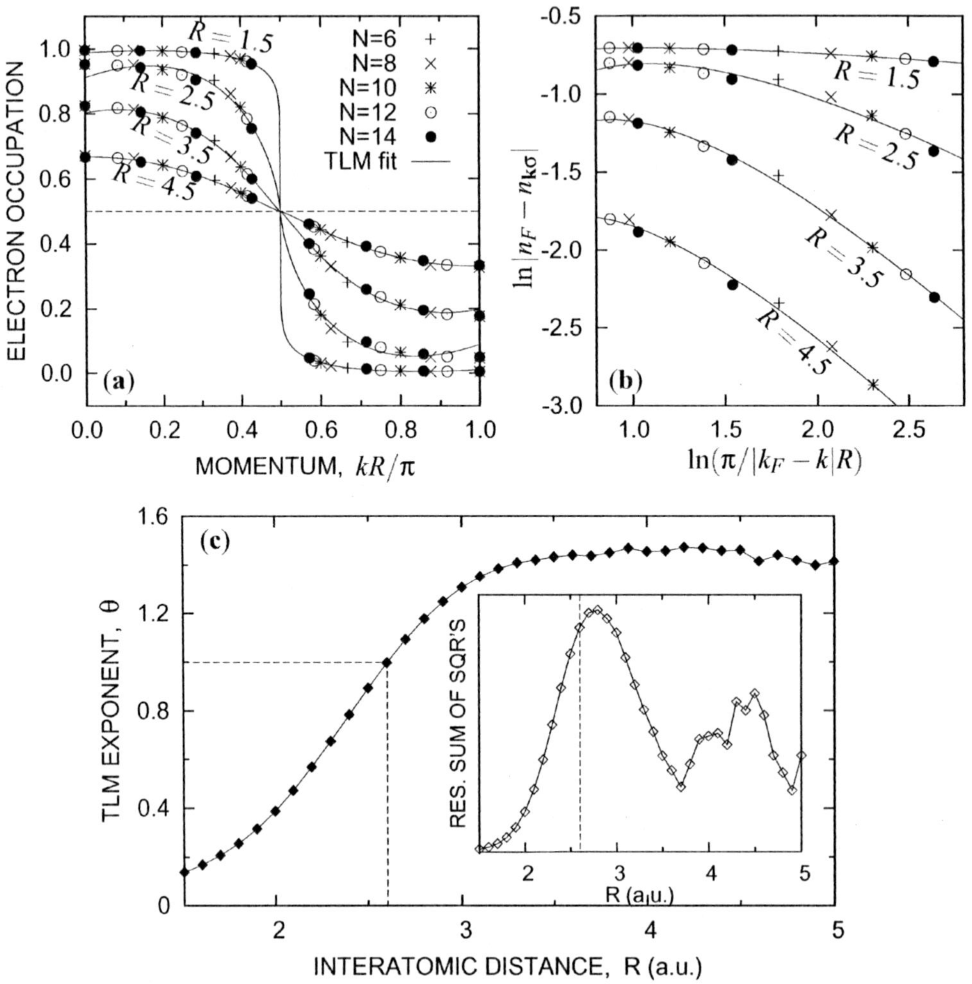

FIGURE 1. Luttinger–liquid scaling for a *half–filled* 1D chain of $N = 6 \div 14$ atoms with *long–range* Coulomb interactions: (a) momentum distribution for electrons in the linear and (b) log–log scale; (c) Tomonaga–Luttinger model exponent θ vs. lattice parameter R (specified in a_0) and (*inset*) the corresponding residual sum of squares. Solid lines in Figs. (a) and (b) represent the fitting of Eq. (3).

of $N = 16$ atoms (*cf.* Figure 3a) the charge is almost uniformly distributed for $R \lesssim 3a_0$, but charge–density waves sets in rapidly in the range $R/a_0 = 4 \div 5$. The corresponding order parameter, defined as $\theta_{\text{CDW}} \equiv N^{-1} \sum_m (-1)^m \langle \Delta n_i \Delta n_{i+m} \rangle$, (where $\Delta n \equiv n - \langle n \rangle$) approaches its maximal value $\theta_{\text{CDW}} = 1/4$ for $R \gtrsim 8a_0$ (*cf.* Figure 3b). The correspondence between the appearance of such charge–order and the system conductivity is discussed in the end of this paper.

The results for the half–filled system with *on–site* Hubbard repulsion only [11], are qualitatively very similar to those displayed in Fig. 1 (when the long–range interactions are included). The critical value of the lattice parameter $R_{\text{crit}} = 2.16a_0$ also does not

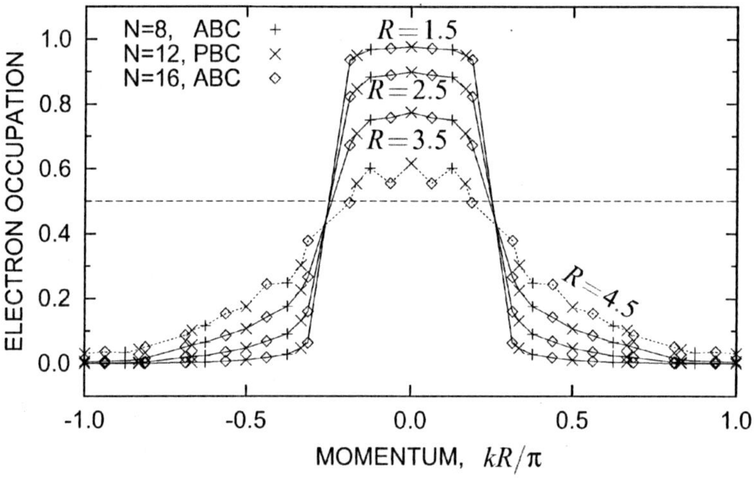

FIGURE 2. Momentum distribution $n_{k\sigma}$ for electrons on a chain of $N = 8 \div 16$ atoms in the *quarter–filled* band case ($N_e = N/2$). Lines are drawn as a guide to the eye only. ABC and PBC denotes the *antiperiodic* and *periodic* boundary conditions, respectively.

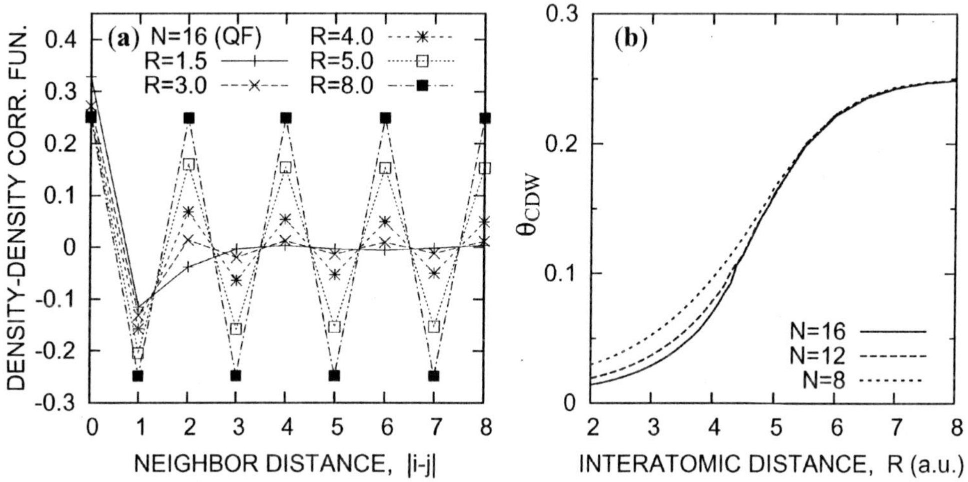

FIGURE 3. Charge–density distribution for the *quarter–filled* ($N_e = N/2$) nanochain: (*a*) density fluctuation correlation function $\langle \Delta n_i \Delta n_j \rangle$ vs. distance $|i - j|$, (*b*) *charge–density wave* order parameter for the alternating density fluctuation (*see* main text for the definition) vs. interatomic distance *R*.

differ drastically from the previous one. This is because such nanoscopic systems may always show a conducting behavior in the large–density limit as the external electron tunnels through a finite potential barrier. Therefore, such half–filled band systems, both with– and without inclusion of the long–range interactions can be regarded as close to the metal–insulator transition, with no contradiction to the infinite–chain RG result by Fabrizio [12]. This discussion is completed by the calculation of the charge–energy gap, as well as of the conductivity, in the next two Sections.

The charge–energy gap

For a further verification, whether the system is metallic or insulating in the Luttinger–liquid like regime presented above, we perform an extrapolation with $1/N \to 0$ of the charge–gap defined (for the *half-filling*) as

$$\Delta E_C(N) = E_G^{N+1} + E_G^{N-1} - 2E_G^{N}, \tag{4}$$

where $E_G^{N_e}$ is the ground-state energy of the system containing N_e electrons. The corresponding numerical results are shown in Fig. 4, where we again use the boundary conditions (*periodic* or *antiperiodic*) that minimize the ground–state energy. The extrapolation with $1/N \to 0$ performed using the 2–nd and the 3–rd order polynomials in $1/N$ provides nonzero value of ΔE_C for any lattice parameter R; only for the lowest examined value $R = 1.5a_0$, $E_G^{N_e}$ reaches zero within the extrapolation error; for other values, it is nonzero. The gap is also significantly smaller than the corresponding Hartree–Fock value in the regime $R \lesssim 4.5a_0$, suggesting some kind of reorganizatio of the dielectric state, e.g. from the Slater– to the Mott–type, as discussed for parametrized models [13]. This hypothesis is verified by estimating the system conductivity in the *next* Section.

The situation becomes completely different when we consider the *quarter–filled* band case $N_e = N/2$ (*cf.* Fig. 5). The parabolic extrapolation with $1/N \to 0$ provides now the value of the charge–gap $\Delta E_C \approx 0$ (within the errorbars) for lattice parameter $R \lesssim 2a_0$. In the range of $R/a_0 = 2.5 \div 4.5$ the gap reaches nonzero values (significantly greater than the corresponding errorbars), but random dispersion of the datapoints suggests instability of the performed extrapolation due to nonanalytic behavior of ΔE_C when the system is close to the metal–insulator transition. For $R \gtrsim 4.5a_0$ the gap smoothly grows to a nonzero value corresponding to the insulating charge–density wave state, identified

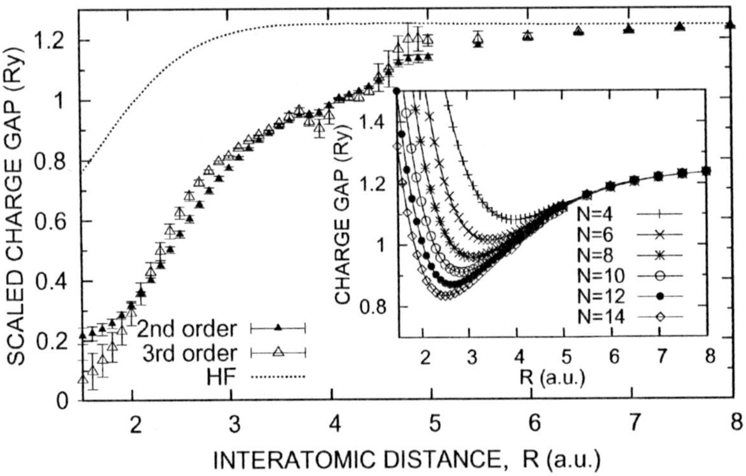

FIGURE 4. Charge–energy gap obtained through finite–size scaling for the chains of $N = 4 \div 14$ atoms. The corresponding Hartree–Fock value (the Slater gap) for an infinite system is also drawn (*dotted* line). The *inset* exhibits the original data used for the scaling.

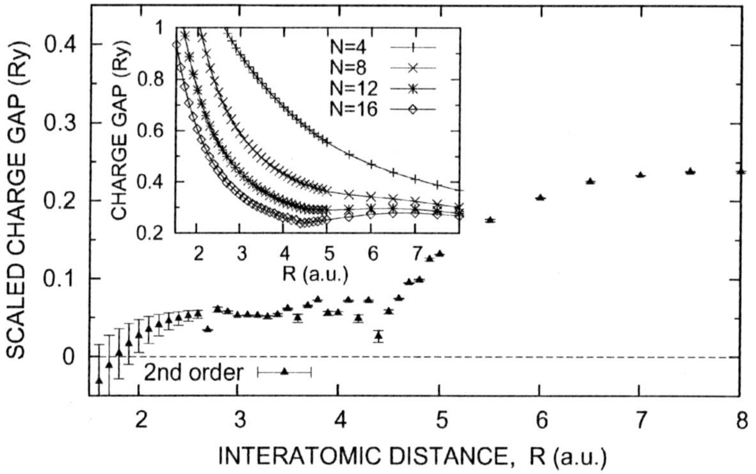

FIGURE 5. Charge–energy gap obtained through finite–size scaling for the chains of $N = 4 \div 16$ atoms in the *quarter–filled* band case. The 2–nd order polynomial has been fitted to perform the extrapolation with $1/N \to \infty$. *Inset* provides the original data used for the scaling.

in the *previous* Section. The more precise position of the MIT point is determined *below*, where we calculate the Drude weight for the system.

Charge stiffness and optical conductivity

We start the discussion with the real part of the optical conductivity at $T = 0$, which is determined by the real part of the linear response to the applied electric field [14], and can be written as $\sigma(\omega) = D\delta(\omega) + \sigma_{\text{reg}}(\omega)$, where the regular part is

$$\sigma_{\text{reg}}(\omega) = \frac{\pi}{N}\sum_{n\neq 0}\frac{|\langle\Psi_n|j_p|\Psi_0\rangle|^2}{E_n - E_0}\delta(\omega - (E_n - E_0)), \tag{5}$$

whereas the Drude weight (the *charge stiffness*) D is given by

$$D = -\frac{\pi}{N}\langle\Psi_0|T|\Psi_0\rangle - \frac{2\pi}{N}\sum_{n\neq 0}\frac{|\langle\Psi_n|j_p|\Psi_0\rangle|^2}{E_n - E_0}, \tag{6}$$

with the kinetic term T as in Eq. (1) and the current operator $j_p = it\sum_{j\sigma}(a^\dagger_{j\sigma}a_{j+1\sigma} - \text{HC})$. Here the states $|\Psi_n\rangle$ in Eqs. (5) and (6) are the eigenstates of Hamiltonian (1) corresponding to the eigenenergies E_n, with boundary conditions that minimize the ground–state energy for a given system size N. Matrix elements $\langle\Psi_n|j_p|\Psi_0\rangle$ are calculated within the Lanczos technique set up by Dagotto [15].

For finite system of N atoms D is always nonzero due to a finite tunneling probability through a potential barrier of finite width. Because of that, the finite–size scaling with

$1/N \to \infty$ has to be performed for D. Here we use the following parabolic extrapolation

$$\ln|D_N^*| = a + b(1/N) + c(1/N)^2, \tag{7}$$

where D_N^* denotes the *normalized Drude weight* $D^* = -(N/\pi)D/\langle\Psi_0|T|\Psi_0\rangle$ for the system of N sites, that provides the value in the range $0 \leqslant D^* \leqslant 1$, and thus can be regarded as an order parameter for MIT.

The results for 1D system of $N = 6 \div 14$ atoms in the *half-filled* band case are shown in Fig. 6a. For the system with long–range interactions, the extrapolated Drude weight becomes significantly grater then zero only for small lattice parameters $R \leqslant 2.6a_0$. The corresponding border for the Hubbard model is $R \leqslant 2.1a_0$ [11]. The limiting values match those for which the Luttinger–liquid exponent cross the critical values $\theta = 1$, corresponding to the metal–insulator boundary. The above results suggests transition–like behavior in such 1D systems at half–filling. However, the optical conductivity $\sigma_{\mathrm{reg}}(\omega)$, drawn in Fig. 6b, shows the isolated Hubbard peak at $\omega \approx U$ and no interband transitions present in conducting systems. Due to this fact, and also because of the nonzero value of the charge–gap for any R (cf. *above*), one should regard both the half–filled systems studied here as the Mott–insulators in the large N limit. Nevertheless, the finite–N results show the conductivity of the nanoscopic chain falls by factor $\sim 10^2$ when the corresponding increase of R is in the range of $40 \div 50\%$, so one can either consider such system as undergoing the transformation from nanoliquid to localized spin system at the half–filling. It would be very intresting to confirm experimentally these results.

The situation becomes again completely different at the *quarter–filling* (QF). The normalized Drude weight of QF systems of $N = 8 \div 16$ atoms, depicted in Fig. 7a, shows a highly–conducting behavior ($D^* \approx 1$) for $R \lesssim 3.5a_0$ and transforms gradually to zero in the range $R/a_0 = 4 \div 5$. Also, the regular part of conductivity $\sigma_{\mathrm{reg}}(\omega)$ (*cf.*

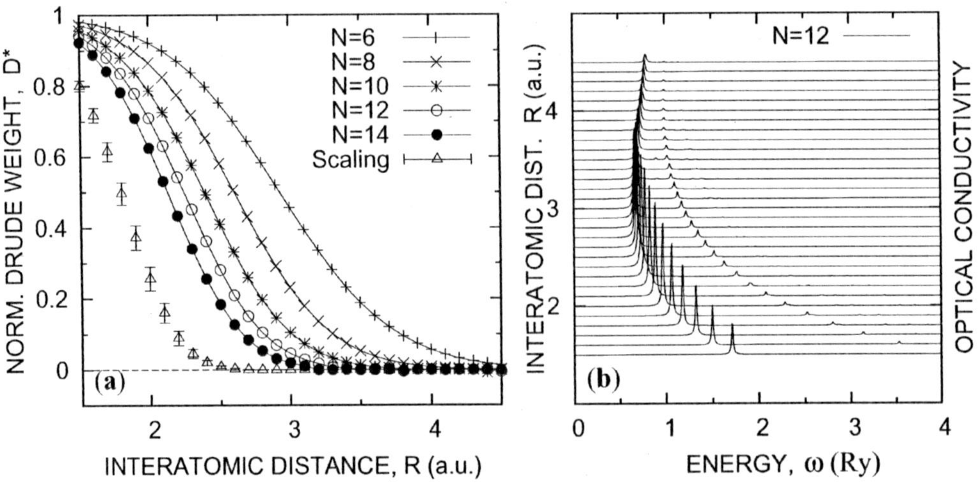

FIGURE 6. Optical conductivity for the nanochain in the *half-filled* band case: (*a*) normalized Drude weight vs. lattice parameter R (specified in a_0) and its values obtained through finite size scaling with $1/N \to 0$; (*b*) regular part of the conductivity, $\sigma_{\mathrm{reg}}(\omega)$ for $N = 12$ atoms.

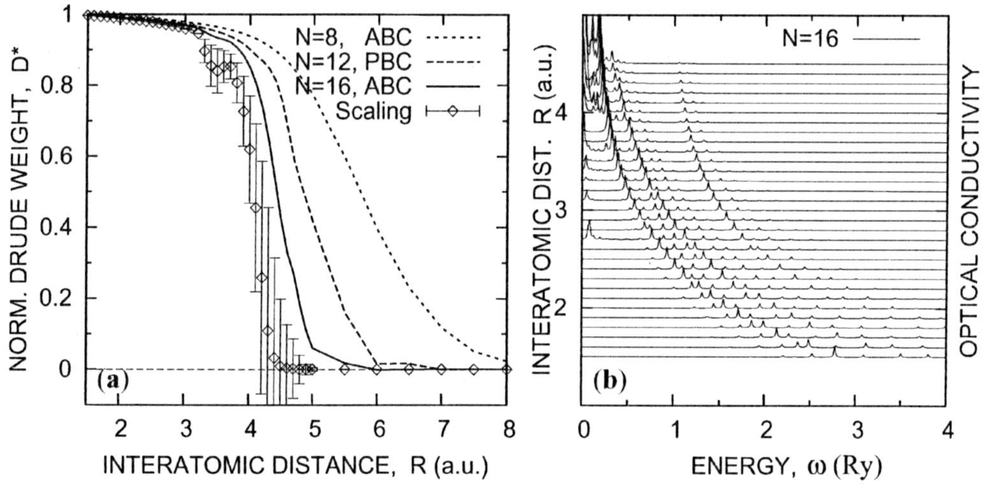

FIGURE 7. Optical conductivity for the *quarter–filled* chain: (*a*) normalized Drude weight vs. lattice parameter R and its values extrapolated with $1/N \to 0$ (the Aitken method has been used to estimate the errors); (*b*) regular part of the conductivity, $\sigma_{\mathrm{reg}}(\omega)$ for $N = 16$ atoms.

Fig. 7b) demonstrates the interband transitions in the metallic range, those vanish at $R \approx 4.5a_0$ (for $N = 16$). Such behavior provides the model case for transformation from a nanometal in the small R range to the charge–ordered system (*cf.* Fig. 3) for larger R.

A BRIEF OVERVIEW

We have presented a fairly complete description of finite 1D chain in the framework of EDABI method, which combines the *exact* diagonalization of many–fermion Hamiltonian in the Fock space with a subsequent *ab initio* readjustment of the single–particle (Wannier) function. The ground–state and dynamical properties have been obtained as a function of the lattice parameter R. Our approach thus *extends* the current theoretical treatments to the *strongly correlated systems* within the parametrized (second–quantized) models by providing the determination of those parameters and, in turn, analyzing the correlated state explicitly as a function of R.

We start from analyzing the situation with one electron per atom (the *half–filled* band case), including the *long–range* Coulomb interactions. The Luttinger liquid type of the electron momentum distribution suggests the *crossover transition* from the metallic to the insulating (spin–ordered) state with the increasing R (the same is about the system without the long–range interactions [11]), but metallic behavior is manifested to much stronger degree when the long–range part of the Coulomb interactions is included). However, the finite–size scaling with $1/N \to 0$, performed on the charge–energy gap shows the insulating nature of the ground state for the large N limit, in agreement with the renormalization–group results for the *infinite* system with two Fermi points [12]. Such an apparently contradictory nature of the nanoscopic systems is confirmed by their transport properties. On one hand, the Drude weight is nonzero in the small R limit (and critical values of R agree with those obtained from the Luttinger–liquid exponent), but the regular part of the optical conductivity exhibits the insulating behavior.

An illustrative example of the nanonscopic system with a clear transformation from the nanometal to the nanoinsulator with charge–density wave order is provided for a similar system in the *quarter–filled* band case (and when including the long–range Coulomb interactions). For that system, the Drude weight is reduced gradually from its maximal value to zero, and the other studied properties evolve appropriately with the increasing lattice parameter R. The intermediate range of R, where the evolution takes place, also shrink rapidly with increasing N, suggesting the zero–temperature transition in the large N limit. To the best of our knowledge, such a transition has not been identified before for 1D system of $S = 1/2$ electrons.

ACKNOWLEDGMENT

We thank to our colleagues, Krzysztof Rościszewski and Maciej Maśka, for discussions about the Lanczos algorithm and the role of boundary conditions for finite systems. We are grateful to the Institute of Physics of the Jagiellonian University for the support for computing facilities used in part of the numerical analysis. The support from the Committee for Scientific Research (KBN) of Poland (Grant Nos. 2 P03B 064 22 and 2P03B 050 23), and from Foundation for Polish Science (FNP) is also acknowledged.

REFERENCES

1. Hohenberg, P.C., Kohn, W., Sham, L.I., in *Adv. Quantum Chemistry*, ed. S.B. Trickey, Academic, San Diego, Vol. 21, pp. 7–26 (1990); Temmerman, W., *et al.*, in *Electronic Density Functional Theory: Recent Progress and New Directions*, ed. J.F. Dobson *et al.*, Plenum, New York, pp. 327–347 (1998).
2. Anisimov, V.I., *et al.*, *Phys. Rev. B* **44**, 943 (1991); Wei, P., Qi, Z.Q., *ibid.* **49**, 10864 (1994).
3. Svane, A., Gunnarson, O., *Europhys. Lett.* **7**, 171 (1988); *Phys. Rev. Lett.* **65**, 1148(1990).
4. Ezhov, S., *et al.*, *Phys. Rev. Lett.* **83**, 4136 (1999); Held, K., *et al.*, *ibid.* **86**, 5345 (2001).
5. Spałek, J., *et al.*, *Phys. Rev. B* **61**, 15676 (2001), and unpublished; Rycerz, A., Spałek, J., *ibid.* **63**, 073101 (2001); **65**, 035110 (2002); Spałek, J., Rycerz, A., *ibid.* **64**, R161105 (2001).
6. Spałek, J., *et al.*, *Acta Phys. Polon. B* **31**, 2879 (2000); **32**, 3189 (2001); Rycerz, A., *et al.*, *ibid.* **34**, 651 (2003); 655 (2003).
7. Rycerz, A., *et al.*, in *Lectures on the Physics of Highly Correlated Electron Systems VI*, ed. F. Mancini, AIP Conf. Proc. Vol. 629, pp. 213–222, New York, (2002); *ibid.* Vol. 678, pp. 313–322, New York (2003).
8. Spałek, J. *et al.*, in *Concepts in Electron Correlation*, Proc. of the NATO Adv. Res. Workshop, eds A.C. Hewson and V. Zlatić, pp. 257–268, Kluwer, Dordrecht (2003).
9. Solyom, J., *Adv. Phys.* **28**, 201 (1979); Voit, J., *Rep. Prog. Phys.* **57**, 977 (1995).
10. Sorella, S., *et al.*, *Europhys. Lett.* **12**, 721 (1990).
11. Rycerz, A., *Ph. D. Thesis*, Jagiellonian University, Kraków (2003). Unpublished.
12. Fabrizio, M., *Phys. Rev. B* **54**, 10054 (1996); see also Voit, Ref. [9].
13. Resta, R., Sorella, S., *Phys. Rev. Lett.* **82**, 370 (1999); Korbel, P., *et al.*, *Eur. Phys. J. B* **32**, 315 (2003).
14. Shastry, B.S., Sutherland, B., *Phys. Rev. Lett.* **65**, 243 (1990); Millis, J.A., Coppersmith, S.N., *Phys. Rev. B* **42**, 10807 (1990).
15. Dagotto, E., *Rev. Mod. Phys.* **66**, 763 (1994); Prelovšek, P., and Zotos, X., in *Lectures on the Physics of Highly Correlated Electron Systems VI*, ed. by F. Mancini, AIP Conf. Proc. Vol. 629, p.161 (2002).

Electronic Raman Scattering in Density Waves

András Ványolos* and Attila Virosztek*†

*Department of Physics, Budapest University of Technology and Economics, 1521 Budapest, Hungary

†Research Institute for Solid State Physics and Optics, PO Box 49, 1525 Budapest, Hungary

Abstract. We calculate the electronic Raman spectrum of pure conventional and unconventional density waves (U)DW in the mean field approximation. The calculation is carried out in the $\mathbf{q} \to 0$ long wavelength limit, retaining only the quasiparticle contribution. In analogy with unconventional superconductivity, the Raman spectra depend strongly on the polarization of the incoming and scattered light, which affects the low frequency power law behavior and the peak position in the spectra, and can be used to identify which type of gap structure is present in the DW. The Coulomb screening is also considered in the RPA, and we find that it can be ignored in the long wavelength limit due to the vanishing average of the Raman vertex.

INTRODUCTION

Electronic Raman scattering has been proven to be a valuable spectroscopic tool contributing for example to the establishment of the d-wave nature of the order parameter in high temperature superconductors (HTSC) [1]. It has also been applied recently in order to investigate the temperature and pressure dependence of the charge density wave amplitude mode in $1T$-$TiSe_2$ [2]. Superconducting and density wave condensates, both unconventional, are believed to be present in the underdoped cuprates [3, 4]. A theoretical analysis of this complex situation with respect to Raman scattering has also been attempted [5].

The recent surge of interest in unconventional density waves (UDW) is mostly due to their potential applicability in the pseudogap phase [6, 7, 8] of HTSC materials. However pseudogap phases, and in general various kinds of hidden order, are detected in other substances as well, like in chalcogenides [9], in heavy fermion materials [10], and in Bechgaard salts [11]. Since UDW's are natural candidates for explaining hidden order due to their momentum dependent gap structure [12], they have been proposed to exist in URu_2Si_2 [13] and in α-$(ET)_2$ salts [14].

The aim of the present paper is to develop a theory of Raman scattering in pure quasi-one dimensional conductors with conventional, or unconventional density wave ground state. Basics of electronic Raman scattering and mean field treatment of density waves are given in Section II. The quasiparticle contribution to the light scattering intensity in various polarizations and gap structures are calculated in Section III. We investigate the effect of Coulomb screening in Section IV, and Section V. is devoted to our conclusions.

CP715, *Lectures on the Physics of Highly Correlated Electron Systems VIII: Eighth Training Course*, edited by A. Avella and F. Mancini

In this paper we use the temperature Green's function formalism in order for further generalizations to be more straightforward [15]. Light coupling to electrons via the vector potential $\mathbf{A}$ can be treated in second-order perturbation theory. The intensity of scattered light in a Raman experiment can be expressed [16] as

$$\frac{d\sigma}{d\omega d\Omega dV} = r_0^2 \frac{\omega_s}{\omega_i} S_{\gamma\gamma}(\mathbf{q},\omega), \tag{1}$$

where $r_0^2 = e^2/mc^2$ is the Thomson radius, $\omega_i, \mathbf{q}_i$ and $\omega_s, \mathbf{q}_s$ are the energies and momenta of the incoming and scattered photon, respectively. Furthermore the energy and momentum transfer to the material are $\omega = \omega_i - \omega_s$ and $\mathbf{q} = \mathbf{q}_f - \mathbf{q}_i$. The generalized structure factor $S_{\gamma\gamma}$ is related to the Raman response through the fluctuation-dissipation theorem

$$S_{\gamma\gamma}(\mathbf{q},\omega) = \frac{1}{\pi}[1+n(\omega)]\mathrm{Im}\chi_{\gamma\gamma}(\mathbf{q},\omega), \tag{2}$$

where $n(\omega)$ is the Bose function. The Raman response of the electron system measures "effective density" fluctuations

$$\chi_{\gamma\gamma}(\mathbf{q},\omega) = i\langle[\tilde{\rho}(\mathbf{q}),\tilde{\rho}(-\mathbf{q})]\rangle(\omega)/V, \tag{3}$$

where

$$\tilde{\rho}(\mathbf{q}) = \sum_{\mathbf{k},\sigma} \gamma_{\mathbf{k}} c^\dagger_{\mathbf{k}+\mathbf{q},\sigma} c_{\mathbf{k},\sigma}, \tag{4}$$

V is the volume of the system, and since we are interested in the $\mathbf{q} \to 0$ behavior of the $\chi_{\gamma\gamma}$ susceptibility, we neglected the $\mathbf{q}$ dependence of the vertex $\gamma_{\mathbf{k}}$. Here $c^\dagger_{\mathbf{k},\sigma}$, $(c_{\mathbf{k},\sigma})$ is the creation (annihilation) operator of an electron with momentum $\mathbf{k}$ and spin σ in the single conduction band $\varepsilon_{\mathbf{k}} = -2t_a \cos ak_x - 2t_b \cos bk_y - 2t_c \cos ck_z$ with $t_a \gg t_b, t_c$. The strength of the scattering is determined by the momentum dependent function $\gamma_{\mathbf{k}}$ called the Raman vertex, which has the form

$$\gamma_{\mathbf{k}} = (\mathbf{e}_i\mathbf{e}_s) + \frac{1}{m}\sum_b \left(\frac{\langle \mathbf{k}|\mathbf{p}\mathbf{e}_s|b\mathbf{k}\rangle\langle b\mathbf{k}|\mathbf{p}\mathbf{e}_i|\mathbf{k}\rangle}{\varepsilon_{\mathbf{k}} - \varepsilon_{b\mathbf{k}} + \omega_i} + \frac{\langle \mathbf{k}|\mathbf{p}\mathbf{e}_i|b\mathbf{k}\rangle\langle b\mathbf{k}|\mathbf{p}\mathbf{e}_s|\mathbf{k}\rangle}{\varepsilon_{\mathbf{k}} - \varepsilon_{b\mathbf{k}} - \omega_s} \right), \tag{5}$$

where b stands for the band index of the electron excited out of the conduction band, and the corresponding states are $|\mathbf{k}b\rangle$ and $|\mathbf{k}\rangle$, respectively. In addition the polarization vectors of the incoming and scattered light are denoted by $\mathbf{e}_i, \mathbf{e}_s$. If the incoming and scattered light frequencies can be neglected in comparison to the optical band gap [17], then the Raman vertex is related to the inverse mass tensor $\gamma_{\alpha\beta}(\mathbf{k}) = m\partial^2\varepsilon_{\mathbf{k}}/\partial k_\alpha \partial k_\beta$ through the relation

$$\gamma_{\mathbf{k}} = \sum_{\alpha,\beta} e_s^\alpha \gamma_{\alpha\beta}(\mathbf{k}) e_i^\beta. \tag{6}$$

The retarded susceptibility of the effective density can be evaluated with analytical continuation from the Fourier transform of the corresponding τ (imaginary time) ordered response $\chi_{\gamma\gamma}(\mathbf{q},\tau) = -\langle T_\tau[\tilde{\rho}(\mathbf{q},\tau)\tilde{\rho}(-\mathbf{q})]\rangle$ in the usual way. The one particle temperature Green's function of the DW using Nambu's notation reads

$$G_\sigma(\mathbf{k},i\omega_n) = -\int_0^\beta d\tau \langle T_\tau[\Psi_\sigma(\mathbf{k},\tau)\Psi^\dagger_\sigma(\mathbf{k})]\rangle e^{i\omega_n\tau}, \tag{7}$$

where the two component spinor field

$$\Psi_\sigma(\mathbf{k},\tau) = \begin{pmatrix} c_{\mathbf{k},\sigma}(\tau) \\ c_{\mathbf{k}-\mathbf{Q},\sigma}(\tau) \end{pmatrix} \tag{8}$$

is used to treat the left and right moving electrons in momentum space in a convenient way. After straightforward calculation using the mean field Hamiltonian of the system [12] one finds

$$G_\sigma^{-1}(\mathbf{k},i\omega_n) = i\omega_n - \xi_{\mathbf{k}}\rho_3 - \Delta_\sigma(\mathbf{k})\rho_1, \tag{9}$$

where ρ_i stand for the Pauli matrices, while the linearized spectrum of the highly anisotropic electron system around the Fermi energy is $\xi_{\mathbf{k}} = \varepsilon_{\mathbf{k}} - \mu = v_F(k_x - k_F) - 2t_b\cos(bk_y) - 2t_c\cos(ck_z)$. In the followings we suppress the spin index of the order parameter $\Delta_\sigma(\mathbf{k})$, since in order to describe a spin density wave (SDW) with polarization vector parallel to the z-axis of the spin space, we can utilize the relation $\Delta_\uparrow(\mathbf{k}) = -\Delta_\downarrow(\mathbf{k}) = \Delta(\mathbf{k})$, while for charge density wave (CDW) $\Delta_\uparrow(\mathbf{k}) = \Delta_\downarrow(\mathbf{k}) = \Delta(\mathbf{k})$ holds. The order parameter is either independent of the momentum, which is the case of a conventional DW, or it can have four different type of wavevector dependence ($\Delta(\mathbf{k}) = \Delta\cos bk_y$, $\Delta(\mathbf{k}) = \Delta\sin bk_y$, $\Delta(\mathbf{k}) = \Delta\cos ck_z$, $\Delta(\mathbf{k}) = \Delta\sin ck_z$) as discussed in detail in Ref. [12].

QUASIPARTICLE CONTRIBUTION

Making use of the Green's function in Eq. (9), the quasiparticle contribution to the Raman susceptibility can be written as

$$\chi_{\gamma\gamma}(\mathbf{q},i\nu_n) = \\ -\frac{T}{V}\sum_{\mathbf{k},\sigma,\omega_m}\mathrm{Tr}\begin{pmatrix}\gamma_{\mathbf{k}} & 0 \\ 0 & \gamma_{\mathbf{k}-\mathbf{Q}}\end{pmatrix} G_\sigma(\mathbf{k},i\omega_m)\begin{pmatrix}\gamma_{\mathbf{k}+\mathbf{q}} & 0 \\ 0 & \gamma_{\mathbf{k}-\mathbf{Q}+\mathbf{q}}\end{pmatrix} G_\sigma(\mathbf{k}+\mathbf{q},i\omega_m+i\nu_n). \tag{10}$$

This formula corresponds to a bubble diagram with self energy corrections due to the order parameter of the condensate. With the aid of this result, we can simply treat the Coulomb screening in the usual random phase approximation (RPA), which will be presented in the next section. After performing the Matsubara sum and the analytic continuation in Eq. (10) we obtain

$$\chi_{\gamma\gamma}(\mathbf{q},\omega+i\delta) = \frac{1}{2V}\sum_{\mathbf{k}}\Bigg[\{f(E_{\mathbf{k}+\mathbf{q}})-f(E_{\mathbf{k}})\}$$
$$\times\Bigg\{\left(\frac{1}{\omega+i\delta+E_{\mathbf{k}}-E_{\mathbf{k}+\mathbf{q}}}-\frac{1}{\omega+i\delta-E_{\mathbf{k}}+E_{\mathbf{k}+\mathbf{q}}}\right)$$
$$\times\left(\left(1+\frac{\xi_{\mathbf{k}}\xi_{\mathbf{k}+\mathbf{q}}}{E_{\mathbf{k}}E_{\mathbf{k}+\mathbf{q}}}\right)\gamma_{\mathbf{k}}^{(1)}+\frac{\Delta_{\mathbf{k}}\overline{\Delta_{\mathbf{k}+\mathbf{q}}}}{E_{\mathbf{k}}E_{\mathbf{k}+\mathbf{q}}}\gamma_{\mathbf{k}}^{(2)}+\frac{\overline{\Delta_{\mathbf{k}}}\Delta_{\mathbf{k}+\mathbf{q}}}{E_{\mathbf{k}}E_{\mathbf{k}+\mathbf{q}}}\gamma_{\mathbf{k}}^{(3)}\right)$$
$$+\left(\frac{1}{\omega+i\delta+E_{\mathbf{k}}-E_{\mathbf{k}+\mathbf{q}}}+\frac{1}{\omega+i\delta-E_{\mathbf{k}}+E_{\mathbf{k}+\mathbf{q}}}\right)\left(\frac{\xi_{\mathbf{k}}}{E_{\mathbf{k}}}+\frac{\xi_{\mathbf{k}+\mathbf{q}}}{E_{\mathbf{k}+\mathbf{q}}}\right)\gamma_{\mathbf{k}}^{(4)}\Bigg\} \quad (11)$$
$$+\{1-f(E_{\mathbf{k}+\mathbf{q}})-f(E_{\mathbf{k}})\}$$
$$\times\Bigg\{\left(\frac{1}{\omega+i\delta+E_{\mathbf{k}}+E_{\mathbf{k}+\mathbf{q}}}-\frac{1}{\omega+i\delta-E_{\mathbf{k}}-E_{\mathbf{k}+\mathbf{q}}}\right)$$
$$\times\left(\left(1-\frac{\xi_{\mathbf{k}}\xi_{\mathbf{k}+\mathbf{q}}}{E_{\mathbf{k}}E_{\mathbf{k}+\mathbf{q}}}\right)\gamma_{\mathbf{k}}^{(1)}-\frac{\Delta_{\mathbf{k}}\overline{\Delta_{\mathbf{k}+\mathbf{q}}}}{E_{\mathbf{k}}E_{\mathbf{k}+\mathbf{q}}}\gamma_{\mathbf{k}}^{(2)}-\frac{\overline{\Delta_{\mathbf{k}}}\Delta_{\mathbf{k}+\mathbf{q}}}{E_{\mathbf{k}}E_{\mathbf{k}+\mathbf{q}}}\gamma_{\mathbf{k}}^{(3)}\right)$$
$$+\left(\frac{1}{\omega+i\delta+E_{\mathbf{k}}+E_{\mathbf{k}+\mathbf{q}}}+\frac{1}{\omega+i\delta-E_{\mathbf{k}}-E_{\mathbf{k}+\mathbf{q}}}\right)\left(\frac{\xi_{\mathbf{k}}}{E_{\mathbf{k}}}-\frac{\xi_{\mathbf{k}+\mathbf{q}}}{E_{\mathbf{k}+\mathbf{q}}}\right)\gamma_{\mathbf{k}}^{(4)}\Bigg\}\Bigg],$$

where

$$\gamma_{\mathbf{k}}^{(1)} = \gamma_{\mathbf{k}}^2+\gamma_{\mathbf{k}-\mathbf{Q}}^2, \quad (12a)$$
$$\gamma_{\mathbf{k}}^{(2)} = \gamma_{\mathbf{k}}\gamma_{\mathbf{k}-\mathbf{Q}}, \quad (12b)$$
$$\gamma_{\mathbf{k}}^{(3)} = \gamma_{\mathbf{k}-\mathbf{Q}}\gamma_{\mathbf{k}}, \quad (12c)$$
$$\gamma_{\mathbf{k}}^{(4)} = \gamma_{\mathbf{k}-\mathbf{Q}}^2-\gamma_{\mathbf{k}}^2. \quad (12d)$$

Here $E_{\mathbf{k}}^2 = \xi_{\mathbf{k}}^2+|\Delta_{\mathbf{k}}|^2$ is the excitation spectrum and $f(E)$ is the Fermi function. Eq. (11) is reduced to the density correlator if we set $\gamma_{\mathbf{k}} = 1$ [12]. The first term in the susceptibility corresponds to the intraband scattering and vanishes in the long wavelength limit, while the second term represents the interband scattering between the two quasiparticle bands and has finite contribution in the $\mathbf{q}\to 0$ limit. Therefore we obtain

$$\mathrm{Im}\chi_{\gamma\gamma}(\mathbf{q}\to 0,\omega>0) = 2\pi\frac{\tanh(\omega/4T)}{\omega^2}\frac{1}{V}\sum_{\mathbf{k}}\delta(\omega-2E_{\mathbf{k}})|\Delta_{\mathbf{k}}|^2(\gamma_{\mathbf{k}}-\gamma_{\mathbf{k}-\mathbf{Q}})^2. \quad (13)$$

Making use of the $\varepsilon_{\mathbf{k}}$ kinetic energy spectrum and Eq. (6) we easily obtain

$$(\gamma_{\mathbf{k}}-\gamma_{\mathbf{k}-\mathbf{Q}})^2 = \left((e_i^x\Gamma^x e_s^x)\sin(ak_F)\sin a(k_x-k_F)+\sum_{\alpha=y,z}(e_i^\alpha\Gamma^\alpha e_s^\alpha)\cos(\delta_\alpha k_\alpha)\right)^2$$
$$= (e_i^x\Gamma^x e_s^x)^2\sin^2(ak_F)\sin^2 a(k_x-k_F)+\sum_{\alpha=y,z}(e_i^\alpha\Gamma^\alpha e_s^\alpha)^2\cos^2(\delta_\alpha k_\alpha)+\text{cross terms}, \quad (14)$$

where $\Gamma^{x,y,z} = 4m\delta^2_{x,y,z}t_{x,y,z}$, $\delta_{x,y,z} = a,b,c$, furthermore the contributions to the susceptibility from the six cross terms vanish because of the symmetry properties of the integral in Eq. (13). As is seen above, the nonvanishing terms corresponding to the three different polarization directions contribute independently to the total intensity of scattered light, thus $\mathrm{Im}\chi_{\gamma\gamma} = \sum_\alpha (e^\alpha_i e^\alpha_s)^2 \chi''_\alpha$, $\alpha = x,y,z$. At zero temperature they are given by

$$\chi''_x = \Gamma^{x2}\frac{2\pi g_0(0)}{\omega W^2}\int_{-\pi}^{\pi}\frac{d(bk_y)}{2\pi}\int_{-\pi}^{\pi}\frac{d(ck_z)}{2\pi}|\Delta_{\mathbf{k}}|^2\mathrm{Re}\sqrt{\omega^2-4|\Delta_{\mathbf{k}}|^2}, \tag{15a}$$

$$\chi''_\alpha = \Gamma^{\alpha 2}\frac{2\pi g_0(0)}{\omega}\int_{-\pi}^{\pi}\frac{d(bk_y)}{2\pi}\int_{-\pi}^{\pi}\frac{d(ck_z)}{2\pi}\mathrm{Re}\frac{|\Delta_{\mathbf{k}}|^2\cos^2\delta_\alpha k_\alpha}{\sqrt{\omega^2-4|\Delta_{\mathbf{k}}|^2}}, \quad \alpha = y,z, \tag{15b}$$

where $g_0(0) = 1/\pi bcv_F$ is the normal state density of states for one spin direction and $W \simeq 4t_a$ stands for the bandwidth. The response functions for finite T are obtained simply by multiplying Eqs. (15) by the factor $\tanh(\omega/4T)$.

In the case of conventional DW, when the gap on the Fermi surface is constant, then Eqs. (15) simplify to

$$\chi''_{x,\mathrm{conv}} = \Gamma^{x2}g_0(0)\left(\frac{2\Delta}{W}\right)^2\frac{\pi\mathrm{Re}\sqrt{x^2-1}}{2x}, \tag{16a}$$

$$\chi''_{\alpha,\mathrm{conv}} = \Gamma^{\alpha 2}g_0(0)\mathrm{Re}\frac{\pi}{4x\sqrt{x^2-1}}, \quad \alpha = y,z, \tag{16b}$$

where Δ is supposed to be real, and $x = \omega/2\Delta$. For polarizations perpendicular to the chain direction x, the Raman intensity has the usual inverse squareroot divergence at 2Δ, and below this threshold there is no transition. It is worth mentioning, that for the chain polarization the divergent peak is suppressed, which is the consequence of the vanishing vertex on the Fermi surface as it's readily seen in Eq. (14).

In the case of UDW it's enough to consider the k_y dependent gap, because similar results can be obtained for the k_z dependent order parameter, with the following relations

$$\chi''_x[\Delta(k_y)] = \chi''_x[\Delta(k_z)], \quad \chi''_y[\Delta(k_y)] = \chi''_z[\Delta(k_z)], \quad \chi''_z[\Delta(k_y)] = \chi''_y[\Delta(k_z)], \tag{17}$$

which can be easily verified using the symmetries of the integrals in Eqs. (15). Furthermore, in principle there are six cases for the susceptibilities depending on the polarization and gap function, but as it can be readily seen from the symmetry of the integrals in Eq. (15), in our model we are actually left with four different responses to evaluate, which are tabulated in Table 1. After straightforward calculation the susceptibilities can be expressed with the complete elliptic integrals of the first and second kind as

TABLE 1. The relationships between the imaginary part of the susceptibilities in the cases when there is a cosine or sine gap structure present in the k_y direction.

Polarization	$\Delta(\mathbf{k}) = \Delta\cos bk_y$		$\Delta(\mathbf{k}) = \Delta\sin bk_y$
$x-x$	$\chi''_{x,\cos}$	$=$	$\chi''_{x,\sin}$
$y-y$	$\chi''_{y,\cos}$	$\neq$	$\chi''_{y,\sin}$
$z-z$	$\chi''_{z,\cos}$	$=$	$\chi''_{z,\sin}$

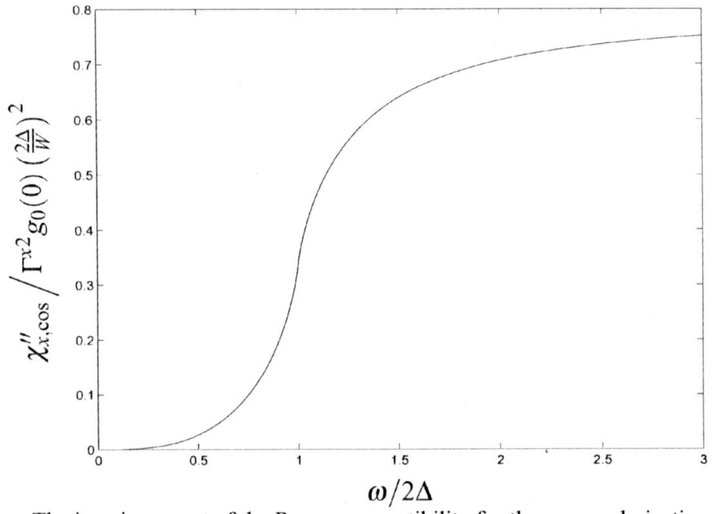

FIGURE 1. The imaginary part of the Raman susceptibility for the $x-x$ polarization at $T = 0, \Delta(\mathbf{k}) = \Delta\cos(bk_y)$.

$$\chi''_{x,\cos} = \frac{\Gamma^{x2} g_0(0)}{3x} \left(\frac{2\Delta}{W}\right)^2 \begin{cases} 2(x^2-1)K(x) - (x^2-2)E(x), & x < 1, \\ x(x^2-1)K(1/x) - x(x^2-2)E(1/x), & x \geq 1, \end{cases} \tag{18a}$$

$$\chi''_{y,\cos} = \frac{\Gamma^{y2} g_0(0)}{3x} \begin{cases} (x^2+2)K(x) - 2(x^2+1)E(x), & x < 1, \\ x(2x^2+1)K(1/x) - 2x(x^2+1)E(1/x), & x \geq 1, \end{cases} \tag{18b}$$

$$\chi''_{z,\cos} = \frac{\Gamma^{z2} g_0(0)}{2x} \begin{cases} K(x) - E(x), & x < 1, \\ x(K(1/x) - E(1/x)), & x \geq 1, \end{cases} \tag{18c}$$

$$\chi''_{y,\sin} = \frac{\Gamma^{y2} g_0(0)}{3x} \begin{cases} (1-x^2)K(x) - (1-2x^2)E(x), & x < 1, \\ 2x(1-x^2)K(1/x) - x(1-2x^2)E(1/x), & x \geq 1, \end{cases} \tag{18d}$$

where the gap maximum Δ is supposed to be real, $x = \omega/2\Delta$, and the corresponding plots of each susceptibility can be seen on Figs. 1-4.

In contrast to conventional DW, in an UDW there are nodes on the Fermi surface, giving rise to arbitrarily small energy excitations. Therefore the scattering intensity is finite for frequencies smaller than the maximum optical gap 2Δ. Thus it is useful to

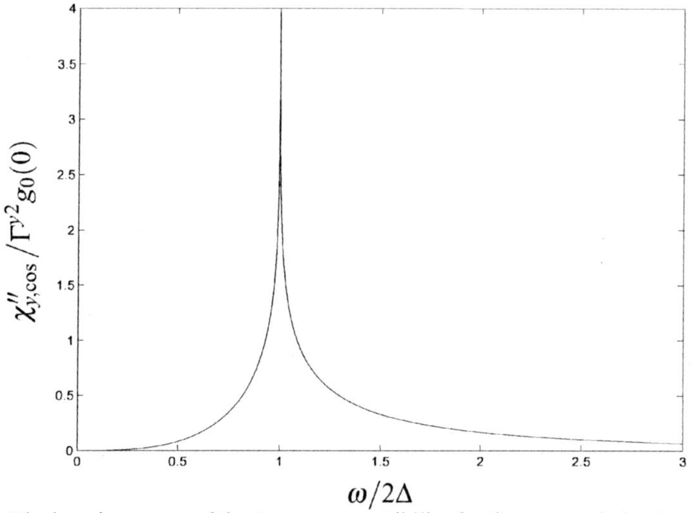

FIGURE 2. The imaginary part of the Raman susceptibility for the $y-y$ polarization at $T=0, \Delta(\mathbf{k}) = \Delta\cos(bk_y)$.

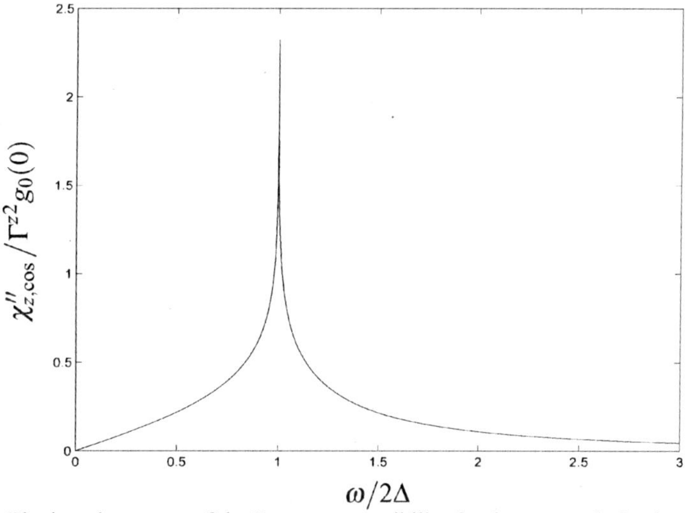

FIGURE 3. The imaginary part of the Raman susceptibility for the $z-z$ polarization at $T=0, \Delta(\mathbf{k}) = \Delta\cos(bk_y)$.

consider the low frequency behavior, and we obtain

$$\chi''_{x,\cos}(\omega \to 0) = \pi\Gamma^{x2}g_0(0)(2\Delta/W)^2x^3/16 + \mathcal{O}(x^5), \tag{19a}$$

$$\chi''_{y,\cos}(\omega \to 0) = 3\pi\Gamma^{y2}g_0(0)x^3/16 + \mathcal{O}(x^5), \tag{19b}$$

$$\chi''_{z,\cos}(\omega \to 0) = \pi\Gamma^{z2}g_0(0)x/8 + \mathcal{O}(x^3), \tag{19c}$$

$$\chi''_{y,\sin}(\omega \to 0) = \pi\Gamma^{y2}g_0(0)x/4 + \mathcal{O}(x^3). \tag{19d}$$

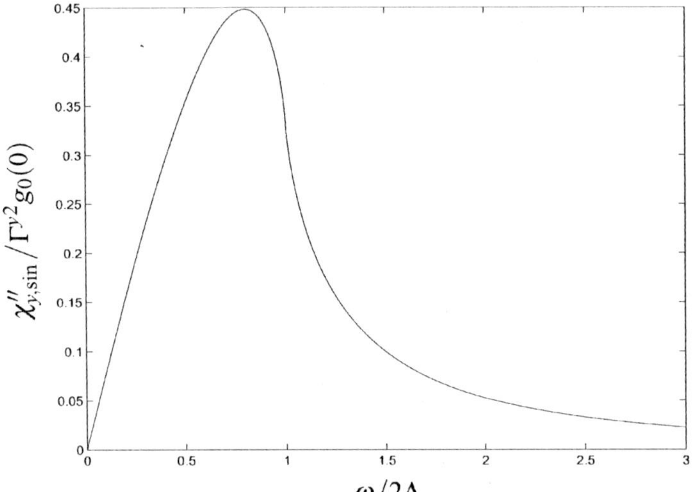

FIGURE 4. The imaginary part of the Raman susceptibility for the $y-y$ polarization at $T=0, \Delta(\mathbf{k}) = \Delta\sin(bk_y)$.

As is seen on Figs. 1-4. and in Eqs. (18-19), the Raman response depends signifi cantly on the polarization directions of the incoming and scattered light. On the one hand, in an UDW with given order parameter the dependence manifests itself (i) in the qualitatively different lineshapes shown on Figs. 1-3 for a cosine gap structure, (ii) quantitatively in the absolute amplitudes of the responses; the intensity of scattered light in the case of chain polarization is expected to be much smaller in magnitude than for perpendicular polarizations due to the $(2\Delta/W)^2 \ll 1$ factor in Eq. (18a) coming from the Raman vertex vanishing on the Fermi surface, (iii) in the low frequency power law behavior with different exponents. On the other hand, if the polarization is fi xed in the y direction, many differences can be observed between the $\chi''_{y,\cos}, \chi''_{y,\sin}$ suscetibilities obtained for a cosine and sine gap (see Figs. 2 and 4): the line shape changes, the divergent peak is suppressed, while the peak position is shifted downwards, and the low frequency behavior alters from ω^3 to ω. All these important features of the Raman response makes the Raman experiment to be a relevant and fruitful probe in identifying the magnitude and symmetry of the energy gap. Similar analysis [18] contributed to the establishment of the d-wave nature of the order parameter in HTSC.

COULOMB SCREENING

In the case when the light produces charge fluctuation in the electron gas, the coupling to the long-range Coulomb forces reduces the scattering rate. Therefore, it is useful to treat these forces separetly. The Coulomb energy operator is

$$H_c = \frac{1}{2V}\sum_{\mathbf{q}} \rho(\mathbf{q})V(\mathbf{q})\rho(-\mathbf{q}), \tag{20}$$

where $V(\mathbf{q}) = 4\pi e^2/q^2$. In the usual RPA approximation the screened Raman susceptibility reads as

$$\chi_{\gamma\gamma}^{\rm sc} = \chi_{\gamma\gamma} - \chi_{\gamma 1}\left(V - V\chi_{11}V + \ldots\right)\chi_{1\gamma} = \chi_{\gamma\gamma} - \frac{\chi_{\gamma 1}\chi_{1\gamma}}{\chi_{11}} + \frac{\chi_{\gamma 1}\chi_{1\gamma}}{\chi_{11}^2}\chi_{11}^{\rm sc}, \qquad (21)$$

where $\chi_{\gamma\gamma}$ is already defined in Eq. (11), $\chi_{11}^{\rm sc} = \chi_{11}/(1+V\chi_{11})$ is the screened density correlator with χ_{11} being the one bubble contribution [12]. Furthermore, the $\chi_{\gamma 1}(\mathbf{q}, i\nu_n)$ and $\chi_{1\gamma}(\mathbf{q}, i\nu_n) = \chi_{\gamma 1}(-\mathbf{q}, -i\nu_n)$ susceptibilities are the Fourier components of the corresponding τ-ordered operators

$$\chi_{\gamma 1}(\mathbf{q}, \tau) = -\langle T_\tau[\tilde{\rho}(\mathbf{q}, \tau)\rho(-\mathbf{q})]\rangle, \qquad (22a)$$
$$\chi_{1\gamma}(\mathbf{q}, \tau) = -\langle T_\tau[\rho(\mathbf{q}, \tau)\tilde{\rho}(-\mathbf{q})]\rangle. \qquad (22b)$$

After straightforward calculations, the explicit expression for $\chi_{\gamma 1}(\mathbf{q}, i\nu_n)$ is the same as in Eq. (11) with the $\omega + i\delta \rightarrow i\nu_n$ substitution and with the following simplifications in the $\gamma_{\mathbf{k}}^{(1-4)}$ coefficients

$$\gamma_{\mathbf{k}}^{(1)} = \gamma_{\mathbf{k}} + \gamma_{\mathbf{k}-\mathbf{Q}}, \quad \gamma_{\mathbf{k}}^{(2)} = \gamma_{\mathbf{k}}, \quad \gamma_{\mathbf{k}}^{(3)} = \gamma_{\mathbf{k}-\mathbf{Q}}, \quad \gamma_{\mathbf{k}}^{(4)} = \gamma_{\mathbf{k}-\mathbf{Q}} - \gamma_{\mathbf{k}}. \qquad (23)$$

Considering the $\mathbf{q} \rightarrow 0$ long wavelength limit, we find that $\chi_{\gamma 1}(\mathbf{q} = 0, i\nu_n) = 0$ independently of the functional form of the Raman vertex $\gamma_{\mathbf{k}}$. Therefore the leading order in $\mathbf{q}$ is the first order, which is easily obtained from Eq. (11) with the above mentioned modifications and reads as

$$\chi_{\gamma 1}(\mathbf{q} \rightarrow 0, i\nu_n) = \frac{2}{i\nu_n V}\sum_{\mathbf{k}}[-f'(E_{\mathbf{k}})]\frac{\xi_{\mathbf{k}}(\mathscr{D}E_{\mathbf{k}})}{E_{\mathbf{k}}}(\gamma_{\mathbf{k}} - \gamma_{\mathbf{k}-\mathbf{Q}})$$
$$+ \frac{1}{2V}\sum_{\mathbf{k}}\tanh(E_{\mathbf{k}}/2T)\left(\frac{1}{i\nu_n + 2E_{\mathbf{k}}} + \frac{1}{i\nu_n - 2E_{\mathbf{k}}}\right)\left(\mathscr{D}\frac{\xi_{\mathbf{k}}}{E_{\mathbf{k}}}\right)(\gamma_{\mathbf{k}} - \gamma_{\mathbf{k}-\mathbf{Q}}), \qquad (24)$$

where $\mathscr{D} = \mathbf{q}\nabla$ is a differential operator. After evaluating the integrals in Eq. (24) we obtain that the first order term vanishes due to the fact, that the Brillouin zone average of the $\gamma_{\mathbf{k}} - \gamma_{\mathbf{k}-\mathbf{Q}}$ combination of the Raman vertex is zero, as it is readily seen from Eq. (14). Thus, the leading order in the $\mathbf{q} \rightarrow 0$ limit is at least q^2 in the $\chi_{\gamma 1}$ susceptibility. The density correlator χ_{11} is $\mathscr{O}(q^2)$ [12], therefore the second and third term in Eq. (21) can be ignored next to the $\chi_{\gamma\gamma} = \mathscr{O}(q^0)$ term. The above calculations lead us to the conclusion that the Raman response is not affected even if we take into account Coulomb screening in RPA, due to the fact, that the "effective density" $\tilde{\rho}$ does not couple to the real density ρ.

CONCLUSIONS

We have investigated theoretically the electronic Raman scattering in quasi-one dimensional interacting electron systems with density wave ground state. Mean field treatment of conventional as well as unconventional density waves in pure systems has been applied in order to determine the Raman intensity in various scattering geometries. We have found distinct, characteristic lineshapes especially in the unconventional situation, depending on the particular momentum dependence of the density wave order parameter. We conclude, that the Raman experiment could serve as a valuable tool in identifying materials supporting unconventional density waves, and in specifying their particular gap structure. We have also considered Coulomb screening, and we found it ineffective due to the negligible coupling of density fluctuations to the Raman vertex in our nearest neighbor tight-binding model.

ACKNOWLEDGMENTS

We have benefi ted from discussions with R. Hackl and B. Dóra. This work was supported by the Hungarian Scientifi c Research Fund under Grants No. OTKA T032162, NDF45172 and T037451.

REFERENCES

1. Devereaux, T. P., Einzel, D., Stadlober, B., Hackl, R., Leach, D. H., and Neumeier, J. J., *Phys. Rev. Lett.*, **72**, 396 (1994).
2. Snow, C. S., Karpus, J. F., Cooper, S. L., Kidd, T. E., and Chiang, T.-C., *Phys. Rev. Lett.*, **91**, 136402 (2003).
3. Benfatto, L., Caprara, S., and Castro, C. D., *Eur. Phys. J. B*, **17**, 95 (2000).
4. Chakravarty, S., Laughlin, R. B., Morr, D. K., and Nayak, C., *Phys. Rev. B*, **63**, 094503 (2001).
5. Zeyher, R., and Greco, A., *Phys. Rev. Lett.*, **89**, 177004 (2002).
6. Jánossy, A., Fehér, T., Oszlányi, G., and Williams, G. V. M., *Phys. Rev. Lett.*, **79**, 2726 (1997).
7. Opel, M., Nemetschek, R., Hoffmann, C., Philipp, R., Müller, P. F., Hackl, R., Tüttő, I., Erb, A., Revaz, B., Walker, E., Berger, H., and Forró, L., *Phys. Rev. B*, **61**, 9752 (2000).
8. Kaminski, A., Rosenkranz, S., Fretwell, H. M., Campuzano, J. C., Li, Z., Raffy, H., Cullen, W. G., You, H., Olson, C. G., Varma, C. M., and Höchst, H., *Nature*, **416**, 610 (2002).
9. Németh, L., Matus, P., Kriza, G., and Alavi, B., *Synth. Metals*, **120**, 1007 (2001).
10. Isaacs, E. D., McWhan, D. B., Kleiman, R. N., Bishop, D. J., Ice, G. E., Zschack, P., Gaulin, B. D., Mason, T. E., Garrett, J. D., and Buyers, W. J. L., *Phys. Rev. Lett.*, **65**, 3185 (1990).
11. Christ, P., Biberacher, W., Kartsovnik, M. V., Steep, E., Balthes, E., Weiss, H., and Müller, H., *JETP Lett.*, **71**, 303 (2000).
12. Dóra, B., and Virosztek, A., *Eur. Phys. J. B*, **22**, 167 (2001).
13. Ikeda, H., and Ohashi, Y., *Phys. Rev. Lett.*, **81**, 3723 (1998).
14. Maki, K., Dóra, B., Kartsovnik, M. V., Virosztek, A., Korin-Hamzić, B., and Basletić, M., *Phys. Rev. Lett.*, **90**, 256402 (2003).
15. Abrikosov, A. A., Gorkov, L. P., and Dzyaloshinski, I. E., *Methods of Quantum Field Theory in Statistical Physics*, Dover, New York, 1963.
16. Klein, M. V., and Dierker, S. B., *Phys. Rev. B*, **29**, 4976 (1984).
17. Abrikosov, A. A., and Genkin, V. M., *Zh. Eksp. Teor. Fiz*, **65**, 842 (1973), [Sov. Phys. JETP **38**, 417 (1974)].
18. Devereaux, T. P., and Einzel, D., *Phys. Rev. B*, **51**, 16336 (1995).

Author Index